Smart Materials for Energy, Communications and Security

NATO Science for Peace and Security Series

This Series presents the results of scientific meetings supported under the NATO Programme: Science for Peace and Security (SPS).

The NATO SPS Programme supports meetings in the following Key Priority areas: (1) Defence Against Terrorism; (2) Countering other Threats to Security and (3) NATO, Partner and Mediterranean Dialogue Country Priorities. The types of meeting supported are generally "Advanced Study Institutes" and "Advanced Research Workshops". The NATO SPS Series collects together the results of these meetings. The meetings are co-organized by scientists from NATO countries and scientists from NATO's "Partner" or "Mediterranean Dialogue" countries. The observations and recommendations made at the meetings, as well as the contents of the volumes in the Series, reflect those of participants and contributors only; they should not necessarily be regarded as reflecting NATO views or policy.

Advanced Study Institutes (ASI) are high-level tutorial courses intended to convey the latest developments in a subject to an advanced-level audience

Advanced Research Workshops (ARW) are expert meetings where an intense but informal exchange of views at the frontiers of a subject aims at identifying directions for future action

Following a transformation of the programme in 2006 the Series has been re-named and re-organised. Recent volumes on topics not related to security, which result from meetings supported under the programme earlier, may be found in the NATO Science Series.

The Series is published by IOS Press, Amsterdam, and Springer, Dordrecht, in conjunction with the NATO Public Diplomacy Division.

Sub-Series

A.	Chemistry and Biology	Springer
B.	Physics and Biophysics	Springer
C.	Environmental Security	Springer
D.	Information and Communication Security	IOS Press
E.	Human and Societal Dynamics	IOS Press

http://www.nato.int/science
http://www.springer.com
http://www.iospress.nl

Series B: Physics and Biophysics

Smart Materials for Energy, Communications and Security

edited by

Igor A. Luk'yanchuk
University of Picardy,
Amiens, France

Daoud Mezzane
Cadi Ayyad University,
Marrakech, Morocco

Springer

Published in cooperation with NATO Public Diplomacy Division

Proceedings of the NATO CCMS Workshop on
Smart Materials for Energy, Communications and Security (SMECS)
Marrakech, Morocco
December 2007

Library of Congress Control Number: 2008931005

ISBN 978-1-4020-8795-0 (PB)
ISBN 978-1-4020-8794-3 (HB)
ISBN 978-1-4020-8796-7 (e-book)

Published by Springer,
P.O. Box 17, 3300 AA Dordrecht, The Netherlands.

www.springer.com

Printed on acid-free paper

PREFACE

Rapid evolution of trade, cultural and human relations provides the qualitative and quantitative enhancement of international collaborations, linking the countries with different economical and technological level. Delocalization of High-Tech industry inevitably leads to development of the material science and engineering researches in emergent countries, requiring transfer of know-how, restructuration of basic research and educational networks.

This book presents the contributions of participants of the Advanced Research Workshop "*Smart Materials for Energy, Communications and Security*" (ARW SMECS; *www.smecs.ferroix.net*), organized in December 2007 in Marrakech in frame of the "NATO - Science for Peace" program. The objective of this event was the attempt to overview several hot topics of material physics related with problems of modern society: transformation and storage of energy, treatment and transmission of information, environmental security issues etc., with the focus of their implementation in Mediterranean Dialogue (MD) countries: Algeria, Egypt, Mauritania, Morocco and Tunisia.

The workshop is an important stage in developing of the research network "*Mediterranean Electronic Materials*" – MEM (*www.reseau-MEM.org*), that has an objective to encourage the inter-Maghreb and Europe-Maghreb collaborative studies in the area of electroactive materials.

Participants of the Advanced Research Workshop "*Smart Materials for Energy, Communications and Security*" , Marrakech, Morocco, December 2007

Started in 1990s, this program was supported by various institutions, including NATO collaborative linkage grant, French and Moroccan foreign ministries and several regional programs. Large number of joint research projects, training exchange programs for experts and young specialists, international meetings and colloquiums were organized in frame of the network MEM.

The important feature of ARW SMECS was its multidisciplinarity that allowed the sharing of the mutual interests and experience of participants. Beyond scientific sessions, the wide Round-Table discussion of contemporary tendencies and tools for transfer of technologies between European (NATO) and MD countries was organized. Special attention was given to amplification of horizontal collaborative links between MD countries and to integration of the young research institution of Mauritania into this structure.

The sections of the book correspond to the principal topics of *ARW SMECS*:

- Materials for transmission, treatment and storage of information, multiferroic, magnetic and superconducting materials for spintronics, electrochromic and nitride-based semiconducting materials for infrared and optoelectronic devices, carbon-based terahertz electronic technologies, memory-capacious materials.
- Energy storage materials and photovoltaic cells
- Piezo and Electroactive materials for detectors, acoustic transducers and gas sensors
- Functional materials for electronics and environmental security, measurement and characterisation techniques

We would like to thank the authors of the articles as well as other participants of ARW SMECS for their highly beneficial contribution to success of workshop that, we believe, will enhance the mutual understanding between all partners of the Mediterranean Dialogue.

We acknowledge here the political, public and research institutions that, together with NATO, helped us to organize the workshop: Cady Ayad University of Marrakech, Office of Naval Research Global, French and Moroccan Foreign Affairs Ministries and French program ACI VOLUBILIS. We especially thank to Ms. Anaïs Sene for the technical help in preparation of this book.

Co-directors of ARW SMECS

Igor A. Luk'yanchuk
Amiens, France

Daoud Mezzane
Marrakech, Morocco

LIST OF CONTRIBUTORS

Mourad AROUS
Laboratoire des Matériaux composites
Céramiques et Polymères
Faculté des Sciences de Sfax
3018 Sfax Tunis
mourad.arous@fss.rnu.tn

Mohammed BENAISSA
National Center of Scientific Research and Technology (CNRST)
52 Omar Ibn El Khattab BP 8027
10102 Rabat-Agdal Moroco
benaissa@cnrst.ma

Abdel BENLHACHEMI
Laboratoire LME Faculté des Sciences
Agadir Université Ibn Zouhr
BP 8106 Cité Dakhla
80000 Agadir Morocco
a.benlhachemi@gmail.com

Lahcen BIH
UFR PCMI Céramiques et Verres
Fac. Sci. et Techniques-Errachidia
BP 509
Boutalamine Morocco
bihlahcen@yahoo.fr

Marie-José CASANOVE
Groupe MC2, CEMES – UPR 8011
BP 94347
31005 Toulous Cedex 04 France
casanove@cemes.fr

Lahcen DAOUDI
Département des Sciences de la Terre
Faculté des Sciences et Techniques
Université Cadi-Ayyad BP 549
Marrakech Morocco
daoudi@fstg-marrakech.ac.ma

Essebti DHAHRI
Laboratoire de Physique Appliquée
Faculté des Sciences de Sfax
BP 802
3018 Sfax Tunis
essebti@yahoo.com

Aldo DI CARLO
Dept. Electronics Engineering
Univ. of Rome "Tor Vergata"
Via Del Politecnico 1
00133 Roma Italy
aldo.dicarlo@uniroma2.it

Lahcen ESSALEH
Université Cadi Ayyad LMCN FSTG
Département de Physique Gueliz
BP 549
Marrakech Morocco
lessaleh@gmail.com

Ahlem KABADOU
Laboratoire de Métallurgie Appliquée
3018 Sfax Tunis
ahlemkabadou@yahoo.fr

Wolfgang KLEEMANN
Laboratorium für Angewandte
Physik Gerhard-Mercator-
Universität
Duisburg
D-47048 Duisburg Germany
kleemann@uni-duisburg.de

Mihai POPESCU
National Institute for Materials
Physics Atomistilor
Str. 105 bis, POB. MG. 7
77125 Bucharest-Magurele
Romania
mpopescu@infim.ro

Mickael PORTNOI
School of Physics, University
of Exeter
Stocker Road
EX4 4QL Exeter UK
m.e.portnoi@exeter.ac.uk

Aline ROUGIER
Laboratoire de Réactivité et de
Chimie
Des Solides University of Picardy
33 rue Saint Leu
80039 Amiens Cedex 01 France
aline.rougier@u-picardie.fr

Ismael SAADOUNE
Université Cadi Ayyad FST ECME
Avenue A. Khattabi BP 549
Marrakech Morocco
saadoune@fstg-marrakech.ac.ma

Anaïs SENE
Laboratory of Condensed Matter
Physics, University of Picardy
33, rue Saint Leu
80039 Amiens Cedex 01 France
anais.sene@gmail.com

Anatoli SIDORENKO
Institute of Electronic Engineering
and Industrial Technologies ASM
Academic 3/3
MD 2028 Kishinev Moldova
anatoli.sidorenko@int.fzk.de

Salaheddine SAYOURI
LPTA Département de Physique
Faculté des Sciences-DM BP 1796
Fès-Atlas Morocco
ssayouri@yahoo.com

CONTENTS

Part I

MATERIALS, TRANSMISSION, TREATMENT AND STORAGE OF INFORMATION

MULTIFERROIC AND MAGNETOELECTRIC MATERIALS FOR SPINTRONICS

WOLFGANG KLEEMANN* AND PAVEL BORISOV
Angewandte Physik, Universität Duisburg-Essen, 47048 Duisburg, Germany

Abstract. Purely voltage controlled multiferroic and magnetoelectric materials promise to fulfil the requirements of minimal heat dissipation in three-dimensional spintronic architectures. Actually, most promising concepts aim at electrically controlling the giant or tunneling magnetoresistance of magnetic multilayer stacks. They are based either on magnetoelectric and/or multiferroic tunnel barriers or on exchange-coupled magnetoelectric pinning layers. The physical principles and the state of the art of these concepts will be discussed for devices involving $La_{0.1}Bi_{0.9}MnO_3$ tunneling barriers and Cr_2O_3 pinning layers, respectively.

Keywords: Spintronics, giant magnetoresistance, tunneling magnetoresistance, magnetic data storage, MRAM, exchange bias, magnetoelectric effect, multiferroics.

1. Introduction

The next-but-one generation of microelectronic devices (45 nm base length) will have to experience three-dimensional (3D) packaging to a large extent. The heat release of the central processing units in computers will become a serious challenge,[1] in particular after the introduction of novel 3D storage devices with current controlled read-write protocols like magnetic random access memories, MRAM.[2] The need of purely electric field-controlled devices with a minimum amount of Joule heating becomes of paramount interest.[1] In this situation the revived interest in magnetoelectric (*ME*) materials[3] finds its utmost pertinence. ME response signifies magnetic control via electric fields and vice versa.[4] It promises to maximize in multiferroic materials close to their ordering temperatures.[5] That is why *ME* multiferroics, i.e. materials undergoing phase transitions

*To whom correspondence should be addressed: Wolfgang Kleemann, Angewandte Physik, Universität Duisburg-Essen, D-47048 Duisburg, Germany. Email: wolfgang.kleemann@uni-due.de

I. A. Luk'yanchuk and D. Mezzane (eds.), *Smart Materials for Energy, Communications and Security.*

into long range magnetic and electric order close or above room temperature are declared the 'holy grail' in materials science,[6] which is still far from being reached.

2. Spintronics

The electronic spin has been 'found' to be functional in electronics only after the discovery of the 'giant magneto-resistance (GMR)' in exchange-coupled *FM* thin films.[7,8] Because of its eminent importance in information technology it was awarded the physics Nobel prize in 2007 to Albert Fert[9] and Peter Grünberg.[10] Its introduction into magnetic hard disk technology has warranted Moore's law to stay valid in this field of technology since 1995. Both magnetic read heads and MRAM cells (Fig. 1) are based on the same principle. Two *FM* thin films (e.g. Co) being separated by a thin non-*FM* conducting layer (e.g. Cu) are allowed to have either parallel or antiparallel directions of their magnetization. A probing current along the triple layer stack will then experience a low and a high resistance, $R_{\uparrow\uparrow} < R_{\uparrow\downarrow}$, respectively. In order to well-define these two switching states, denoted as '0' and '1', respectively, the magnetization of one layer is 'pinned' by an *AF* overlayer. Here the so-called exchange bias effect due to interlayer exchange-induced unidirectional anisotropy[11] becomes functional. The other ('free') *FM* layer is soft magnetic and reacts by spin reversal due to external magnetic fields being as low as 1–3 mT.

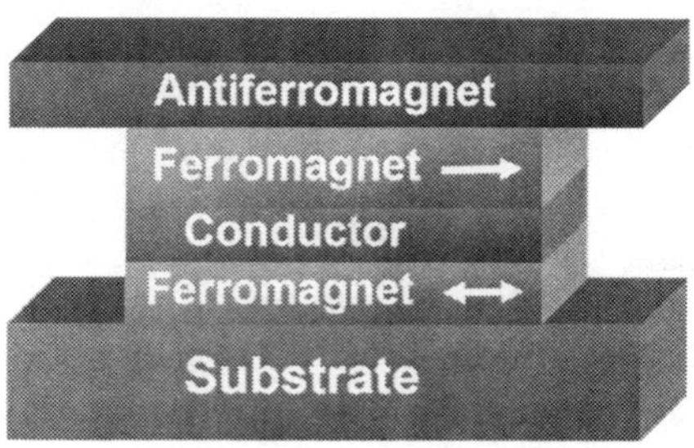

Figure 1. Schematics of a Magnetic Random Access Memory (MRAM) cell.

3. Multiferroics and magnetoelectrics

The world of magnetically and electrically polarizable materials is schematically shown in Fig. 2.[12] The diagram shows the location of the *multiferroic* materials within a small overlap zone of (anti)ferromagnetic (*AF* and *FM*, respectively) and ferroelectric (*FE*) materials. They are conventionally defined as materials revealing two (or more) magnetic and electric order parameters, e.g. magnetization,

M*,** and electric polarization, ***P. On the other hand, *ME* materials are conventionally supposed to reveal a *linear ME* effect allowing for cross-linked electric and magnetic field-induced contributions, $\Delta\boldsymbol{M} = \mu_0\alpha_{EM}\boldsymbol{E}$ and $\Delta\boldsymbol{P} = \alpha_{ME}\boldsymbol{H}$, where $\alpha_{EM} = \alpha_{ME}$ in single phase magnetoelectrics by definition. From a more general point of view also magnetoelectrics with higher order coupling may be considered.[3] *ME* materials may have magnetic or electric or even no long-range order at all as indicated by the black-hatched circle in Fig. 2. The overlap of the *ME* circle with the multiferroic region designates the material class of '*ME multiferroics*' (central black triangle). Both in Fig. 2 and in nature this region is very small and therefore sometimes referred to as '*the holy grail*'.[6] Accepted members of this family are scarce. Most of them are typical low-temperature materials like the NiCl boracites $Ni_3B_7O_{13}Cl$ (*FM* and *FE* Curie temperatures $T_{CM} = 29$ K and $T_{CE} = 398$ K, respectively[13]). One of the rare high temperature multiferroic systems, $BiFeO_3$ (*AF* $T_N = 625$ K[14] and $T_{CE} = 1{,}083$ K[15]), unfortunately offers only comparably small linear *ME* coupling either in extremely strong magnetic fields[16] or in strained thin films.[17] In another promising multiferroic material, $BiMnFeO_3$ ($T_{CM} = 100$ K[18] and $T_{CE} = 440$ K[19]) linear *ME* coupling is forbidden at all by symmetry. This case contrasts with the *AF* ($T_N = 308$ K[20]), but non-polar hexagonal crystal Cr_2O_3, the first-ever investigated[21] and theoretically well-understood[4,22] *ME* material.

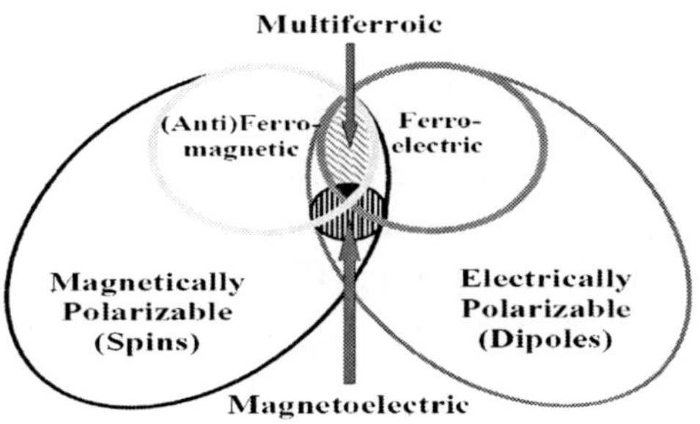

Figure 2

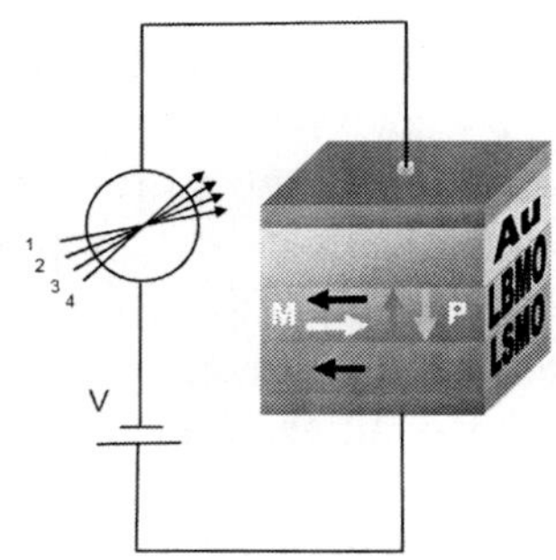

Figure 3

Figure 2. Magnetic and electric polarizable materials and their subsets of ordered, multiferroic and magnetoelectric materials.[12]

Figure 3. Schematics of a 4-bit tunneling magneto- (TMR) and electroresistance (TER) scheme of a trilayer device LSMO/LBMO/Au, where the TMR is low (high) for parallel (antiparallel) LSMO and LBMO magnetizations ***M***, and the TER is low (high) for up and down directed polarization ***P*** in the LBMO layer, respectively.[23] Four different tunneling currents, 1–4, schematically denote the 4-bit states.

4. Multiferroic 4-bit data storage

The prominent multiferroic material $BiMnO_3$ has recently been proposed to become the heart of future spintronic spin-filtering devices, which have the capacity to establish a novel 4-bit logic.[23] Actually, the solid solution $La_{0.1}Bi_{0.9}MnO_3$ (LBMO, $T_{CM} = 90$ K, $T_{CE} > 400$K) has been intercalated as a thin film of 2 nm thickness between layers of *FM* half-metallic $La_{0.9}Sr_{0.1}MnO_3$ (LSMO; $T_{CM} = 250$ K) and metallic Au, respectively. As depicted schematically in Fig. 3 tunneling currents of spin polarized electrons are then filtered very effectively in dependence on the directions of either the magnetization ***M*** or the polarization ***P***. A specific merit of this device is the virtual independence of the magnetically and electrically switchable resistivities thanks to the *absence* of any *ME* coupling. Unfortunately its use as a 4-bit memory is limited due to the low *FM* Curie temperature, $T_{CM} < 100$ K, but future practical applications are envisaged nevertheless.

5. Magnetoelectric field controlled data storage

The use of antiferromagnets in spintronics lies in their capacity to pin the magnetization of an attached *FM* layer via the 'exchange bias' effect.[11,24] In a first attempt we have proposed to introduce this effect into spin valve technology as depicted in Fig. 4.[25] Two *FM* layers with perpendicular magnetic anisotropy ('*FM*1' and '*FM*2'; magnetization arrows up or down) are intercalated by a thin *ME* tunneling layer ('*ME*'). Owing to their different thicknesses they have different coercive fields under magnetic switching, Hence, they show an enhanced tunneling magneto-resistance *R* in an intermediate field regime, where the magnetizations are antiparallel (arrows). By mutually shifting the hysteresis curves of both layers via a magnetic extra moment induced by an external voltage in the single-domained *AF* layer '*ME*' (up and down arrows, respectively), the corresponding *R* vs. *H* curves will be shifted to lower and higher fields, respectively. A suitably designed trilayer will then be capable of switching the magnetization of *FM II* by an external voltage, but in the absence of any magnetic field from 'spin up' to 'spin down'. This causes the resistivity to change from 'low' ('0') to 'high'('1').

We have tested the *ME* hysteresis shift of a perpendicularly anisotropic *FM* multilayer of (Co 0.3 nm/Pt 1.2 nm)$_3$ and found it disappointingly small within the exchange bias system Cr_2O_3(0001)/Pt(0.7 nm)/(Co/Pt)$_3$/Pt 3.1 nm.[26] After single-domaining the *AF* Cr_2O_3 substrate by '*ME* cooling'[3] in antiparallel freezing fields $E_{fr} = -300$ kV/m and $\mu_0 H_{fr} = 0.5$ T from room temperature to 250 K, the hysteresis was observed to shift by only 4.5 mT, which is much too small for practical purposes. However, the idea sketched in Fig. 4 is still acknowledged to be one of the most promising for *ME* spintronics. It is world-wide pursued by various research groups in attempts to find the ultimate material combination for this principle, e.g. on the basis of electric domain switching in exchange biasing $BiFeO_3$,[27] but still a functional device has not yet been presented.

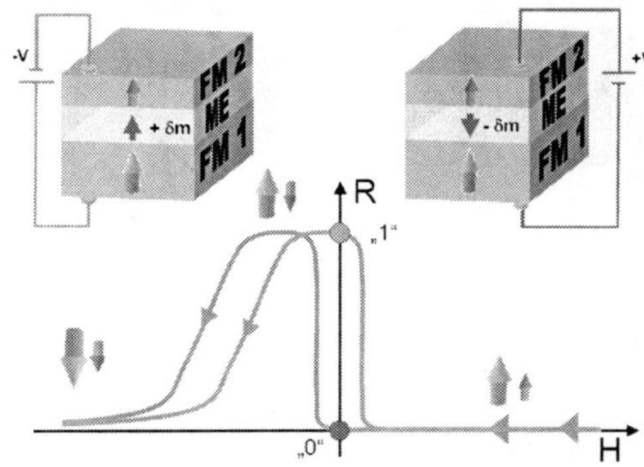

Figure 4. Magnetoelectric random access memory cell[25] in its two switching stages and their associated resistance characteristics, *R* vs. *H*. At $H = 0$ the 2-bit resistance values are marked by circles and denoted by '1' and '0', respectively.

After our discovery of *magnetoelectric switching* of the exchange bias in magnetoelectric *AF*-*FM* heterostructures[28] we have proposed yet an alternative concept. Figure 5a shows the hysteresis curves of an exchange bias system Cr_2O_3(0001)/Pt(0.5 nm)/Co(0.35 nm)/Pt(3 nm) after *ME* cooling from 350 to 297 K in parallel (right) and antiparallel (left) freezing fields $\mu_0 H_{fr} = 0.3$ T and $E_{fr} = \pm 300$ kV/m.[29] They are completely separated from each other and shifted to the right and to the left of the ordinate axis $H = 0$, respectively. Different spin structures of the field-induced different *AF* single domains (axial spin order $\leftarrow \rightarrow \leftarrow \rightarrow$ and $\rightarrow \leftarrow \rightarrow \leftarrow$, respectively[20]) are at the origin of this surprising behavior. In consequence, at the interface to the *FM* the free energy of the system is differently minimized by the competing exchange, Zeeman and *ME* energy contributions upon cooling to below T_N.[28] In particular, the stable zero-field configurations of the *FM* magnetization, $M < 0$ and $M > 0$ (arrows in Fig. 5a), are switched as well. Note that this kind of magnetization switching needs only switching of an electric field across the insulating *ME* component, hence, no energy consuming stationary current.

The latter argument promises a novel data storage concept based on so-called *ME* random access memory (*ME*RAM, Fig. 5b) and related *ME*-XOR logic cells.[30,31] As in conventional MRAM cells, the logic states '0' and '1' are defined by the magnetoresistances 'low' and 'high' of *FM* bilayers with parallel

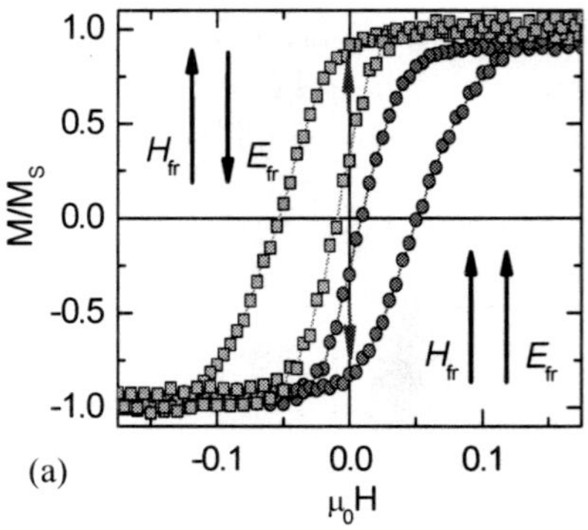

(a)

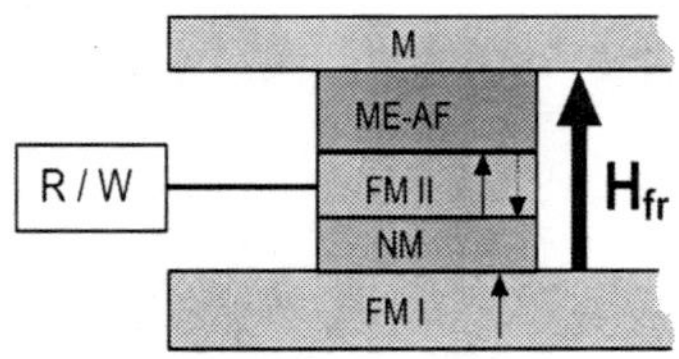

(b)

Figure 5. (a) Exchange bias switched hysteresis loops and zero-field magnetization (blue arrows) of a heterostructure Cr_2O_3(0001)/Pt(0.5 nm)/Co(0.35 nm)/Pt(3 nm) under different *ME* field cooling protocols (black arrows; see text). (b) *ME*-RAM cell using a *ME*-switched free layer *FM II*, a pinned hard magnetic layer *FM I* with stray field H_{fr}, a metallic cap electrode *M*, and a non-magnetic conducting or insulating intermediate layer *NM* for sensing the GMR or the TMR, respectively (see text).

and antiparallel magnetization directions, respectively. As shown in Fig. 5b the *ME*RAM cell uses an electrically switched free layer *FM II* in conjunction with a hard magnetic thick layer *FM I*, which simultaneously provides a magnetic stray field, H_{fr}, being active during *AF* domain freezing upon *ME* cooling from above T_N. *M* is a metallic cap electrode and *NM* is a non-magnetic intermediate layer. Its electric nature, either being conducting or insulating, decides if the GMR or the TMR of the spin valve *FM I/NM/FM II* is sensed via the read/write line *R/W* and the bottom electrode, *FM I.*

Attempts are presently underway[32] to downscale the lateral and vertical dimensions of the *ME*RAM cell to the present microelectronic standard (90 nm and below). Apart from adapting the *ME* switching times to below 100 ns, another drawback of the *ME*RAM design will be overcome. Switching will be possible without intermediate heating to above T_N provided the switching fields will surpass the critical value $|H_{fr}E_{fr}|_{cr}(300\text{ K}) \approx 50$ kOe kV/cm.[33] In a magnetic freezing field of $|\mu_0 H_{fr}| = 10$mT, a freezing voltage $|V_{fr}| = 2$ V will then suffice to switch the *AF* state of a Cr_2O_3 layer of thickness 40 nm. Presently the main challenge is still the growth of properly insulating (0001) oriented thin films of Cr_2O_3, which are grown in a low pressure oxygen atmosphere e.g. by electron beam evaporation.[34]

6. Future developments

ME controlled spintronic components like *ME*RAM and *ME*XOR are promising to overcome the Joule heating challenge of future generations of MRAM, since they avoid stationary currents for producing magnetic switching fields. Surprisingly, up to now the archetypical material Cr_2O_3[21] has been most promising for realistic applications close to room temperature. However, for widespread applications alternative *ME* materials with Néel temperatures around 400 K are needed. There is some hope that solid solutions like $(Cr_{1-x}Fe_x)_2O_3$ might fill this gap provided that the *ME* properties of Cr_2O_3 do not get lost.

The direct use of the *ME* effect as envisaged e.g. in Fig. 4 is still pursued in many research projects, however, the benefit of multiferroic magnetoelectrics for spintronics is still beyond reach. Extremely large *ME* effects have been found in incommensurate magnets like $TbMnO_3$[35] and $Ni_3V_2O_8$,[36] however, their transition temperatures are well below room temperature thus restricting device applications yet to visions.[37] The promising properties of non-*ME* multiferroics like $BiMnO_3$ in spin filtering devices[23] also still have to overcome the break-even of the room temperature barrier. Nevertheless, in view of the present worldwide feverish research activities in this field surprises are not excluded at all ('The best is still to come!'[38]).

ACKNOWLEDGMENTS

Financial support by the *Deutsche Forschungsgemeinschaft* via *Sonderforschungsbereich 491* ('Magnetic Heterolayers: Structure and electronic transport') is gratefully acknowledged.

References

1. A. Vukovic, Power density challenges of next generation telecommunication networks, *Electronics Cooling* **9**:34–40 (2003).
2. A. Moser, K. Takano, D.T. Margulies, M. Albrecht, Y. Sonobe, Y. Ikeda, S. Sun, and E.E. Fullerton, Magnetic recording advancing into the future, *J. Phys.* D **35**:R157–R167 (2002).
3. M. Fiebig, Revival of the magnetoelectric effect, *J. Phys. D* **38**:R123–R152 (2005).
4. L.D. Landau and E.M. Lifshitz, *Electrodynamics of Continuous Media* (Pergamon, Oxford, 1960).
5. W.F. Brown, R.M. Hornreich, and S. Shtrikman, Upper bound on the magnetoelectric susceptibility, *Phys. Rev.* **168**:574–577 (1968).

6. A.M. Kadomtseva, Yu.F. Popov, A.P. Pyatakov, G.P. Vorob'ev, A.K. Zvezdin, and D. Viehland, Phase transitions in multiferroic $BiFeO_3$ crystals, thin-layers, and ceramics: enduring potential for a single phase, room-temperature magnetoelectric 'holy grail', *Phase Trans.* **79**: 1019–1042 (2006).
7. P. Grünberg, R. Schreiber, Y. Pang, M.B. Brodsky, and H. Sowers, Layered magnetic structures: evidence for antiferromagnetic coupling of Fe layers across Cr interlayers, *Phys. Rev. Lett.* **57**:2442–2445 (1986).
8. M.N. Baibich, J.M. Broto, A. Fert, F. Nguyen Van Dau, F. Petroff, P. Eitenne, G. Creuzet, A. Friederich, and J. Chazelas, Giant magnetoresistance of (001)Fe/(001)Cr magnetic superlattices, *Phys. Rev. Lett.* **61**:2472–2475 (1988).
9. http://nobelprize.org/nobel_prizes/physics/laureates/2007/fert-lecture.html
10. http://nobelprize.org/nobel_prizes/physics/laureates/2007/grunberg-lecture.html
11. W.H. Meiklejohn and C.P. Bean, New Magnetic Anisotropy, *Phys. Rev.* **102**:1413–1414 (1956).
12. W. Eerenstein, N.D. Mathur, and J.F. Scott, Multiferroic and magnetoelectric materials, *Nature* **442**:759–765 (2006).
13. J.-P. Rivera, H. Schmid, J.M. Moret, and H. Bill, Measurement of the linear magnetoelectric effect in nickel-chlorine boracite $Ni_3B_7O_{13}Cl$, *Int. J. of Magnetism*, **6**:221–220 (1974).
14. J.R. Teague, R. Gerson, and W.J. James, Dielectric hysteresis in single crystal $BiFeO_3$, *Solid State Commun.* **8**:1073–1074 (1970).
15. P. Fischer, M. Polomka, I. Sosnowska, and M. Szymanski, Temperature dependence of the crystal and magnetic structures of $BiFeO_3$, *J. Phys.* C **13**:1931–1940 (1980).
16. Yu.F. Popov, A.K. Zvezdin, A.M. Kadomtseva, V.A. Murashov, G.P. Vorob'ev, Linear magnetoelectric effect and phase transitions in bismuth ferrite, $BiFeO_3$, *JETP Lett.* **57**:69–73 (1993).
17. J. Wang, J.B. Neaton, H. Zheng, V. Nagarajan, S.B. Ogale, B. Liu, D. Viehland, V. Vaithyanathan, D.G. Schlom, U.V.W aghmare, N.A. Spaldin, K.M. Rabe, M. Wuttig, and R. Ramesh, Epitaxial $BiFeO_3$ multiferroic thin film heterostructures, *Science* **299**:1719–1722 (2003).
18. C.H. Yang, T.Y. Koo, S.H. Lee, C. Song, K.B. Lee, and Y.H. Jeong, Orbital ordering and enhanced magnetic frustration of strained $BiMnO_3$ thin films, *Europhys. Lett.* **74**:348–354 (2006).
19. T. Kimura, S. Kawamoto, I. Yamada, M. Azuma, M. Takano, and Y. Tokura, Magnetocapacitance effect in multiferroic $BiMnO_3$, *Phys. Rev.* B **67**:180401(R) (2003).
20. G.T. Rado and V.J. Folen, Observation of the magnetically induced magnetoelectric effect and evidence for antiferromagnetic domains, *Phys. Rev. Lett.* **7**:310–311 (1961).
21. D.N. Astrov, The magnetoelectric effect in antiferromagnets, *JETP* **11**:708–709 (1960).
22. I.E. Dzialoshinskii, On the magnetoelectrical effect in antiferromagnets, *Sov. Phys. JETP* **10**:628–629 (1959).
23. M. Gajek, M. Bibes, S. Fusil, K. Bouzehouane, J. Fontcuberta, A. Barthélémy, and A. Fert, Tunnel junctions with multiferroic barriers, *Nature Mater.* **6**:296–302 (2007).
24. J. Noguès and I.K. Schuller, Exchange bias, *J. Magn. Magn. Mater.* **192**:203–232 (1999).
25. Ch. Binek, A. Hochstrat, X. Chen, P. Borisov, and W. Kleemann, Electrically controlled exchange bias for spintronic applications, *J. Appl. Phys.* **97**:10C514 (2005); C. Binek and B. Doudin, Magnetoelectronics with magnetoelectrics, *J. Phys.: Cond. Matter* **17**:L39–L44 (2005).
26. Ch. Binek, P. Borisov, X. Chen, A. Hochstrat, S. Sahoo, and W. Kleemann, Perpendicular Exchange bias and its control by magnetic, stress and electric fields, *Eur. Phys. J.* B**43**:197–201 (2005).

27. H. Béa, M. Bibes, F. Ott, B. Dupé, X.-H. Zhu, S. Petit, S. Fusil, C. Deranlot, K. Bouzehouane, and A. Barthélémy, Mechanisms of exchange bias with multiferroic $BiFeO_3$ epitaxial thin films, *Phys. Rev. Lett.* **100**:017204 (2008).
28. P. Borisov, A. Hochstrat, X. Chen, W. Kleemann, and Ch. Binek, Magnetoelectric switching of exchange bias, *Phys. Rev. Lett.* **94**:117203 (2005).
29. P. Borisov, A. Hochstrat, X. Chen, and W. Kleemann, Multiferroically composed exchange bias systems, *Phase Trans.* **79**:1123–1134 (2006).
30. X. Chen, A. Hochstrat, P. Borisov, and W. Kleemann, Magnetoelectric exchange bias systems in spintronics, *Appl. Phys. Lett.* **89**:202508 (2006)
31. International patent application PCT/EP2006/002892.
32. P. Borisov, A. Hochstrat, V.V. Shvartsman, and W. Kleemann, Magnetoelectric Cr_2O_3 for spintronic applications, *Integrated Ferroelectrics* (in print).
33. T.J. Martin and J.C. Anderson, Antiferromagnetic domain switching in Cr_2O_3, *IEEE Trans. Magn.* **2**:446–449 (1966).
34. P. Borisov, A. Hochstrat, V.V. Shvartsman, W. Kleemann, T. Eimüller, and A. Fraile-Rodriguez, Thin Cr_2O_3 films for magnetoelectric data storage deposited by reactive e-beam evaporation, *Ferroelectrics* (in print).
35. N. Hur, S. Park, P.A. Sharma, J.S. Ahn, S. Guha, and S.-W. Cheong, Electric polarization reversal and memory in a multiferroic material induced by magnetic fields, *Nature* **429**: 392–395 (2004).
36. G. Lawes, A.B. Harris, T. Kimura, N. Rogado, R.J. Cava, A. Aharony, O. Entin-Wohlman, T. Yildirim, M. Kenzelmann, C. Broholm, and A.P. Ramirez, Magnetically driven ferroelectric order in $Ni_3V_2O_8$, *Phys. Rev. Lett.* **95**:087205 (2005).
37. A.B. Harris and G. Lawes, Ferroelectricity in incommensurate magnets, In: *The Handbook of Magnetism and Advanced Magnetic Materials*, Vol. 4 – *Novel Materials*, H. Kronmüller, S.P.S. Parkin (Eds.) (Wiley, New York, 2007), pp. 2428–2460.
38. Editorial, *Nature Mater.* **6**:1 (2007).

LARGE AREA NB NANOLAYERS WITH ADVANCED SUPERCONDUCTING PROPERTIES AS A BASE FOR SUPERCONDUCTING SPINTRONICS

VLADIMIR ILICH ZDRAVKOV*, ROMAN ANDREEVICH MORARI AND ANATOLI SERGEEVICH SIDORENKO
Institute of Electronic Engineering and Industrial Technologies ASM, MD2028 Kishinev, Moldova

Abstract. High quality Nb nanoscale films of large area (7 × 80 mm^2) and constant thickness were deposited by DC-magnetron sputtering in commercial "Leybold Z400" vacuum system. Homogeneity and proper thickness of the Nb layer provided by the target-holder movement during the DC sputtering using specially constructed arrangement based on controllable DC motor with a gear. Rutherford backscattering spectrometry was used for precise thickness measurements. The increase of superconducting critical temperature (>1.5 K for films with comparable thickness) and of superconducting coherent length (up to 30–35%) in comparison with films prepared by common technique is reported.

Keywords: Superconductivity, spintronics, thin films, magnetron sputtering, nano-size Nb layers.

1. Introduction

Niobium superconducting thin films (d_{Nb} = 5 ÷ 15 nm) and multilayers are the base of various superconducting electronic devices. For example, single Nb films are used for different applications such as mixers[1] and detectors[2] of electromagnetic radiation. Nb film with a large value of the electron mean-free path is necessary to reach the intermediate frequency bandwidth larger than 10 GHz for the diffusion-cooled hot-electron bolometer mixer.[3]

The design of spintronic superconducting devices based on the hybrids is the object of intense investigations for last decades.[4] The investigation of proximity

*To whom correspondence should be sent: V. I. Zdravkov, Institute of Electronic Engineering and Industrial Technologies ASM, MD2028 Kishinev, Moldova

effect in Superconductor/Normal metal (S/N) and Superconductor/Ferromagnet (S/F) nano-size layered structures require technological approach yields high quality superconducting films with constant thicknesses and advanced superconducting properties. The common material for superconducting electronics, Nb, has as an advantage relatively high temperature of superconducting transition, T_c 9.2 K. Unfortunately, Nb has enhanced getter capability whereas adsorbed gases intensively affect the superconducting properties especially for nanoscale films. On the other hand, S/F superconducting structures with Nb nano-layers demonstrate main interesting physical phenomena based on space oscillation of the order parameter due to proximity effect.[5–7] Reliable producing of the most applicable range of superconducting Nb layer thicknesses, 5–15 nm (keeping $d_{Nb} \approx \xi_{Sc}$ – superconducting coherence length) with reproducible high quality and T_c close to the bulk value is a challenge for constructing superconducting devices based on proximity effect.[4,5] The efficient technological approach for fabrication of the set of the samples with equal superconducting layers was demonstrated in.[8] The disadvantage of this approach is relatively low T_c of Nb layer (6.4 K for single Nb layer with $d_{Nb} = 31$ nm).[8]

The increase of T_c for deposited Nb films is possible under deeper vacuum conditions inside the system, by optimizing the substrate temperature or deposition rate.[9,10] The protection cup and buffer layers from neutral material (Al_2O_3 or Si for example) could be used to avoid impurity penetration in Nb layer. From the other hand, the experimental setup of vacuum system and the necessity to avoid the interdiffusion with other hybrid components often restrict Nb layer's quality improvement.

2. Methodology

The samples were prepared by magnetron sputtering on a flame-polished glass substrates or on commercial (111) silicon substrates kept at room temperature. The base pressure in the "Leybold Z400" vacuum system was about 2×10^{-6} mbar; pure argon (99.999%, "Messer Griesheim") at pressures of 8×10^{-3} mbar was used as a sputter gas. The targets, 75 mm in diameter, from Nb, Si were pre-sputtered a few times for 3–5 min to remove contaminations from the targets surface as well as to reduce the residual gas pressure in the chamber during the pre-sputtering of Nb. As the next step we deposited silicon buffer layer with RF magnetron to obtain clean interface for the subsequent niobium layer. To provide homogeneity and proper thickness of the Nb layer the target-holder was moved during the DC sputtering using specially constructed arrangement based on controllable DC motor with a gear. With this setup we were able to achieve the average growth rate of the Nb layer 1.3 nm/s (the steady-state deposition rate would be about 4 nm/s) which is close to the optimal value for magnetron deposition technique.[9] To prevent the deposited samples from degradation in ambient atmosphere as well as from influence of contaminated Si surface we protected them with ~5 nm thick silicon cap and buffer layers.

Figure 1. The design of fabricated nano-sized layered structure based on Nb-film.

Figure 1 demonstrates the design of fabricated nanosize layered structure. The resulting long samples (length = 80 mm, width = 7 mm) were cut equidistantly and subsequently perpendicular to the long side of the sample onto 2.5 mm wide strips for resistance measurements using a diamond cutter. Thus each long sample is the origin of the series of mentioned above strips. We identified the samples by the id of the long sample (S15, S16 …) and the number of each 2.5 mm strip subsequently produced by the cutting of the long sample. Aluminum wires 50 μm in diameter were further attached to the strip by an ultrasonic bonder for four-probe resistance measurements. The standard DC four-probe method was used, applying a sensing current of 10 μA with alternating polarity. The magnetic measurements were performed in "Oxford Instruments" superconducting magnet system. The Rutherford backscattering spectrometry (RBS) gave us the possibility to determine the absolute thickness of the layers at the level of 1 nm with an accuracy of 0.03 nm. Details of the RBS measurements are described in.[8] The Auger-electron analysis was performed on PHI 680 Auger Nanoprobe system using small-angle 250 V Ar^+ ion beam sputtering, considering 1 nm/min rate.

3. Results and discussion

The thickness measurements and the elementary analysis were performed by RBS and Auger-electron spectroscopy after T_c and H_{c2} (T) detection. We did not detect impurities inside the Nb layer in the range of RBS and Auger-electron spectroscopy accuracy (1.5 ÷ 2 at%). The Auger data presented in Fig. 2 demonstrate also the effective protection of the Nb layer by Si buffer and cap layers against contamination of Nb from the Si-substrate and from outer surface of the structure.

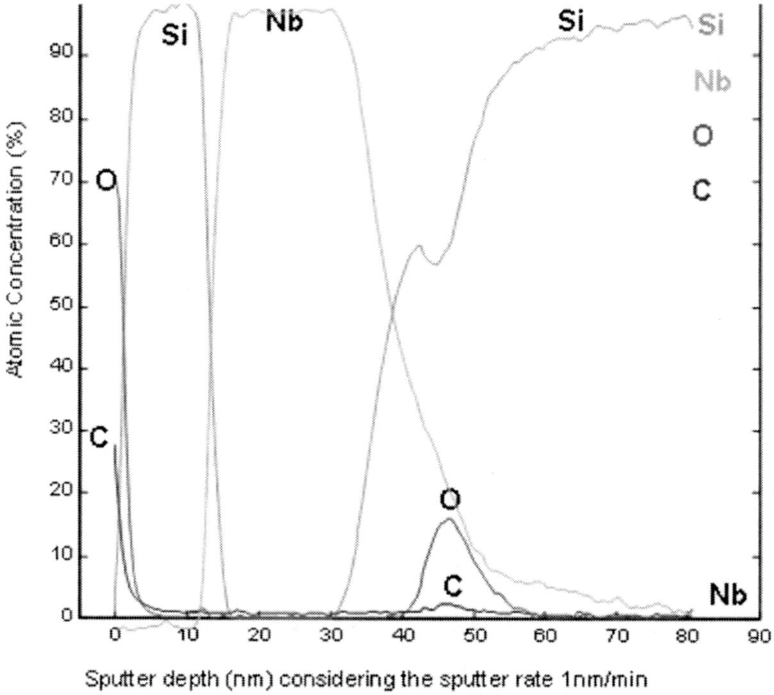

Figure 2. The data of Auger-analysis of the sample with 11 nm thick Nb layer. (Nb thickness was determined by RBS.)

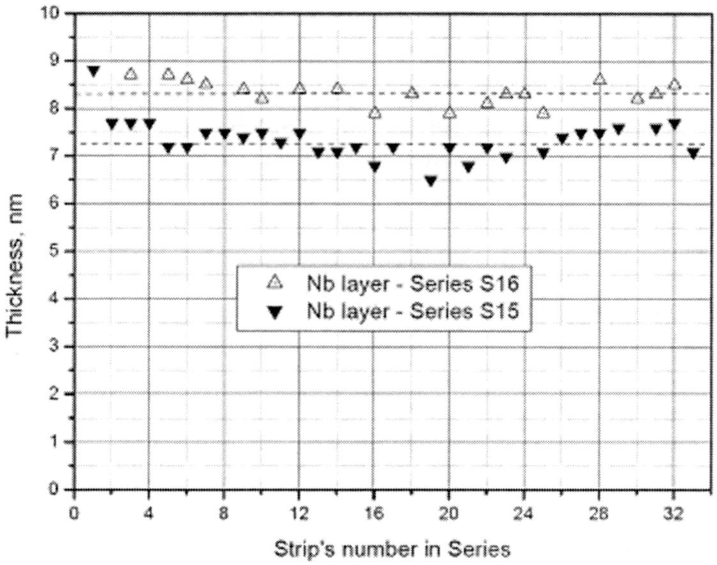

Figure 3. Thicknesses of strips cut from samples S15 and S16 derived from RBS data. Strip's number in series correspond the cutting sequence.

Figure 3 presents Nb layer thickness determination for the series S15 and S16. The dashed lines in Fig. 3 are the averaged values of Nb films thicknesses, 7.3 and 8.3 nm, for each long sample (S15 and S16). The accuracy of RBS method for such range of thicknesses is within 0.4–0.5 nm. The numbers within series correspond to cut sequence of strips for the long sample. Thickness deviation did not exceed 0.5 nm from strip to strip for each series (except one maximum and one minimum value) and it is still in the accuracy of the measurements range. Anyway, this value of thickness deviation corresponding to 0.1–0.15 K difference in T_c from strip to strip is quite suitable for proximity effect investigation and for microelectronic devices implementation.

The residual resistivities (ρ_N) of Nb films were 16.5; 12.6; 8.9 μΩ cm for $d_{Nb} \approx 7; 9; 13$ nm respectively. The critical temperatures of the samples with Nb thicknesses from about 5.5 up to 100 nm are presented in Fig. 4. The critical temperature for films with thickness 5.5–10 nm (5.6–7.5 K) are close to ones detected for the best of Nb thin films of the same thickness range (see for

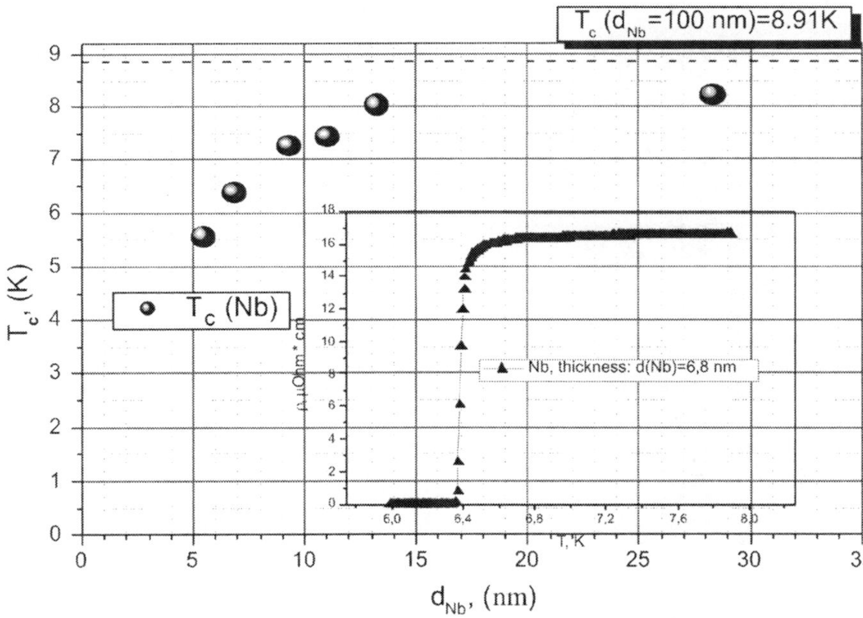

Figure 4. The critical temperature of the samples with Nb layer thickness from 5.5 nm up to 100 nm. *Inset:* Typical superconducting transition for Nb film. The critical temperature for Nb film with thickness 6.8 nm is 6.37 K (criteria $0.5R_N$).

example[10,11]), but our films were formed in much simpler equipment and relatively rough vacuum conditions (~2×10^{-6} mbar). It is worth to mention the other important advantage of presented technique: the possibility to form the large area film in single deposition run and to combine it with other layers without the vacuum interruption.

The shape of typical superconducting transition is presented in the inset of Fig. 4. The critical temperature (0.5 R_N criteria) of the sample with Nb thickness of 6.8 nm is 6.37 K. The width of transition (criteria 0.9 R_N-0.1 R_N, R_N is the resistance in normal state before transition) is 0.05 K. These characteristic values have advantage in comparison with Nb films with similar T_c but with the thickness of 31 nm deposited by common technique.[8]

We investigated also the superconducting properties of the samples in magnetic field to estimate the ξ_{Sc} and to derive d_S/ξ_{Sc} ratio – some of the main values which determine the proximity effect and spin-valve effect in S/F structures.[5,6]

Figure 5 demonstrates the temperature dependence of the upper critical magnetic field applied parallel to the Nb layer for the same sample as for Fig. 2. Extrapolation of the parallel critical magnetic field to 0 K yields the values $H^{\|}_{c2}(0) = 13 - 13.5$ T. The Ginsburg-Landau superconducting coherence length at 0 K $\xi_{GL}(0)$ was calculated using the formula from[12]:

$$\xi_{GL}(0) = \frac{\sqrt{12}\Phi_0}{2\pi \times d_s H^{\|}_{c2}(0)} \tag{1}$$

where Φ_0 is the flux quantum, d_S is the thickness of superconducting layer.

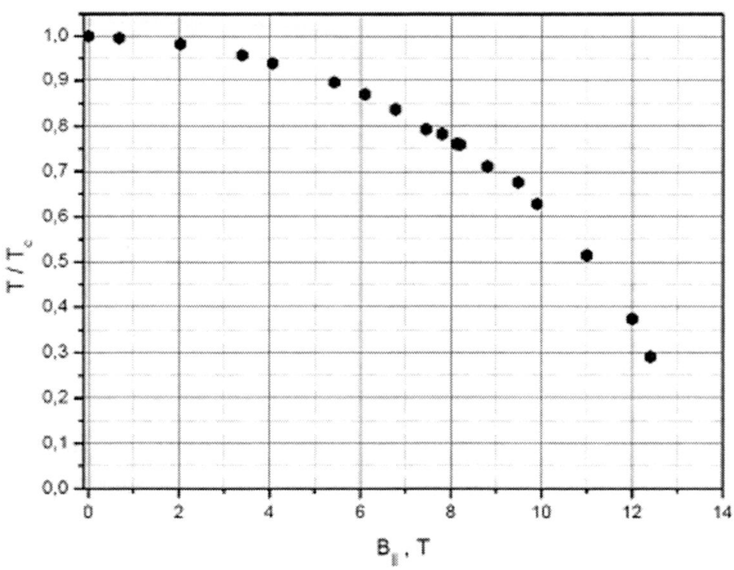

Figure 5. Dependence of the upper critical magnetic field applied parallel to the Nb film plane from the temperature, normalized by T_C.

Using $d_S \approx 6.8$ nm and $H_{c2}^{\|}(0) \approx 13.5$ T one can derive $\xi_{GL}(0) = 12.4$ nm. The significant increase of this value in comparison with $\xi_{GL}(0) = 9.3$ nm for Nb films obtained by common technique[8] is the evidence of the higher quality of the films, prepared by developed technique.

Taking in mind that $\xi_S = (2/\pi)\xi_{GL}(0)$[8] one can derive for the sample the ratio $d_S/\xi_{Sc} \approx 0.9$ being quite promising for spintronic application based on proximity effect in S/F multilayers.[5]

4. Conclusions

The presented technological approach yields significant improvement of superconducting properties of large area nanoscale Nb films in comparison with common methods based on DC-magnetron deposition. The thickness deviation of Nb layer along the sample with length of 80 mm does not exceed 5–6% for all strips and not more than 3–4% from strip to strip. This value is in general within the accuracy of the thickness determination method (RBS). The increase of superconducting critical temperature (>1.5 K for films with comparable thickness) and superconducting coherent length (30–35%) opens the possibility of proximity effect investigation and spintronic applications based on large area superconducting films with thicknesses of a few nanometers. The ratio $d_S/\xi_{Sc} \approx 0.9$ for the Nb films with thickness in the range of 5.5–10 nm, being achieved by our technique, is very promising for spintronic applications based on proximity effect in S/F multilayers.

ACKNOWLEDGEMENTS

The authors are grateful to Professor Carmine Attanasio, Jaan Aarts, Siegfried Horn, Renhard Tidecks, Valerii Ryazanov for long-term fruitful collaboration, to V. Oboznov for stimulating discussion, to S. Gsell for precise RBS analysis, to E. Nold for Auger analysis of the samples. This work was supported through the BMBF project MDA01/007, collaborative Grant RFBR-Moldova "Investigation of nanostructures superconductor/ferromagnet – base element of superconducting spintronics" and the collaborative grant "Belorussia-Moldova".

References

1. B. S. Karasik, M. C. Gaidis, W. R. McGrath, B. Bumble, and H. G. LeDuc, Appl. Phys. Lett. **71**, 1567 (1997).
2. A. Semenov, A. Engel, H.-W. Hübers, K. Il'in, and M. Siegel, SPIE's 47th Annual Meeting, Hawaii, August, 2002; Millimeter and Submillimeter Detectors for Astronomy, Proceedings of SPIE 4855, edited by T. G. Phillips and J. Zmuidzinas (California Institute of Technology, Pasadena, 2003), p. 249.
3. D. E. Prober, Appl. Phys. Lett. **62**, 2119 (1993).
4. A. I. Buzdin, Rev. Mod. Phys. 77, 935 (2005).
5. L. R. Tagirov, Physica C **307**, 145–163 (1998).
6. V. Zdravkov, A. Sidorenko, G. Obermeier, S. Gsell, M. Schreck, C. Müller, S. Horn, R. Tidecks, L. R. Tagirov, Phys. Rev. Lett. **97**, 057004 (2006).
7. V. V. Ryazanov, V. A. Oboznov, A. Yu. Rusanov, V. A. Veretennikov, A. A. Golubov, J. Aarts, Phys. Rev. Lett. **86**, 2427 (2001).
8. A. S. Sidorenko, V. I. Zdravkov, A. Prepelitsa, C. Helbig, Y. Luo, S. Gsell, M. Schreck, S. Klimm, S. Horn, L. R. Tagirov, and R. Tidecks, Annalen der Physik (Leipzig-Berlin) **12**, 37 (2003).
9. N. N. Iosad, T. M. Klapwijk, S. N. Polyakov, V. V. Roddatis, and E. K. Kov'ev, P. N. Dmitriev, IEEE Trans. Appl. Superconduct. **9**(2), 1720, June 1999.
10. Th. R. Lemberger, I. Hetel, J. W. Knepper, and F. Y. Yang, Phys. Rev. B **76**, 094515 (2007).
11. A. Sidorenko, C. Sürges, T. Trappmann, and H. v. Löneysen, Phys. Rev. B **53**, 11751 (1996).
12. Z. Radovi´c, L. Dobrosavljevi´c-Gruji´c, A. I. Buzdin, and J. R. Clem, Phys. Rev. B **38**, 2388 (1988).

AMORPHOUS CHALCOGENIDE MATERIALS WITH SMART MEMORY

MIHAI POPESCU*
National Institute R&D of Materials Physics, Department of Low Dimension Materials, Bucharest-Magurele, P.O. Box. MG. 7, Romania

Abstract. Intelligent machines using multi-level memory chips are at the ends of our fingers. The first prototypes of smart memories are under construction. The science behind the devices is growing up very rapidly. New materials with multi-level storage are discovered and tested. The applications of the smart memories are extensive and in the same time rewarding for the welfare of the human kind.

Keywords: Chalcogenide, phase-change materials, smart memory.

1. Introduction

The major sciences emerged in the last century are: Physical Science striving for understanding of the structure of matter through quantum mechanics; the Life Sciences that are trying to understand the structure of cells and the mechanisms of life through biology and genetics, and Information Science, that lays at the basis of communication through the development of computational means.

Microelectronics and its recent advancement, nanoelectronics, embrace all the major sciences and touch every aspect of the human life: food, energy, transportation, communication, health and exploration of new spaces.

With microelectronic devices we monitor the food safety, the pollution, we produce electricity, control the vehicles, transmit information, we help to cure or enhance the human body through artificial senses or biomaterial manipulation.

The progress is impressive. Nevertheless, the microelectronics is far to be able to imitate Nature in terms of integration density, functionality and performance. For example, a state-of-the-art low power Pentium processor consumes nearly

* To whom correspondence should be sent: Email: mpopescu@infim.ro

I. A. Luk'yanchuk and D. Mezzane (eds.), *Smart Materials for Energy, Communications and Security.*

twice as much power as a human brain, while it has 1,000 times fewer transistors than the number of cells in the brain.

The electronics goes step-by-step to the core of the solid or condensed matter, at the nano-scale exploiting the huge possibilities of the atomic networks.

One of the very important class of materials that is foreseen to be the key factor for the forthcoming smart computers in the informatics is the class of chalcogenides: materials that have in their composition at least one chalcogen. The chalcogens are sulphur, selenium, and tellurium. The chalcogenides, in their amorphous, glassy or, generally, in their non-crystalline form are flexible materials, able to change their properties under the action of many external factors, as e.g. light, other radiations, fields, particles, heat, pressure etc.[1] This flexibility allows to use chalcogenide in a large number of applications. Nearly 44 effects found and investigated in chalcogenide materials wittness the huge prospective of chalcogenides.[2]

Compounds such as GeTe, Sb_2Te_3, or the alloys as $Ge_2Sb_2Te_5$, undergo dramatic changes in optical and electrical properties through a rapid phase change between polycrystalline and amorphous states. This phenomenon has already been exploited in rewritable DVDs and in some electric, non-volatile memories. They are also candidates for a variety of reconfigurable applications in RF circuitry, antennae, analog circuits, and even as reconfigurable wiring harnesses.

In this paper we show how the chalcogenide materials can be put at work for making smart computer memories, that operate with more than two logical values.

2. Multi-level switching: a way towards smart memories

Computers are characterized by two fundamental attributes:

Operation is based on binary logic (storage and manipulation of data occurs through conversion to binary strings).

Operation is sequential in the manner described by John von Neumann. The development of a computational function is a step by step process. Computer programs are simply line-by-line instructions that describe a sequence of steps to be implemented.

The silicon computers became insufficient when the complexity of computation increases. Certain computations, functions and tasks cannot be implemented on conventional silicon computers (e.g. factoring of large numbers or sorting large databases).

Data storage is accomplished in the chalcogenide memory cell by a thermally induced phase change between amorphous and polycrystalline states in a thin film (similar to those used in rewritable CD and DVD optical disks). This rapid, reversible structural change in typical alloys as e.g. Ge-Sb-Te alloy, results in a change in material resistivity that is measured during the read operation. The memory technology uses a short electrical pulse to achieve the amorphous state (high resistance RESET state) and needs a bit longer current pulse to convert the amorphous state to the polycrystalline state (low resistance SET state). The portion of the alloy film near a bottom resistive electrode changes its state as a result of joule heating during the programming pulse.[3]

In the multi-level memory, the memory recording and erasing make use of a new mechanism: nano-dimensional configurations with specific structure are triggered in the bulk of the material at a controlled rate.

As current pulses are applied in the smart regime, minute nano-crystalline regions form, and the volume fraction of such crystalline phases increases with each current pulse. Crystallisation can occur through nucleation/growth upon the application of a current pulse. The microcrystallites generated by a sequence of pulses form temporarily a coherent sequence of states.

The nanocrystallites are distributed randomly throughout the chalcogenide material. As they grow, a percolation path results, which is a continuous high-conductivity pathway across the material between the contacts. Once percolation occurred, the material exits from the amorphous smart regime and enters into crystalline smart regime. The material can be reset in the amorphous state.[4]

The increment of crystallization that occurs upon applications of a current pulse is dictated by the energy accumulated by the action of a number of pulses on the material.

3. The single-level and the multi-level switching

The chalcogenide material, based on Ge-Sb-Te alloy with different doping elements can be operated essentially in two ways: (a) electrically; (b) optically. The reversible phase-change from amorphous state (high resistance, low reflectivity) to crystalline state (low resistance, high reflectivity) is used in current applications.[5]

In the classical electrical switching the amorphous state basculates into the crystalline state through crystallized continuous filaments. This switching is reversible because the crystallized regions can be amorphized by a suitable current pulse (Fig. 1).

Figure 2 shows the electrical switching in a smart, multi-level memory. The change in the resistivity of the smart memory as a function of the programming voltage is shown.

For multi-level memory materials the change from one conduction state to different steps of conduction depends on the number of pulses that knock the chalcogenide film.

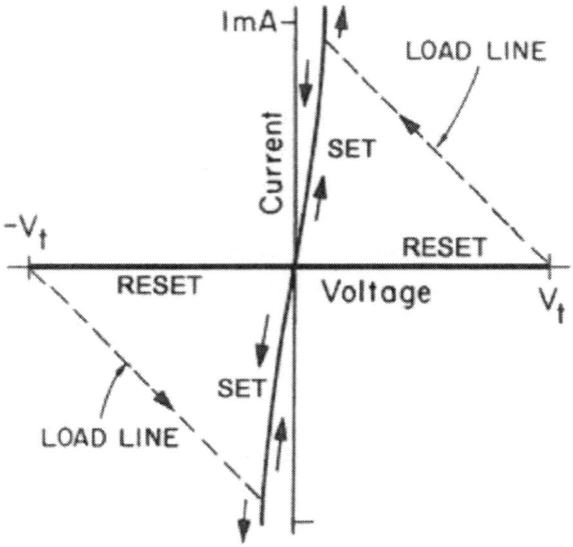

Figure 1. Electrical switching in a single-level memory material.

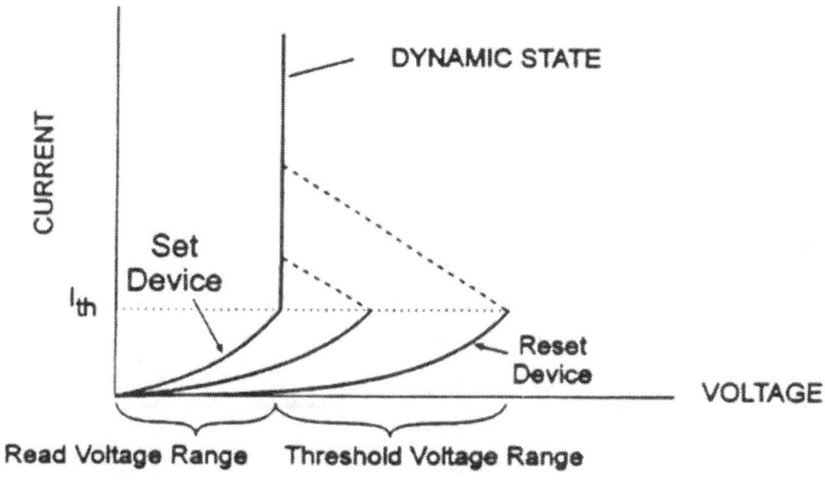

Figure 2. Multi-level memory recording.

For every step of resistance, the energy of ten laser pulses is accumulated.

In the multi-level optical switching the reflectivity is changed in steps, as a function of the number of laser pulses applied to the material.

Figure 3 shows the structural basis of the multi-level memory.[5]

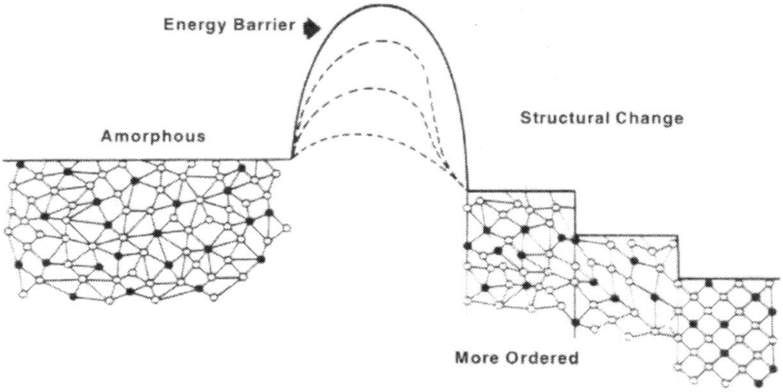

Figure 3. The structural modifications of the amorphous phase-change material with multi-level configurations, during stepwise recording.[3]

The initial amorphous material changes its state of atomic-scale order towards different degrees of ordering characterized by different electrical resistivity.

The multi-level electrical switching in 24 levels is represented in Fig. 4, while the multi-level optical switching in 11 levels is shown in Fig. 5.

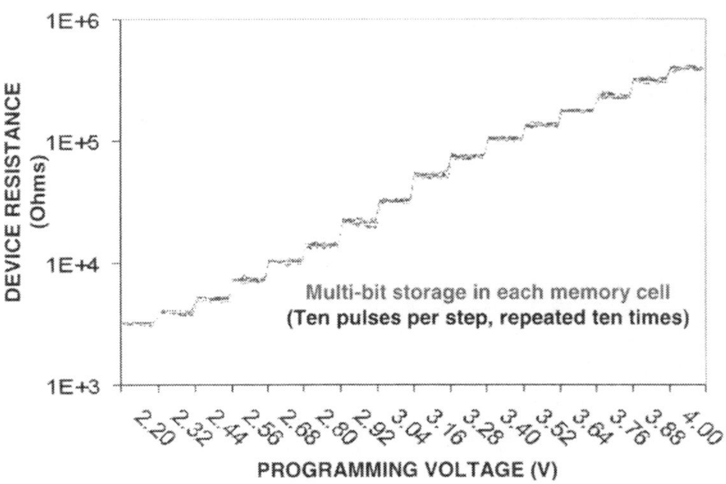

Figure 4. Resistivity modification during multi-level switching.[3]

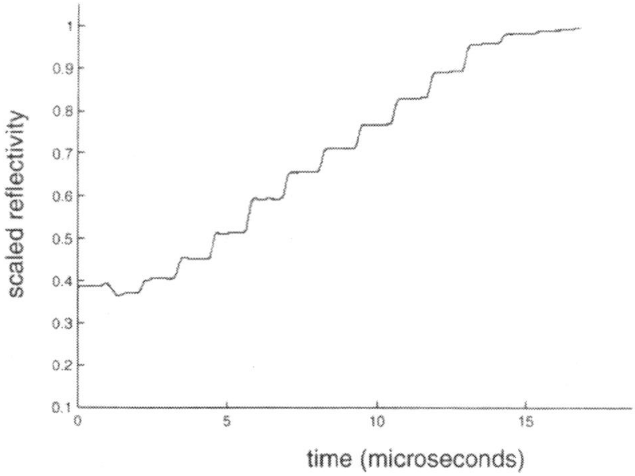

Figure 5. Reflectance modification during multi-level switching.[4]

4. Future trends in fundamentals

The mechanism of the multi-level switching is now under continuous investigation. It is important to know exactly how different parameters of the current laser pulses influence the structure at the atomic scale during phase changes from amorphous to crystalline and back.

The density of crystalline nuclei, structure, size and morphology are not known in every detail. The dynamics of the change is also challenging. Figure 6 shows a possible scheme of the transformation occurring in the material, by gradual transformation of the material during multi-level switching.

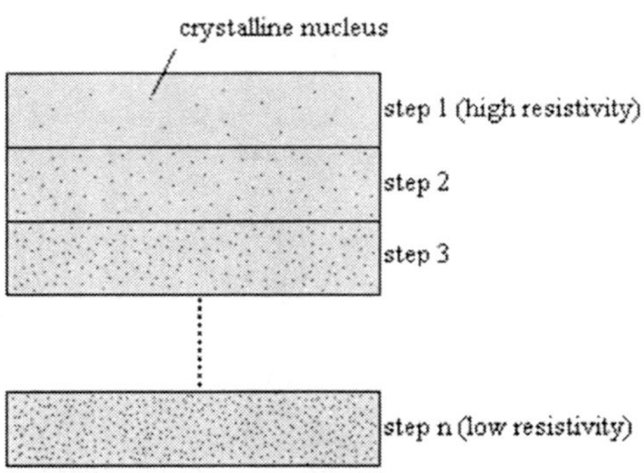

Figure 6. Atom-scale mechanism of multi-level switching.

5. Future trends in applications

5.1 NON-BINARY STORAGE

In non-binary data storage, one multi-level memory device can be programmed to store any one of the three or more numerical values. Each distinct numerical value corresponds to a distinct structural state that exhibits distinct resistance or distinct reflectivity. The information is stored in a non-volatile manner.

Consecutive integral values can be assigned to different ordering states. The energy relative to the reset state required to store an integer is proportional to that integer.

5.2 ENCRYPTION

The encryption capability of the smart device originates from the non-uniqueness of the relation between the structural state of the active material and the information stored in the device. The information content of a particular structural state depends on the number of states included in the non-binary storage protocol. If the number of storage states and the energy increments separating the energy states are not known, the knowledge of the structural state of the device provides no insight about the information value assigned to the structural state. To identify the structural state it is necessary to expose the device to different radiations but these radiations will destroy the stored information. Thus, the stored information is secured.

5.3 NON-BINARY ARITHMETICS

The multilevel, non-binary storage capability provides a natural basis for calculations in non-binary arithmetic systems. The smart computer operates in a non-binary fashion. Decimal (i.e. base 10) operation may be accomplished using a 10 states protocol in which ten current pulses are used to traverse the energy threshold of the device.

5.4 ULTRA-HIGH DATA DENSITY: A WAY OPENED TO 3D MEMORIES

In the mid 1980s Stan Ovshinsky and his colleagues from Energy Conversion Devices, USA, successfully developed the technology for preparation of the first 3D all-thin-film memory devices.

It is a continuous need for high-density, low cost memories.

The smart memories, in principle, can work on very small areas (several nanometers), within the chalcogenide material because every crystalline or amorphous nucleus in the matrix can serve as nano-memory device. The degree of order and the size of the nucleus could be controlled. This means that recording of information on point-like regions is, in principle, possible.

5.5 NEURAL NETWORK (FUZZY) COMPUTERS

Neural network device is possible if the data can be stored in a large number of points with rich connections between points. This is exactly the advantage of smart memory principle in the ultrahigh density of information stored in the 3D mode.

5.6 SMART CARDS AND OTHER TAMPER-PROOF SECURE INFORMATION HOLDERS

The smart memory devices can be used in a range of secure memory storage applications. Tamper-proof information storage is accomplished by subdividing the critical total integrated current into a series of current pulses. Information is recorded by applying the desired number of sub-critical pulses and the data are later read out by applying further sub-critical pulses until the conductivity transition is observed.

The stored data are physically represented by an extremely small volume of crystalline material within the memory cell. Therefore, the forensic methods such as examination by electron microscopy or X-ray diffraction will be not successful at determining the stored information. The electrical interaction of the cell is destructive.

An important application of the invulnerability of the tamper-proof information storage is the electronic cash.

6. Conclusions

Intelligent machines using multi-level memory chips are at the ends of our fingers. The first prototypes of smart memories are under construction. The science behind the devices is growing up very rapidly. New materials with multi-level storage are discovered and tested.

The advancement of the nano-scale and angström-size technologies[6] will lead to an accelerated progress of the new multi-level memory devices, of large practical applications.

References

1. M. Popescu, Non-Crystalline Chalcogenides, Kluwer Academic (Solid State Science and Technology Library, Volume 8), The Netherlands, 2000.
2. M. Popescu, Chalcogenides – Past, Present, Future, *J. Non-Cryst. Solids*, 352, 887 (2006).
3. S. R. Ovshinsky, Applications of non-crystalline materials, pp. 729–779, Chapter 12 in Insulating and Semiconducting Glasses, Kluwer Academic, The Netherlands, 1999.
4. T. Ohta, S. R. Ovshinsky, Photo-Induced Metastability in Amorphous Semiconductors, Chapter 18, Wiley VCH, Weinheim, 2003.
5. Y. Nara, Keynote talk in ISOM'03 Japan, 2003.
6. M. Popescu, *J. Optoelectron. Adv. Mater.* 8(2), 755 (2006).

MAGNETIC REFRIGERATION: APPLICATION TO THE ELECTRON DOPED MANGANITES

ESSEBTI DHAHRI[1*], MOEZ BEJAR[1], SAFA OTHMANI[1], ANOWAR TOZRI[1] AND EL KEBIR HLIL[2]
[1]*Laboratoire de Physique Appliquée, Faculté des Sciences de Sfax, B.P. 802, Sfax 3018, Tunisie*
[2]*Laboratoire de cristallographie CNRS, Institut Néel, Département MCMF B.P. 166, 38042 Grenoble Cedex 9, France*

Abstract. Magnetic refrigeration is a revolutionary, efficient, environmentally friend technology, which is on the threshold of commercialization. The magnetic rare earth materials are utilized as the magnetic refrigerants in most cooling devices. Recently, some data on the magnetocaloric effect in hole-doped manganites are reviewed. It is shown that the variation of interaction exchange energy, taking place under the effect of a magnetic field in the vicinity of the phase transformation, provides a significant contribution to the change of magnetic entropy. In this paper we report different results found for electron-doped manganites $La_{1-x}Ce_xMnO_3$ (x = 0.3, 0.4 and 0.5). These manganites present a large magnetic entropy change induced by low magnetic change, which is beneficial for the household application of active magnetic refrigerant materials.

Keywords: Magnetic refrigeration, magnetocaloric effect, electron-doped manganites, perovskite, double-exchange.

1. Introduction

Recently, studies of perovskite manganite oxides $Ln_{1-x}A_xMnO_3$ (where Ln: rare-earth, A: alkaline element) have revealed that these compounds exhibit colossal magnetoresistance. This is of scientific interest and has potential technological importance in magnetic recording, magnetic switches and magnetic sensors.[1–3] These oxides have been also identified as possible cathode materials for Solid

*To whom correspondence should be sent: E. Dhahri Laboratoire de Physique Appliquée, Faculté des Sciences de Sfax, B.P. 802, Sfax 3018, Tunisie: Email: essebti@yahoo.com

Oxide Fuel Cells (SOFCs), which represent an emerging technology for clean, reliable and flexible power production.[4,5] On the other hand, these materials are largely used as potential candidate for magnetic refrigeration technology.[6–8] This technology has attracted much attention due to its potential impact on energy savings and environmental friendliness over conventional gas refrigeration.[9,10] The magnetic refrigeration is based on the magnetocaloric effect (MCE), which is defined as the heating or cooling (i.e., the temperature change) of a magnetic material due to the application or removal of a magnetic field.[11] The aim of recent studies on MCE is to look for the most useful magnetic materials having large entropy change at low magnetic field for a possible magnetic refrigeration application from low to the sub room temperature. A big attention was paid to study the interplay between structure, magnetic and transport properties in hole-doped manganites. However, there are few results concerning the physical properties of electron-doped manganites.[12,13] In the present paper, we report our recent results on the magnetocaloric effect in cerium doped lanthanum manganites $La_{1-x}Ce_xMnO_3$ (x = 0.3, 0.4 and 0.5). The large entropy change obtained around T_C and the relative cooling power (RCP) values affirm that these compounds could be considered as potentials candidates for magnetic refrigeration at room temperature.

2. Experimental

Powder samples of $La_{1-x}Ce_xMnO_3$ (x = 0.3, 0.4 and 0.5) are synthesized by the sol-gel method in order to obtain a high-purity and homogenous powder. Stoichiometric amounts of La_2O_3, CeO_2 and MnO_2 precursors, with nominal purities >99.9%, were dissolved in concentrated nitric acid resulting light solution. Suitable amounts of citric acid and ethylene glycol, as coordinate agent, were added in the reactor and a complete homogonous brown gel was achieved after heating at 333 K. The gel was heated to give a black-brown powder at 443 K, which was calcinated at 1,373 K over night. Finally the remaining black powder was pressed and sintered at 1,673 K.

3. Data and results

Identification of the phase and structural analysis, carried out by X-ray diffraction technique, were discussed in previous work.[14] We have found that the Rietveld refinements of the X-ray diffraction data reveal the presence of two phases. The first one, corresponding to the $LaCeMnO_3$ phase, crystallizes in the orthorhombic structure with the *Pnma* space group. The second phase, with the *Fm3m* space group, was related to the presence of unreacted CeO_2 amount.

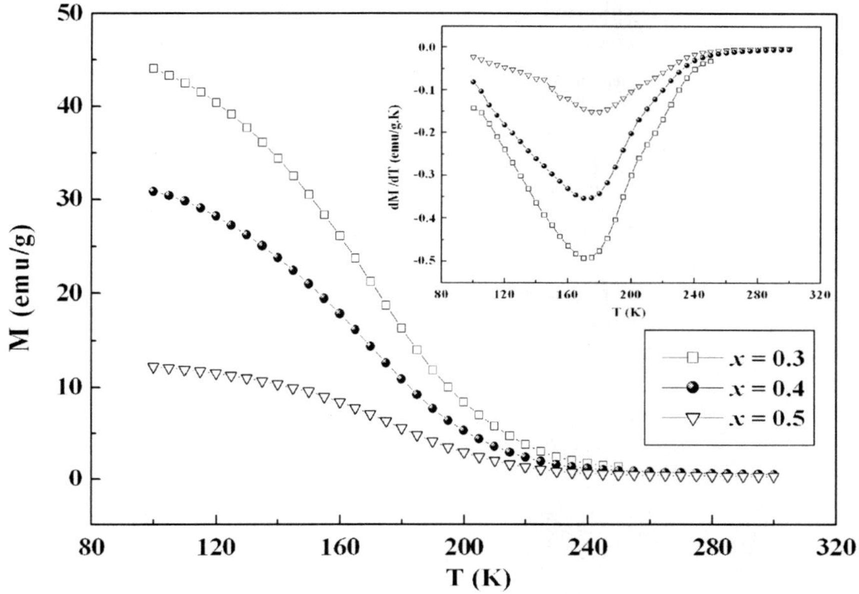

Figure 1. Temperature dependence of the magnetization in a magnetic field $\mu_0H = 0.5T$ for $La_{1-x}Ce_xMnO_3$ samples ($x = 0.3$, 0.4 and 0.5).

The magnetization *M* as a function of the temperature at different applied magnetic field was measured using a Foner magnetometer (FON) equipped with a superconducting coil. Figure 1 shows the temperature dependence of the magnetization (M(T)) of the $La_{1-x}Ce_xMnO_3$ samples ($x = 0.3$, 0.4 and 0.5). The Curie temperature (T_C) of these samples was deduced from the dM/dT curves.

The M(T) curves reveal a decreasing of the magnetization when increasing the *x* content. For the $La_{0.7}Ce_{0.3}MnO_3$ sample, the Mn^{3+}/Mn^{2+} ratio is equal to 7/3 for which the double-exchange interaction is maximal. Then, when increasing *x*, the Mn^{3+}/Mn^{2+} ratio decreases leading to a decrease of the double exchange interaction, which explain the reduction of the magnetization and the Curie temperature. Figures 2a, b and c show the magnetic applied field dependence of the isothermal magnetization M(T, μ_0H) for $La_{1-x}Ce_xMnO_3$ samples. The magnetic entropy change ΔS_M (Fig. 2) was deduced from the M (T, μ_0H) curves using the following equation[15]:

$$\Delta S_M\left(\frac{T_1+T_2}{2}\right)=\frac{1}{T_2-T_1}\left[\int_0^{\mu_0H} M\left(T_2,\mu_0H\right)\mu_0dH - \int_0^{\mu_0H} M\left(T_1,\mu_0H\right)\mu_0dH\right] \quad (1)$$

(a)

(b)

(c)

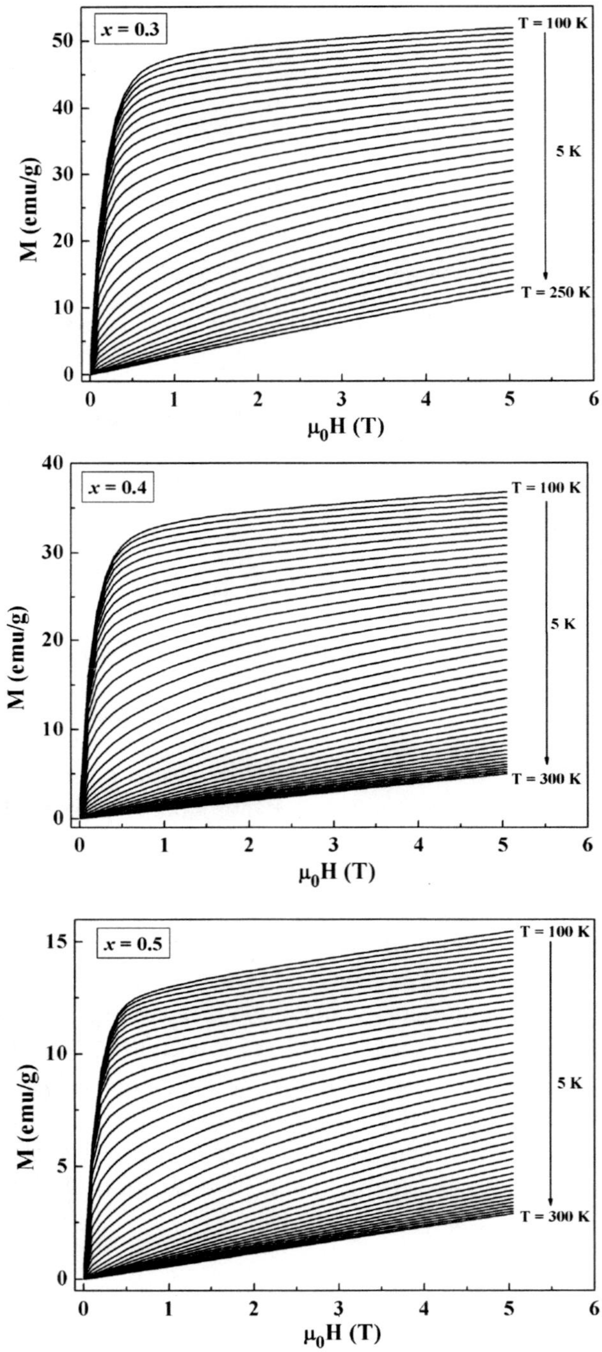

Figure 2. Isothermal magnetization for $La_{1-x}Ce_xMnO_3$ samples measured at different temperature for: (a) $x = 0.3$, (b) $x = 0.4$ and (c) $x = 0.5$.

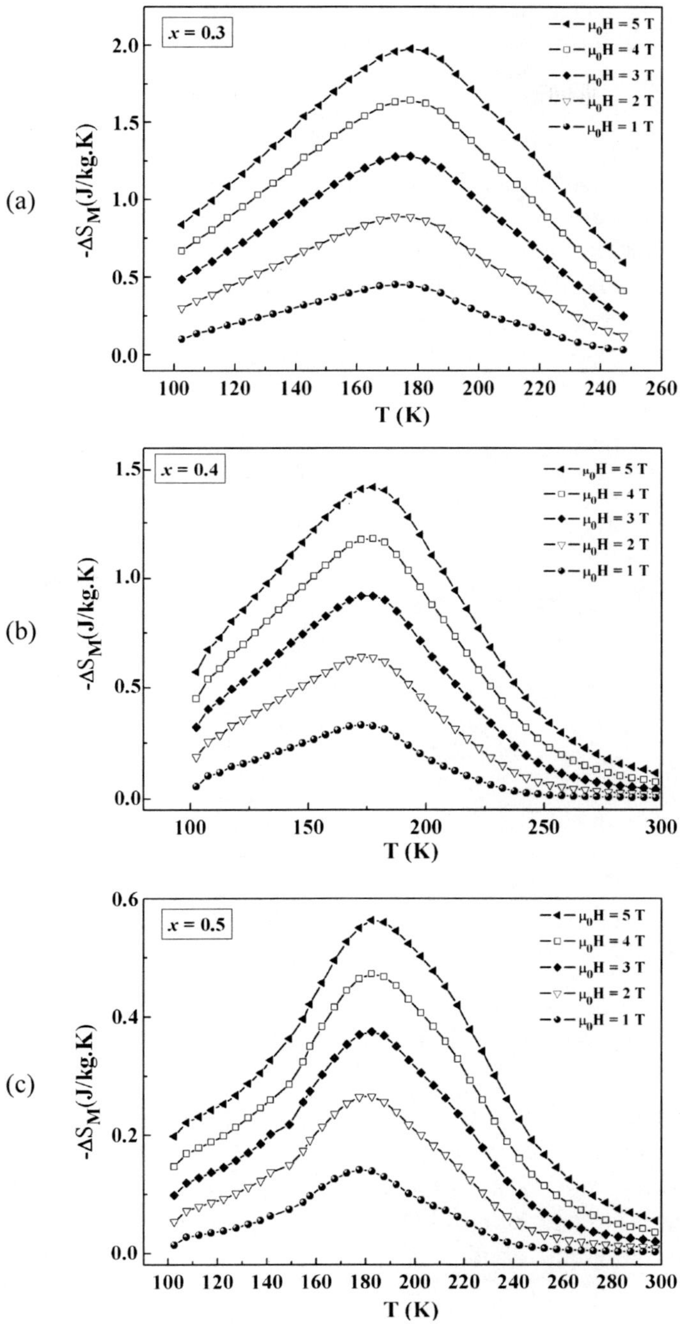

Figure 3. Temperature dependence of the magnetic-entropy change ($-\Delta S_M$) for different magnetic fields for: (a) $x = 0.3$, (b) $x = 0.4$ and (c) $x = 0.5$.

From Fig. 3, we can conclude that all samples exhibit a maximum entropy change ($-\Delta S_{Mmax}$) at their Curie temperature T_C. On the other hand, these curves reveal the strong relation between ($-\Delta S_{Mmax}$) and the magnetic applied field μ_0H. Indeed, for the $La_{0.7}Ce_{0.3}MnO_3$ sample ($-\Delta S_{Mmax}$) increases from 0.45 to 1.976 J/kg K when μ_0H increases from 1 to 5 T, respectively. These values are very close to those found in the case of hole-doped manganites $La_{0.65}Nd_{0.05}Ba_{0.3}Mn_{0.6}O_3$[16] and $La_{0.65}Nd_{0.05}Ca_{0.3}Mn_{0.9}Fe_{0.1}O_3$.[17] So, we can deduce that the substitution of lanthanum in $La_{1-x}A_xMnO_3$ manganites by monovalent, divalent (hole-doped) or tetravalent element (electron-doped) leads to large magnetic entropy change. Figure 4 shows the variation of the magnetic entropy change ($-\Delta S_M$) as a function of temperature for all samples for an applied magnetic field μ_0H of 1T.

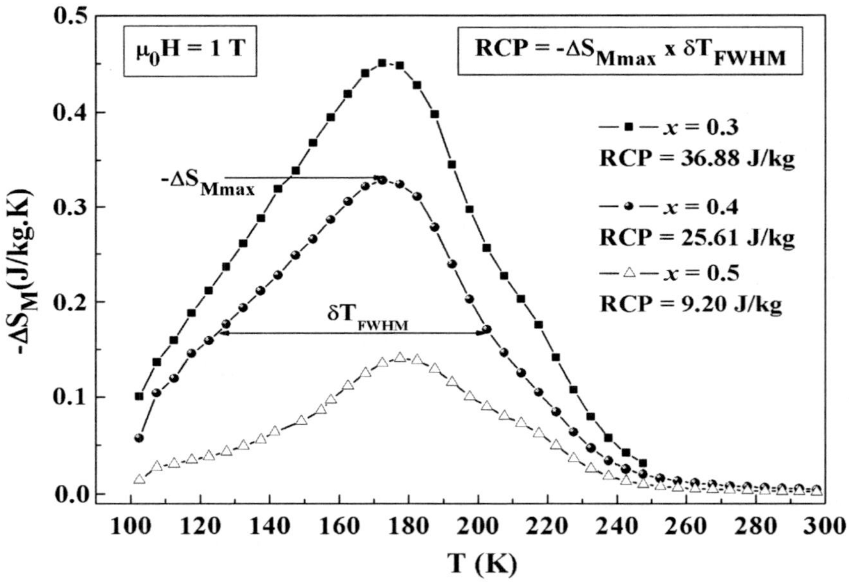

Figure 4. Temperature dependence of the magnetic-entropy change ($-\Delta S_M$) for $La_{1-x}Ce_xMnO_3$ samples (x = 0.3, 0.4 and 0.5) at an applied magnetic field μ_0H=1T.

We can deduce the decrease of the maximum entropy change ($-\Delta S_{Mmax}$) when increasing the *x* content. This can be attributed to the double exchange interaction between Mn^{2+} ($t^3_{2g}\ e^2_g$) and Mn^{3+}($t^3_{2g}\ e^1_g$) ions. This interaction is maximal for Mn^{3+}/Mn^{2+} ratio of 7/3 corresponding to *x* content of 0.3. We have calculated the RCP values for all samples using the following equation[18]:

$$RCP = -\Delta S_{Mmax} \times \delta T_{FWHM} \quad (2)$$

where δT_{FWHM} is the Full Width at Half-Maximum deduced from Fig. 4. Since RCP presents important criteria for selecting magnetic refrigerants, the electron-doped $La_{1-x}Ce_xMnO_3$ compounds can be considered as a potential candidate for magnetic refrigeration. We can mention that the RCP decreases from 85.44 to 20.14 J/kg when increasing x from 0.3 to 0.5. This result indicates that the tetravalent doping in lanthanum manganites system weakens the DE interaction thereby reduces T_C and magnetization of the system.

An attempt to the theoretical modeling of the magnetocaloric effect was done based on the Landau theory for transition phases. The influence of the magnetoelastic coupling on the magnetic entropy change in manganites has been studied. Thus Gibbs free energy can be written as[15]:

$$G(M,T) = G_0 + 1/2\,AM^2 + 1/4\,BM^4 - M\,\mu_0 H \quad (3)$$

where A and B are temperature dependant parameters containing the elastic and magnetoelastic terms of free energy.[12] From the condition of equilibrium $(\partial G/\partial M) = 0$, the magnetization can be described as[15]:

$$AM + BM^3 - \mu_0 H = 0 \ \text{So} \ \mu_0 H/M = A + BM^2 \quad (4)$$

The nature of the magnetization could be determined using the Benerjee criterion.[19] Therefore, we plot $\mu_0 H/M$ versus M^2 curves. The slope denotes whether the observed magnetic transition is of first (negative slope) or second order (positive slope). The variation of $\mu_0 H/M$ versus M^2 (Fig. 5) presents a positive slope, which confirms the transition to be of second order.

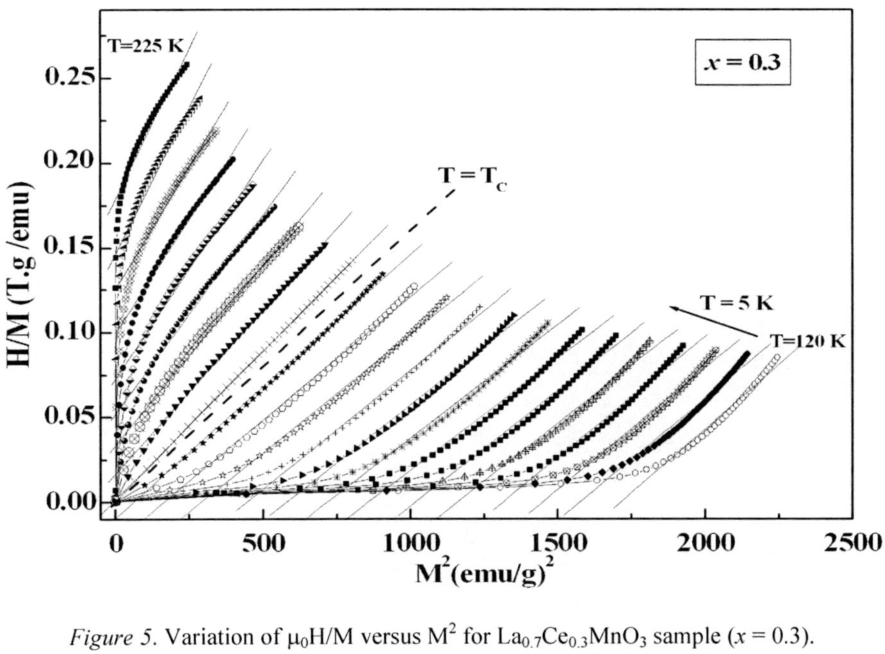

Figure 5. Variation of $\mu_0 H/M$ versus M^2 for $La_{0.7}Ce_{0.3}MnO_3$ sample ($x = 0.3$).

From the linear region of these curves, we have deduced the values of A and B coefficients. The temperature dependence of the coefficient A (Fig. 6) reveals a linear behavior for all samples. For each sample, the intersection of this linear curve with the abscissa axis determines the value of the Curie temperature T_C. On the other hand, the temperature dependence of the parameter B (Fig. 7) is found to be highly non linear.

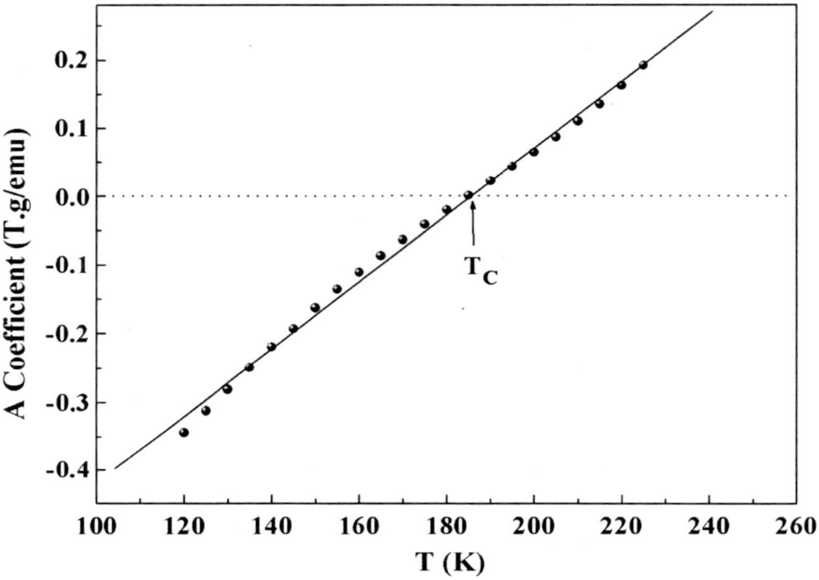

Figure 6. The temperature dependence of the coefficient A for $La_{0.7}Ce_{0.3}MnO_3$ sample ($x = 0.3$).

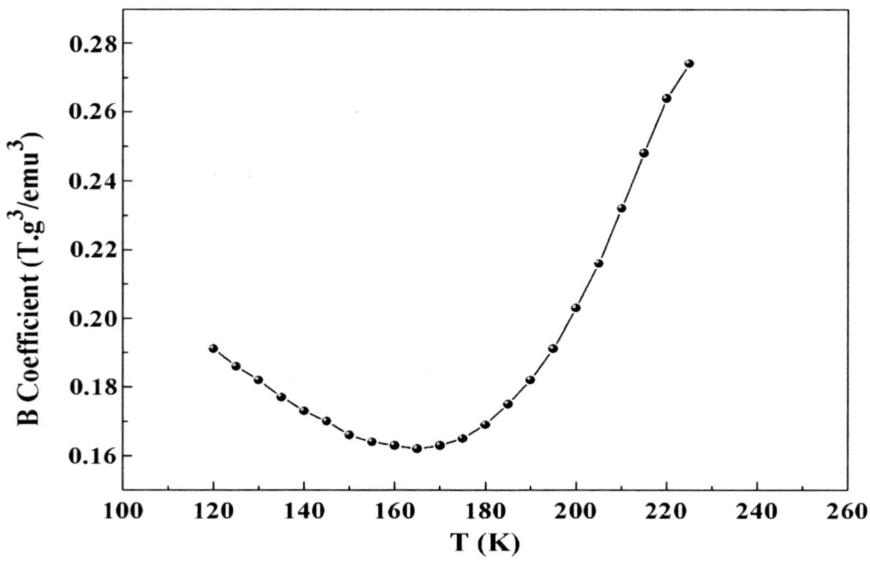

Figure 7. The temperature dependence of the coefficient B for $La_{0.7}Ce_{0.3}MnO_3$ sample ($x = 0.3$).

Amaral et al.[20] found the B coefficient, for La manganites, to be strongly dependent on temperature and it also includes the elastic and magnetoelastic terms. They suggest that the nature of the parameter B takes a crucial role in determining ($-\Delta S_M$). Using the parameters A and B, extracted from the data, the temperature dependence of the magnetic entropy change ($-\Delta S_{Mmax}$) can be calculated differently from the following equation:

$$\Delta S_M(T, \mu_0 H) = -1/2\, A'M^2 - 1/4\, B'M^4 \tag{5}$$

where A' (T) and B' (T) are the temperature derivatives of the expansion coefficients.

This equation represents the ΔS_M variation obtained from Gibb's free energy. We have plotted in Fig. 8 the variation of ΔS_M obtained from Equation (4) (continuous line).

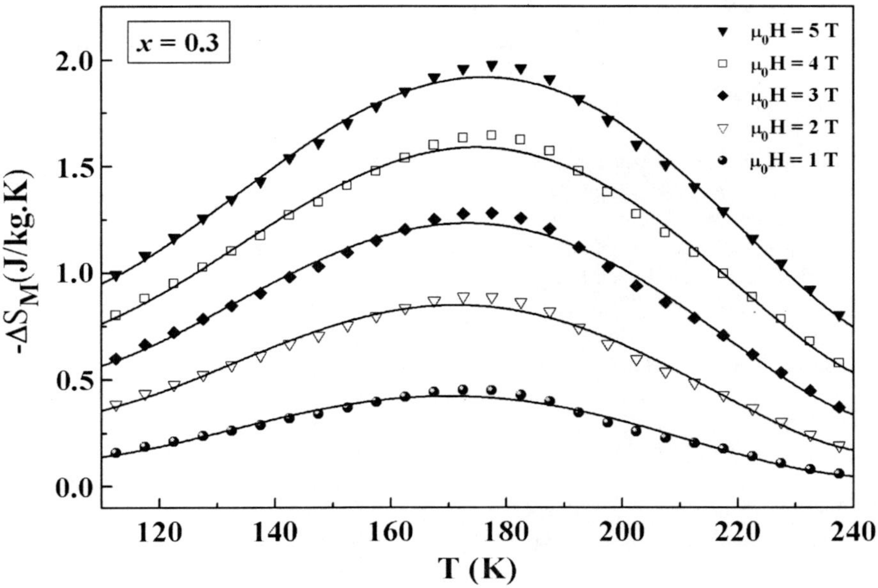

Figure 8. Experimental and theoretical magnetic entropy change for $La_{0.7}Ce_{0.3}MnO_3$ sample.

A comparison of the two ($-\Delta S_M(T)$) curves, obtained with two different ways, shows a clear correspondence between the two curves and, consequently, between the two methods. The present analysis indicates the importance of magnetoelastic coupling and electron interaction in understanding the magnetocaloric properties of lanthanum manganites. A similar behavior was observed with Amaral et al.[15] when comparing the two methods.

4. Conclusion

A series of electron-doped manganites of the type $La_{1-x}Ce_xMnO_3$ (x = 0.3, 0.4 and 0.5) have been prepared by sol-gel method. We have investigated the effect of tetravalent substitution in the magnetic and magnetocaloric properties. The magnetic and magnetocaloric studies reveal a decrease of the magnetization M and the magnetic entropy change ΔS_M when increasing the x content. This behavior can be explained by the double exchange mechanism, which is maximal for $x = 0.3$ (Mn^{3+}/Mn^{2+} ratio of 7/3). We have calculated the ΔS_M values using two different methods and we have found a perfect concordance between the two values. The temperature dependence of ΔS_M indicates a decrease of the maximum entropy change ΔS_{Mmax} when increasing x content. Finally, we have found that our samples present an important value of ΔS_{Mmax} and a large value of the relative cooling power RCP, which are very close to those reported in literature. Thus, our samples can be considered as active and inexpensive magnetic refrigerants.

References

1. G. Venkataiah, D.C. Krishna, M. Vithal, S.S. Rao, S.V. Bhat, V. Prasad, S.V. Subramanyam, and P. Venugopal Reddy, Phys. B 357 (2005) 370.
2. M.B. Salamon and M. Jaime, Rev. Mod. Phys. 73 (2001) 583.
3. E. Dhahri, K. Guidara, A. Cheikh-Rhouhou, J.C. Joubert and J. Pierre, Phase. Trans. 66 (1998) 99.
4. M. Doriya, Solid State Ionics 152/153 (2002) 383–392.
5. H. Tu and U. Stimming, J. Power Sources 127 (2004) 284–293.
6. M. Bejar, R. Dhahri, F. El Halouani and E. Dhahri, J. Alloys. Compd. 414 (2006) 31.
7. S. Das and T.K. Dey, Mat. Chem. Phys. 108 (2008) 220–226.
8. M.H. Phan, H.X. Peng, S.C. Yu, N.D. Tho, H.N. Nhat and N. Chau, J. Magn. Magn. Mat. 316 (2007) e562.
9. K.A. Gschneidner, Jr. and V.K. Pecharsky, Ann. Rev. Mater. Sci. 30 (2000) 387.
10. W. Dai, B.G. Shen, D.X. Li and Z.X. Gao, J. Magn. Magn. Matter. 218 (2000) 25–30.
11. E. Warburg, Ann. Phys. 13 (1881)141.
12. K. Bajaj, J. Jesudasan, V. Bagwe, D.C. Kothari and P. Raychaudhuri, J. Condens. Matter. 1088 (2007) 8984.
13. G. Garbarino, C. Acha, C. Martin, A. Maignan and B. Raveau, Phys. Rev. B. 70 (2004) 014414.
14. S. Othmani, E. Dhahri and M. Bejar, J. Ferroelectrics, in press.
15. M. Foldcaki, R. Chahine, B.R. Gopal and T.K. Bose, J. Magn. Magn. Matter. 150 (1995) 421.
16. W. Chen, L.Y. Nie, W. Zhong, Y.J. Shi, J.J. Hu, A.J. Li and Y.W. Du, J. Alloys Compd. 395 (2005) 23.
17. Z.M. Wang, G. Ni, Q.Y. Xu, H. Sang and Y.W. Du, J. Magn. Magn. Mater. 234 (2001) 371.
18. K.A. Gschneidner, Jr. and V.K. Pecharsky, Ann. Rev. Mater. Sci. 30 (2000) 387.
19. S.K. Banerjee, Phys. Lett. 12 (1964) 16.
20. J.S. Amaral, M.S. Reis, V.S. Amaral and T.M. Mendonça, J. Magn. Magn. Matter. 290 (2004) 686.

TOWARDS ELECTROCHROMIC DEVICES ACTIVE IN THE IR REGION

ALINE ROUGIER[1*], KATHERINE SAUVET[1,2] AND LAURENT SAUQUES[2]
[1]*Laboratoire de Réactivité et Chimie des Solides, UMR6007, 33 Rue St Leu 80039 Amiens, France*
[2]*Délégation Générale de l'Armement CEP (LOT)*
16 bis, avenue Prieur de la Côte d'Or, 94114 Arcueil cedex, France

Abstract. Electrochromic devices (ECD), also described as optical batteries, are able to change their optical properties under an applied voltage. Largely investigated in the visible domain, applications for spacecraft thermal control and military camouflage have recently emphasized the need for ECD active in the infrared region, IR, and particularly in the atmospheric transmission windows called the MW band (mid-wavelength 3–5 μm) and LW band (long wavelength 8–12 μm). Most of all-solid state devices, active in the visible, are based on WO_3 electrochromic layer. Thus, prior to the ECD study, the optical properties of WO_3 thin films were investigated in the IR domain. Strong links between the IR optical characteristics, namely the modulation of reflectance, and the conditions of deposition (substrate nature, duration of deposition, pressure) were established. Contrast in reflectance, ΔR ($\Delta R = R_{deinserted} - R_{inserted}$), as high as 70% in the MW and as high as 30% in the LW were achieved for WO_3 thin films deposited on Au substrate in 45 mTorr, cycled in H_3PO_4 liquid electrolyte. Finally, coupling simulation and experimental data, configurations of ECD active in the IR domain are discussed.

Keywords: Electrochromism, thin films, tungsten oxide, Mid-Infrared, Reflectance modulation, devices, simulation.

*To whom correspondence should be sent: A. Rougier, Laboratoire de Réactivité et Chimie des Solides, UMR6007, 33 Rue St Leu, 80039 Amiens, France; Email: aline.rougier@u-picardie.fr

I. A. Luk'yanchuk and D. Mezzane (eds.), *Smart Materials for Energy, Communications and Security.*

1. Introduction

Since their discovery by Deb in the late sixties,[1] electrochromic materials, which are able to change their optical properties under an applied voltage, have received significant attention. Among transition metal oxides, WO_3 has outperformed the others[2] and thus is widely used in current technologies. Commercialization of EC devices, also described as optical batteries, was first reported in the late eighties, by Gentex, for rear-view mirrors.[3] Since then, few applications including helmet visors or glasses have been reported whereas the most challenging one remains the development of smart windows for use in building and automobile.[2,4] Till recently, most of the applications were turned towards the visible region whereas the infrared, IR, domain was much less addressed.[5–6] In the military field, the preparation of coatings for vehicles and persons, which would be able to blend into their surrounding and, therefore become invisible to an infrared imager, is highly desirable.[7] In this respect, thanks to their capability of controlled modulation of their optical properties, electrochromic devices are promising candidates for tuned emissivity. When designing materials for camouflage applications, the following requirements need to be considered: variation of 80–40% of the infrared emissivity, fast switching, long durability.

Aiming at building an ECD active in the IR region, this paper will first investigate the properties in the IR region of the most utilized and studied in the visible electrochromic material, namely WO_3. Following the optimization of the optical properties (i.e. contrast in reflectance, ΔR) of Radio Frequency Sputtered WO_3 thin films, in respect of the conditions of deposition, their incorporation in the ECD device will be discussed in terms of configuration, multilayer stack or assembly of half cell, choice of the counter electrode and electrolyte…. Besides, the experimental approach will be coupled to a modeling one based on the simulation of optical indexes of each layer.

2. Methodology

2.1 THIN FILM PREPARATION

All thin films were deposited at room temperature on different type of substrates, namely FTO: F doped SnO_2, Au, and Si using reactively Radio Frequency Sputtering (referred here after as RFS-films) technique. WO_3 thin films were grown from a metallic W target, at 250 W with a 90 mm target-substrate distance in a 15 or 45 mTorr environments consisting of 15 sccm (80%) Ar and 3.75 sccm (20%) O_2.

NiO thin films were grown from a Ni target, at 150 W with a 60 mm target-substrate distance in a 15 mTorr environment consisting of 95 sccm (95%) Ar and 5 sccm (5%) O_2. Ta_2O_5 thin films were grown at 250 W from a Ta_2O_5 target with a 60 mm target-substrate distance in a 15 mTorr environment consisting of 15 sccm (95%) Ar and 0.790 sccm (5%) O_2. Depending on the film nature, the deposition time varied from 5 to 240 min, corresponding to film thickness in the 25–750 nm range as determined by profilometry or by optical measurement.

2.2 THIN FILM STRUCTURAL AND TEXTURAL CHARACTERIZATION

The structure of the films was examined by means of X-Ray diffraction (XRD). The data were recorded on a θ-2θ configuration using a Philips PW1470 diffractometer and CuK_{α} radiation $\lambda(Cu_{K\alpha}) = 1.5418$ Å. The film thickness was measured both with a spectroscopic ellipsometer (Sentech) over the 350–800 nm visible range and by profilometry using a Dektak instrument.

2.3 THIN FILM ELECTROCHEMICAL CHARACTERIZATION

All the electrochemical measurements were carried out at room temperature in a two- or three-electrode cell of conventional design using a large platinum foil as counter electrode and filled with an aqueous solution as electrolyte either (i) of diluted phosphoric acid (H_3PO_4 0.1 M) or (ii) of potassium hydroxide (KOH 1 M) when using WO_3 or NiO thin films as working electrode, respectively. The experiments were performed with a VMP automatic cycling/data recording system (Biologic S.A., Claix, France).

2.4 THIN FILM OPTICAL CHARACTERIZATION

The optical indexes, namely the refractive index (n) and the absorption index (k) in the infrared region were calculated using the Film Wizard simulation software,[8] requiring previous measurements of the film thickness, reflectance and transmittance.

For the optical properties in the 2.5–25 μm range, both the reflectance and the transmittance were measured using an Infrared spectrophotometer SOC-100 HDR (Surface Optics Corporation), giving the hemispherical directional reflectance (HDR) and hemispherical directional transmittance (HDT) values. For the HDR and HDT measurements, the thin film was irradiated with a beam at 10° and 0° angle of incidence, respectively. The infrared source was a black body heated at 700°C. A gold specular reference ($R \approx 99\%$ in IR) and the air were used as reference to determine the HDR and the HDT values, respectively.

3. Data and result

Upon electrochemical cycling using a Pt/H_3PO_4 0.1 M/WO_3 cell, WO_3 films are reduced as a result of the injection of protons and electrons: $WO_3 + xe^- + xH^+ \leftrightarrows H_xWO_3$.[2] In the visible domain, the proton insertion is associated with a color change from a transparent to a blue state for the reduced state H_xWO_3 and a contrast in transmission in the 3–5 range.[9] The modulation of the reflectance, ($\Delta R = R_{deinserted} - R_{inserted}$) was recorded after applying both insertion (–1.9 V) and deinsertion (+1 V) voltages during 180 s. The [–1.9 V – 1 V] voltage window was earlier optimized, as the one corresponding to the highest contrast in reflectance for WO_3films cycling in domains ranging from –2.1 to 1.6 V.[10] As gathered in Table 1, the contrast in reflectance is strongly dependant of the deposition conditions, namely, the substrate nature, FTO or Au, the duration of deposition (i.e. the film thickness) and the pressure inside the chamber, either 15 or 45 mTorr. From Table 1 and in respect of applications, films deposited using different conditions of deposition will have to be considered. Indeed, the category of films showing the highest contrast in reflectance in the MW band differs from the one showing the highest contrast in reflectance in the LW. The various optical behaviors were earlier linked to the difference in film morphology, composition and density for WO_3 films that are X-ray amorphous (i.e. featureless X-Ray pattern).[11] In relation with a lower density of 6.18 as compared to 6.53 for films deposited at 15 mTorr, films deposited on Au at 45 mTorr exhibit the highest contrast in reflectance. The other deposition conditions being fixed, the film thickness revealed to be a key factor. Thicker films show higher contrast in reflectance in the LW domain, which remains however largely

TABLE 1. Contrast in reflectance ($\Delta R = R_{deinserted} - R_{inserted}$) in the MW and LW bands for Radio Frequency Sputtered WO_3 thin films deposited in various conditions (pressure, substrate nature, duration of deposition) and cycled in H_3PO_4 0.1 M liquid electrolyte.

Chamber pressure (Ar_15sccm/ O_2_3.75sccm)	Substrate	Duration of deposition (min)	Thickness (nm)	ΔR (%)	
				MW Band	LW Band
15 mTorr	FTO	45	200	30	5
	FTO	110	500	20	27
45 mTorr	FTO	120	320	25	25
	FTO	240	750	25	50
	Au	120	320	73	15
	Au	240	750	43	34

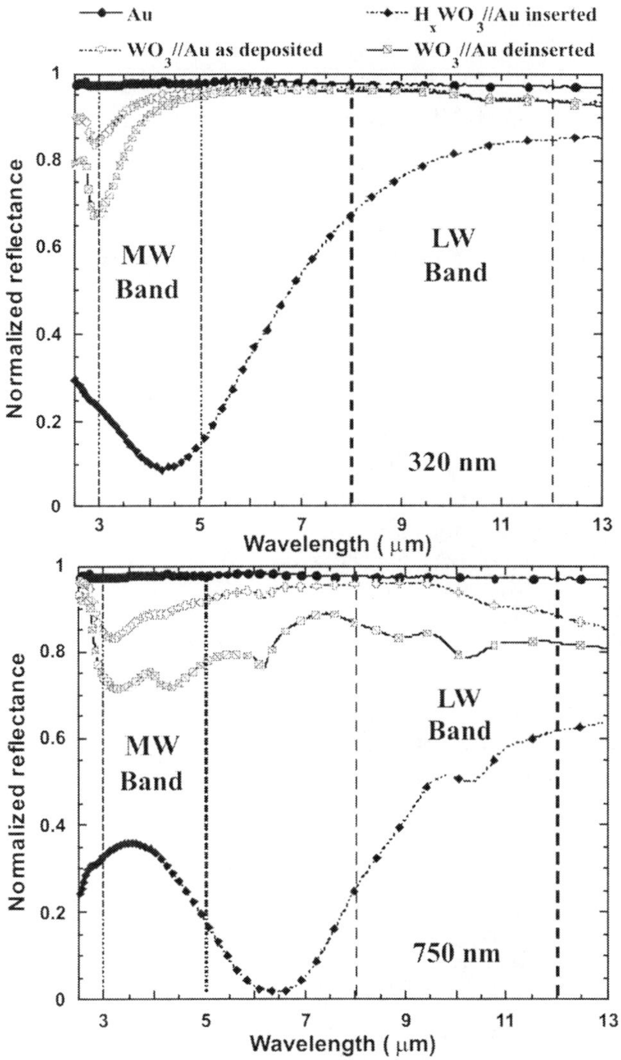

Figure 1. Evolution of the contrast in reflectance, ΔR, in the MW and LW bands for normalized reflectance of a Au blank substrate of 320 and 750 nm RFS-WO_3 thin films in various states: as deposited, inserted at −1.9 V, deinserted at +1 V.

smaller than the one observed in the MW domain. A maximum value of ΔR of 30 % is reached in the 8–12 μm domain as compared to the one of 73% in the 3–5 μm domain, for 750 and 320 nm thick films, respectively (Fig. 1).

Interestingly, the promising property of WO_3 thin films were confirmed using two CEDIP infrared imagers equipped with an InSb detector for the MW band and a microbolometer for the LW band, respectively. The system consisted in heating at 60°C a copper foil supporting the films and a "simulated black body", based on a piece of silicon substrate painted in black. IR imagers recorded digit levels expressed in terms of either luminance ($L = \varepsilon\sigma T^4$) or temperature. For the latter, it corresponded to an apparent temperature and not to the real one as the emissivity of the films was different from the one of the black body ($\varepsilon = 1$). A decrease in the apparent temperature from 59°C, for the inserted state, to 24°C, for the deinserted one, was recorded for 320 nm WO_3 film deposited on Au at 45 mTorr.[11] This variation in apparent temperature of 35°C was associated to a variation of emissivity, $\Delta\varepsilon$, of 0.6–0.8. (The emissivity value was calculated using the reflectance variation as for non transparent material $\varepsilon = 1\text{-}R$.) This range of emissivity values covers a larger domain than the $\Delta\varepsilon = 0.4$ required for application (as mentioned in the introduction). However, this variation concerns a single layer of the device, suggesting smaller variations of the full device.

Having optimized the single layer of WO_3, the next step was its incorporation in the full device. ECD, often described as optical battery, are based on a stack of five layers usually laminated in between a glass or a flexible polyester foil substrate (Fig. 2[12]). The central part of the device consists of three layers, which are the electrochromic layer, the ion storage layer that can either be of neutral color or exhibits complementary electrochromic properties and the electrolyte. The role of the latter is to ensure the ion transport from one electrochromic layer to the other one. The preferred ions are usually Li^+ or H^+ in respect of their small size an expected high mobility. For ECD devices utilized

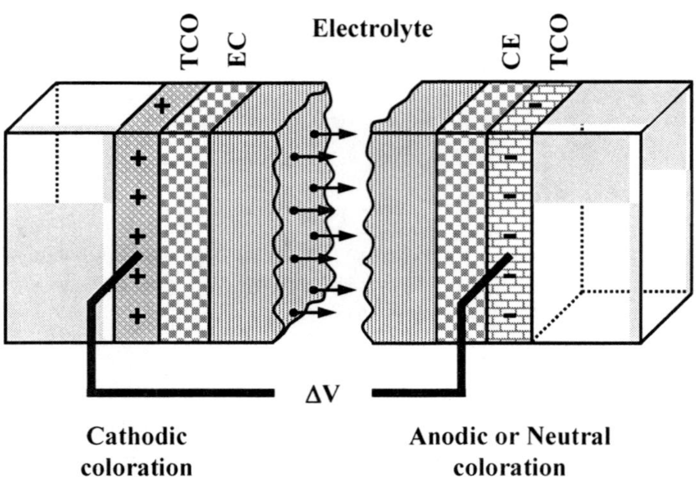

Figure 2. Basic design of an electrochromic device, indicating transport of positive ions under the action of an electric field.[12]

in the transmission mode, the central part is positioned between two conductive and transparent layers. Regarding Transparent Conducting Oxides, TCO, In_2O_3:Sn, ITO, is the state of art with a transmission close to 85% in the visible domain and a conductivity of 1,000–5,000 S/cm.[13] However in respect of a lower cost and availability on large area substrate, ITO is often replaced by FTO. Considering ECD for IR applications, therefore working in the reflection mode, TCO cannot be used as most of them are n-type conductor with no transparency in the IR domain. Thus, different options are possible. Among them, one requires the development of new transparent conducting materials in the IR domain (Fig. 3 – ❺*b*), or their replacement by other type of electronic conductor such as a gold grid deposited on a IR transparent substrate (Fig. 3 – ❺*a*). For the former, p-type conductors are currently investigated in our group with promising IR transparency for delafossite $CuInO_2$ with however limited conductivity.

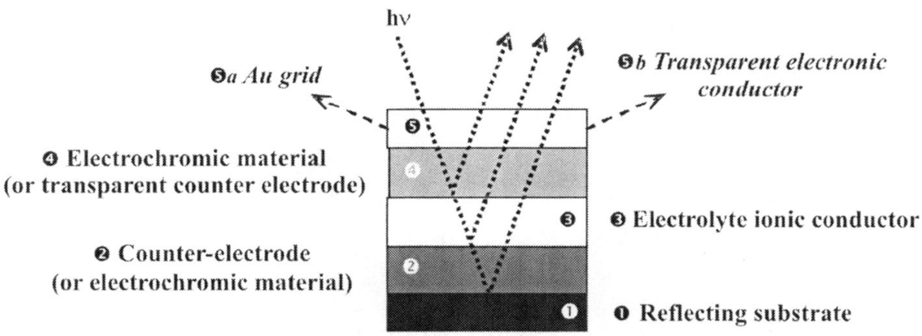

Figure 3. Basic design of electrochromic device (ECD) in reflection mode. Two options, namely a Au grid (Fig. 3 – ❺*a*) or a transparent conductor layer (Fig. 3 – ❺*b*) are considered for the top layer ensuring the electronic contact.

Regarding the configuration of the ECD, two options were evaluated. The first one was to investigate all-solid state devices, which have the advantage of being deposited in a single run. However, the stability of each layer in respect of the environment of deposition of the full stack needs to be carefully checked. The second option was to prepare device by assembling half cells. This approach gives more freedom on the choice of the materials. However, poor electrical contact between the two half cells when assembled may be encountered. In addition as a preliminary step of our experimental work, simulation of the films optical data, more precisely of the optical indexes n and k using the Film Wizard software was performed. The whole simulating process was based on the simulation of the optical indexes of each single layer including the substrate, prior to their incorporation in the multi-layer stack for the determination of the contrast in reflectance of the full device (Fig. 4). For each layer, different states, namely as-deposited, inserted, deinserted as well as different film thickness, ti (or deposition pressures, Pi, for WO_3) were considered.

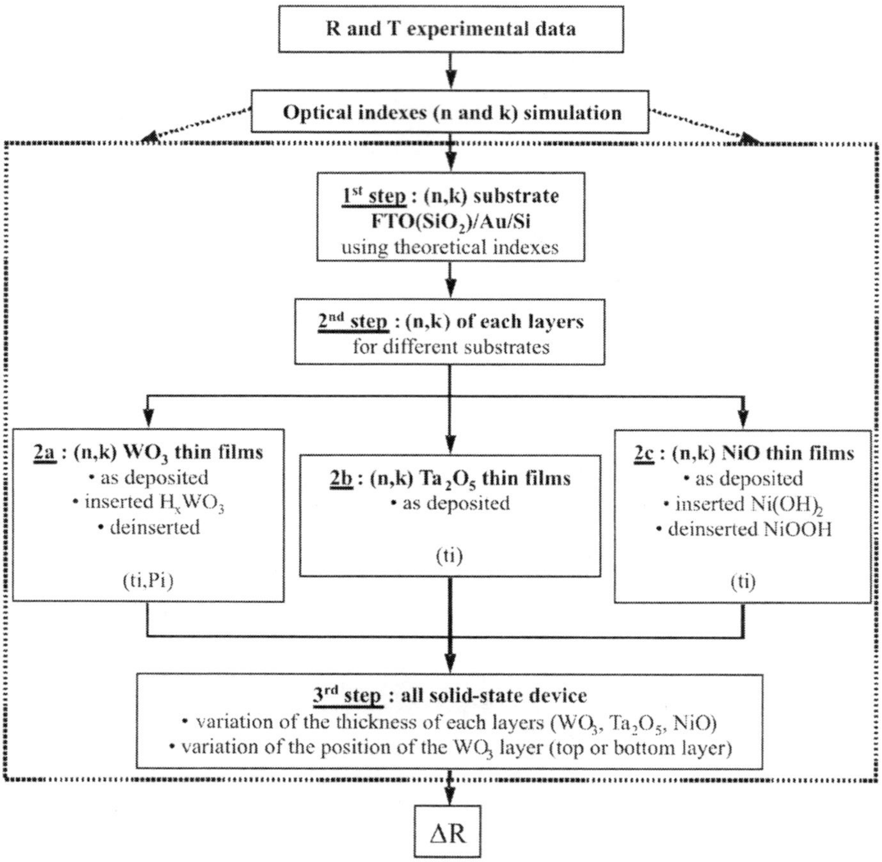

Figure 4. Chart of the calculations of the contrast in reflectance based on the simulation of the optical indexes, n, and k using the simplex method.

Regarding the first option of all-solid-state device and focusing on the three-layer central part, proton conductor electrolytes were favored in respect of faster switching time. RFS-Ta_2O_5 thin films show a transmittance of more than 70% in the IR domain. Due to its large used in ECD for the visible, NiO was chosen as an example of counter electrode. NiO is an anodic electrochromic material switching in the visible from a transparent to a brownish state upon oxidation.[14] RFS-NiO thin films show a contrast of reflectance of about 10% in the full IR domain when cycled in KOH electrolyte, following the $Ni(OH)_2 + OH^- \rightarrow NiOOH + H_2O + e^-$ reaction.

The simulation of the optical indexes (n, k) requires the input of the experimental Reflection (R) and Transmission (T). A very systematic process, starting with the simulation of the optical indexes of the substrate was applied. The simulated n and k values of the substrates are then used for the simulation of the ones of the active electrochromic and electrolyte layers. The optimization method, referred to as the "Modified simplex" or "global modified simplex", enables to evaluate after x iterations the difference between experimental and model points of R and T (RMSE, Root Mean Squared Error). After simulation, the good agreement between the fitted and experimental R and T values of thin films on different substrates is illustrated by the low RMSE value. The contrast in reflectance for entire ECDs is then evaluated using the simulated optical indexes values of each layer.

Having determined the various optical indexes, two types of ECD configurations, depending on the position of the WO_3 electrochromic active layer, were considered. In the first one, WO_3 layer corresponds to the top of the device whereas in the second one it corresponds to the bottom layer. The latter requires that the above electrolyte, counter electrode and conducting layers are IR transparent. For both MW and LW bands, higher contrast in reflectance were calculated for the top configuration as compared to the bottom one (Table 2). Besides, in the top configuration, an increase in film thickness above 400 nm, is associated with a decrease in the contrast in reflectance in the MW whereas an opposite trend is observed for the LW band, i.e. increase in contrast with increasing thickness (Fig. 5).

In respect of the second option and the assembly of half cell, current investigations show promising properties by using PANI as counter electrode[15] and a lab-synthesized electrolyte based on a polymer membrane. Their electrochromic properties in the IR region will be reported in a forthcoming paper.

TABLE 2. Contrast in reflectance ($\Delta R = R_{deinserted} - R_{inserted}$) in the MW and LW bands for electrochromic device based on the $WO_3/Ta_2O_5/NiO/Au$ (top) and $NiO/Ta_2O_5/WO_3/Au$ (bottom) configurations.

	Top		Bottom	
	MW	LW	MW	LW
ΔR (%)	51.8	33.9	36.9	1.9

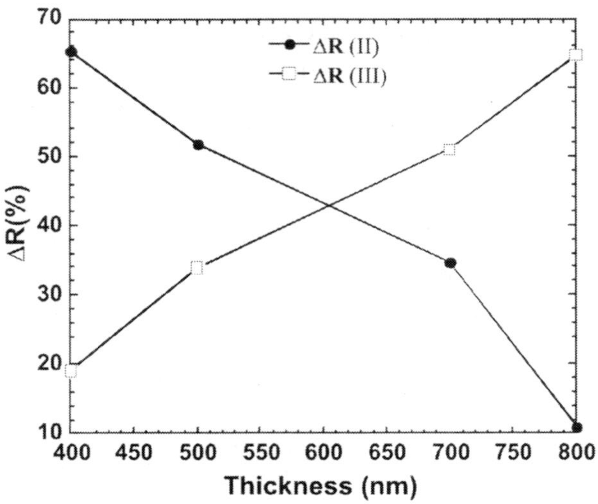

Figure 5. Evolution of the simulated contrast in reflectance (ΔR) in the MW and LW bands as a function of the WO_3 film thickness in WO_3 (?)/Ta_2O_5 (100 nm)/NiO (60 nm)/Au top configuration.

4. Conclusion

This paper reports the extension of the electrochromic properties of materials and devices to the IR region. Detailed investigation of the typical electrochromic material, WO_3 is reported. Depending on the conditions of deposition, contrasts in reflectance as high as 70% in the MW or 50% in the LW are obtained. Simulation of the contrast in reflectance of the full device is proposed using the simulation of the optical indexes of each layer. Deposition of ECD multilayer, using the Radio Frequency Sputtering technique, is currently in progress.

ACKNOWLEDGEMENTS

The authors wish to thank O. Durand (Thales) for X-ray Reflectometry measurements and the determination of the film density, Thierry Dubois (DGA) for IR camera measurements and J-P Bonnet (LRCS) for fruitful discussion and help on PANI deposition.

References

1. S.K. Deb, *Philosophical Magazine* 27, 801 (1973).
2. C.G. Granqvist, *Nature Materials* 5, 89–90 (2006).
3. http://www.gentex.com
4. X. Fanton (St Gobain), *Science* (verre) 9(5) (October 2003).
5. C. Trimble, M. deVries, J.S. Hale, D.W. Thompson, T.E. Tiwald, J.A. Woollam, *Thin Solid Films* 26, 355–356 (1999).
6. A. Bessière, L. Beluze, M. Morcrette, V. Lucas, B. Viana, J.-C. Badot, *Chemistry Materials* 15, 2577 (2003).
7. P. Chandrasekar, B.J. Zay, T. McQueeney, G.C. Birur, V. Sitaram, R. Menon, M. Coviello, R.L. Elsenbaumer, *Synthetic Metals* 155, 623 (2005).
8. FilmWizard, Optical Thin Film Software, version 6.4.2, SCI (Scientific Computing International).
9. A. Rougier, F. Portemer, A. Quédé and M. ElMarssi, *Applied Surface Sciences* 153(1), 1–9 (1999).
10. K. Sauvet, A. Rougier and L. Sauques, *Solar Energy Materials and Solar Cells* 92(2), 209–215 (2008).
11. A. Rougier, K. Sauvet and L. Sauques, *Ionics*, doi:10.1007/s11581-007-0191-y (2008).
12. G.A. Niklasson and C.G. Granqvist, *Journal of Materials Chemistry* 17, 127–156 (2007).
13. H.L. Hartnagel, A.L. Dawar, A.K. Jain, and C. Jagadish, Semiconducting Transparent Thin Films, Institute of Physics Publishing, Bristol, UK (1995).
14. I. Bouessay, A. Rougier, P. Poizot, J. Moscovici, A. Michalowicz and J.-M. Tarascon, *Electrochimica Acta* 50, 3737–3745 (2005).
15. B.P. Jelle and G. Hagen, *Journal of Applied Electrochemistry* 29, 1103–1110 (1999).

NITRIDE SEMICONDUCTORS INVESTIGATED AT A NANOSCALE

MOHAMMED BENAISSA
CNRST-UATRS, 52 Omar Ibn El Khattab,
B.P. 8027, 10102 RabatAgdal, Morocco

Abstract. The present review deals with two important issues that inhibited a rapid development of III-nitrides devices. Namely; p-type doping of GaN and GaN quantum dots on dislocations free surfaces. The first issue was governed by an auto-compensation phenomenon that limited high p-type doping, while the second was affected by threading dislocations that would act as trap centers for carriers preventing an efficient electrical injection. In both cases, the overall electronic behavior is essentially governed both by the intrinsic structure, strain-state and chemistry of these nanoscopic objects, in addition to the nature of interfaces therein. Transmission electron microscopy (TEM) was extensively used to provide local atomic imaging, electron diffraction and spectroscopy with high spatial-resolution and energy-resolution offering therefore numerous possibilities for locally investigating and characterizing the electronic, optical, chemical, and structural properties. This review starts with a brief introduction to basic properties of III-nitride semiconductors, with a description of growth techniques used for the elaboration of studied samples and followed by a recall of electron microscopy and its associated techniques. Particular attention is then paid to detailed investigations regarding p-type doped GaN samples grown with metalorganic vapor phase deposition and GaN QDs grown using molecular beam epitaxy. These results are presented in two separate chapters. A conclusion summarizing main points and a look towards the future is made at the end.

Keywords: III-nitrides, GaN, p-type, quantum dots, epitaxy, HRTEM, EELS, dislocation, strain-state, nano-analysis

1. Introduction

Wide bandgap III-nitrides, including (Al,Ga,In)-N, have seen enormous success in their development especially in the latest stages of the 20th century. Although with a dislocation density in the order of 10^8–10^{10} cm^{-2} III-nitrides are already the building block of many viable devices. Today blue/violet light-emitting diodes (LEDs) and laser diodes (LDs) based on (Al,In,Ga)-N have been successfully commercialized. Blue/green LEDs have already found their market in full-color

I. A. Luk'yanchuk and D. Mezzane (eds.), *Smart Materials for Energy, Communications and Security.*

LCD displays and traffic lights, while blue LDs will tend to replace red lasers in the current CD/DVD read/write systems. The wide bandgap of GaN makes this material suitable not only for light emitting sources but also for high-temperature applications. GaN and its alloys have the potential to form high power electronics such as transistors and UV solar-blind photodetectors. Due to the polar nature of the Ga-N bond, GaN does not possess inversion symmetry. Thus, when GaN is subject to an alternating electric field, the induced polarization is not symmetric. This property of GaN can be used in non-linear optics applications such as second-harmonic generation.

In blue LEDs, the active region consists of one or more InGaN quantum wells sandwiched between n-type and p-type GaN-based cladding layers. By adjusting the relative InN-GaN fraction in the InGaN quantum well, the blue light emission is obtained. In the present review, two important issues that inhibited a rapid development of devices will be addressed, namely; (i) understanding the difficulties in obtaining a relatively good p-type material, and (ii) the fabrication of GaN quantum dots (QDs). As far as the first issue is concerned, GaN doped with Mg is the common used p-type material but the behavior of the free hole concentration (saturation at [Mg] around 10^{18} cm^{-3} followed by a decrease around 10^{19} cm^{-3}) needed a detailed study. While in the second issue, covering blue InGaN QWs with GaN QDs has proven to convert the blue light into white. In order to avoid excitons trapping situation due to threading dislocations, that would act as trap centers for carriers preventing an efficient electrical injection, we attempted the fabrication of GaN QDs on dislocations free surfaces. As can be observed, both issues need a light to be shed on the structural properties at a nanoscale. Indeed, the overall electronic behavior of GaN quantum dots and p-type GaN is essentially governed by the intrinsic structure, strain-state and chemistry of these nanoscopic objects, in addition to the nature of interfaces therein. Therefore, the local and quantitative information about microstructure on a nanometer scale is indispensable getting a thorough understanding of structure-property relationships of these heterostructures. For such a purpose, transmission electron microscopy (TEM) is ideally suited since it is an analytical tool that provides local atomic imaging, electron diffraction and spectroscopy offering therefore numerous possibilities for locally investigating and characterizing the electronic, optical, chemical, and structural properties.

2. Particularities of III–V nitrides

The properties of group III-nitride semiconductors were first investigated in the seventies, when it was demonstrated that the large direct bandgap of nitride semiconductors makes them ideal candidates for visible and ultraviolet light emitters.[1] The first demonstration of stimulated emission from GaN at 2 K was reported by Dingle et al.[2] in 1971. However, the lack of successful p-type doping impeded the development of III-nitrides for more than a decade. Further progress towards GaN semiconductor devices ensued in the eighties with the incorporation of an AlN buffer layer between the sapphire/GaN epitaxial interface resulting in a significant improvement of the quality of GaN layers[3,4] Amano and Akasaki were the first to demonstrate p-type doping of GaN films, using Mg as an acceptor, setting the base for the production of the first p-n junction LEDs.[5] The growth of InGaN alloy was an additional key element for the development of a nitride laser diode. Shuji Nakamura and coworkers at Nichia Laboratories introduced blue LEDs using InGaN heterostructures.[6] Finally, the same researchers achieved firstly pulsed operation and secondly continuous-wave operation (c.w) of nitride laser diodes.[7–11]

2.1 CRYSTAL STRUCTURE

2.1.1 Crystallography

III–V nitride semiconductors – AlN, GaN, InN and their alloys – crystallize in two phases, namely; wurtzite (phase α) and zinc-blende (phase β or sphalerite) structures. These crystallographic phases differ only in the stacking sequence of the polyhedra: sequence ...ABAB... along the [0001] axis for the wurtzite phase and ...ABCABC... for the zinc-blende phase along the [111] axis (Fig. 1). In both structures each atom is tetrahedrally coordinated.

For III–V nitride semiconductors, the wurtzite structure is thermodynamically more stable. The Bravais lattice of the wurtzite structure is composed of two hexagonal closed packed sublattices, which are shifted with respect to each

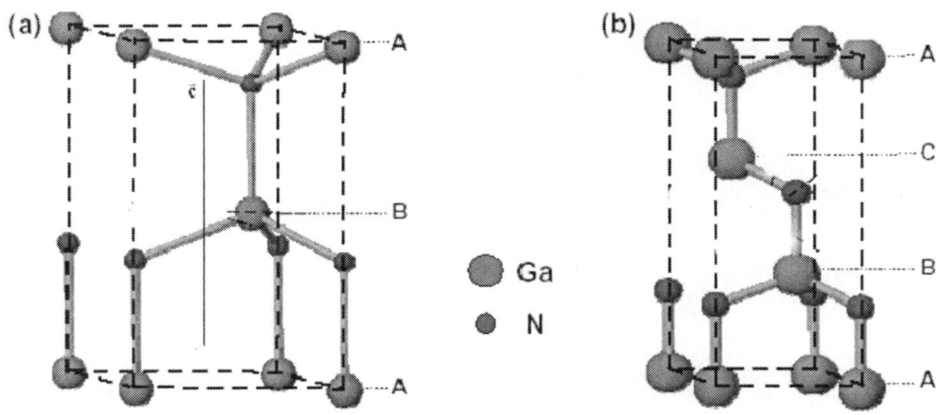

Figure 1. Schematic representation of (a) wurtzite and (b) zinc-blende phases.

other by 3/8[0001]. The associated space group is $P6_3mc$ (C^4_{6V}). In the unit cell the atomic position of the III metal is (2/3,1/3,0) while for nitrogen is (2/3,1/3,u). The parameter u is the ratio between the III metal-N bond length which is aligned along the [0001] axis and the **c** lattice parameter (see Fig. 1a). For an ideal wurtzite lattice all tetrahedral bonds have the same length and the parameter $u = u_0$ is equal to 3/8. The lattice constants [a, c] (nm) of AlN and GaN are respectively [0.311, 0.498] and [0.316, 0.529].

By contrast, the Bravais lattice for the zinc-blende structure is composed by two face-centered cubic sublattices, the second shifted by 1/4[111] with respect to the first (see Fig. 1b). The associated space group is $F\text{-}43m$ (T^2_d). The lattice constants a (nm) of AlN and GaN are respectively 0.438 and 0.452 nm.[12]

2.1.2 Polarity

Wurtzite structures have no centre of inversion and are so called polar materials. Due to this absence of an inversion point, the directions [0001] and [000-1] are not equivalent and need to be defined. We have chosen to define as [0001] direction the vector that begins from a III-metal atom (Al or Ga) and points to a N atom and is aligned along the c polar axis.[13] According to this definition, a Ga-polarity (or Ga-face) layer exists when the growth direction follows the [0001] direction, while we have a N-polarity (or N-face) layer if the [0001] direction is opposite to the growth direction (Fig. 2).

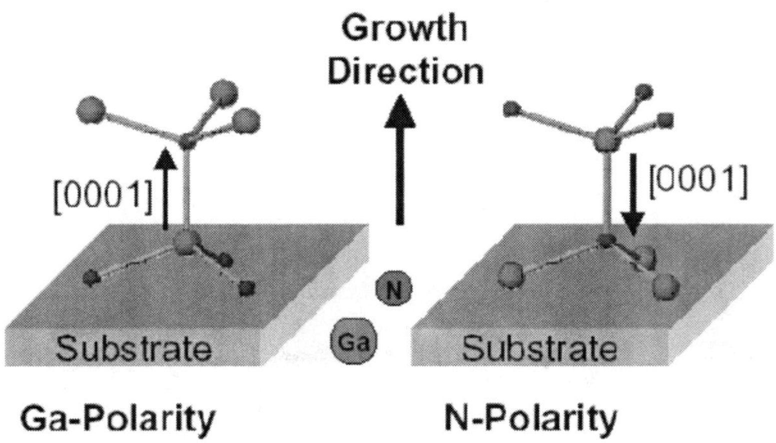

Figure 2. Polarity definition.

The polarity is one of the key parameters in the growth of III-nitride semiconductors because it influences both the surface and bulk properties of the layers. Several techniques have been used to determine and study the polarity of wurtzite nitride structures. In this review, Convergent Beam Electron Diffraction (CBED) technique will be used to determine the polarity. Let recall that CBED is a TEM associated technique.

2.2 GROWTH

2.2.1 Bulk crystals

The production of bulk GaN crystals is difficult because of the high melting point of GaN (2,600 K) and the extremely high nitrogen pressures involved in their growth. Hence, there is currently not a commonly-available single-crystal substrate wafer for nitride epitaxy. Nevertheless, GaN single crystals are grown at UNIPRESS (Poland) at temperatures of 1,400–1,700°C and at nitrogen pressure of 12–20 kbar.[14] These crystals grow as hexagonal platelets with a size of about 1 cm. Although they present excellent structural properties, such a growth method is not scalable for an industrial production. In fact, these substrates are of great interest for evaluation of homoepitaxial GaN or structures with high crystalline quality.

2.2.2 Substrates

Essentially all III-nitride films are grown heteroepitaxially. For the growth of wurtzite nitrides, the most commonly used substrates are sapphire (Al_2O_3), silicon carbide (SiC) and silicon (Si) on which a buffer layer of GaN, AlN or AlGaN needs to be deposited.

Although sapphire has become the most common substrate for nitride LDs and LEDs fabrication,[10] it is far from being an optimum substrate for AlN and GaN growth. The biggest advantage of sapphire is the fact that it is widely available at a low cost. During growth, the c-axis of the AlN (or GaN) and the sapphire are co-linear while the other two main axes are rotated by 30° around the c-axis. For this case, the lattice mismatch between GaN, AlN and Al_2O_3 is given by the relationship: $(\sqrt{3}\, a_{GaN;AlN} - a_{sapphire})/a_{sapphire}$ and it is equal to 16.1% and 13.3% respectively. A detailed study describing the orientation relationships in the III–V nitride/sapphire system can be found in the literature.[15] A comparison of sapphire and SiC as possible substrates shows that SiC has advantages in terms of lattice parameters and coefficients of thermal expansion. Furthermore, SiC has good thermal and electrical conductivity, and both n- and p-type bulk crystals are commercially available. However, SiC is much more expensive compared to sapphire, and the crystalline quality of SiC wafers must still to be improved.

On the other hand, silicon is a low cost substrate and can be found in the form of large wafers. The lattice mismatch is, however, larger than that of sapphire, and the difference in the thermal expansion results in a high density of cracks in the epilayers. To accomplish integration of III-nitride LEDs or LDs with Si electronics, these devices will have to be grown locally on an already processed Si chip. Therefore considerable research work is currently underway with the interest of reducing costs and integrating the III-nitride optoelectronics with Si technology.

In an effort to grow high-quality III–V nitride thin films, many crystal growth techniques and substrate types and orientations have been tried. Growth techniques include metalorganic vapor phase epitaxy (MOVPE), molecular beam epitaxy (MBE), hydride vapor phase epitaxy (HVPE), pulsed laser deposition (PLD) and high-pressure synthesis (HPS). MOVPE and MBE are by far the most extensively used techniques to grow III-nitride epilayers. Indeed, the heteroepitaxial growth of GaN by MOVPE produces GaN layers which, despite huge densities of dislocations, allow the fabrication of highly efficient optoelectronic devices. Henceforth, a new technology in heteroepitaxy of GaN, the epitaxial lateral overgrowth (ELO) has produced GaN layers in which the density of dislocations has been reduced by several orders of magnitude.[16] The development of nitrides by MBE has been hindered during several years by the lack of an efficient nitrogen source. This problem being solved, MBE has recently demonstrated state-of-the-art quantum well and quantum dot heterostructures, and 2D electron gas heterostructures.[16,17]

3. Electron microscopy

3.1 SAMPLE PREPARATION

For TEM experiments, the data quality is at least directly proportional to the quality of the examined specimen. Sample preparation is hence a crucial step toward a rigorous description of microstructural features. In general, a TEM sample needs to be electron transparent. The requirement for electron transparency is a function of the accelerating voltage of the electrons and the average atomic number of the investigated material. Typically, the thickness range of a TEM specimen may vary from few to 50 nm. All the studied samples presented in this review were prepared using the standard technique of mechanical polishing and ion milling using Ar^+ at 3–5 keV. In order to explore both the <1,1,-2,0> and the <1,-1,0,0> zone-axes within the same sample, the sample is cut perpendicular to <-5,1,4,0> axis. This latter was very useful for quantitative HRTEM as will be discussed later on.

3.2 DISLOCATIONS

The dislocations (with **b** as Burger's vector) in the wurtzite structure are the same as in the hcp one.[18] In order to characterize dislocations present in our samples, we have applied the invisibility criterion. The invisibility criterion postulates that if the product **g.R** is equal to zero, then we won't see any contrast because the diffraction planes giving rise to the image contrast are parallel to **R**, where **R** is the displacement field. Since **g.R** is proportional to **g.b**, we can apply this criterion to identify the Burgers vector of a dislocation using TEM diffraction contrast.

Further on, we will refer to the perfect dislocations with Burgers vector **b** = 1/3<11-20> as **a**-type dislocations, those with **b** = <0001> as **c**-type and finally those with **b** = 1/3<11-23> as **a+c** dislocations. The above perfect dislocations, **a**, **b** and **a+c**, show edge, screw and mixed character respectively.

3.3 CBED

As we have already mentioned, the polarity of a nitride layer is an important parameter that influences the surface and bulk properties. In this review, CBED technique was used to determine the polarity of our samples. The absolute polarity of a sample can be determined by comparing experimental patterns with simulated ones. For the case of wurtzite GaN layers, it was found that for CBED patterns taken along the <01-10> zone axis, the (0002) and (000-2) diffaction disks are greatly asymmetric. Simulated CBED patterns were produced using EMS software.[19]

3.4 HRTEM IMAGING

The main part of this review consists of extracting quantitative information from HRTEM images. Thinned specimens were studied using a JEOL 2010F TEM equipped with a field emission gun, operated at 200 keV and having a point to point resolution of 0.19 nm. This information concerns the interface quality, the chemical composition and the strain-state of different nitride nanostructures. In general, HRTEM images are not directly interpreted but image simulations are needed in order to get the exact atomic structure information of the experimental image. Again, the multislice method within the EMS software[19] is used to realize HRTEM image simulations.

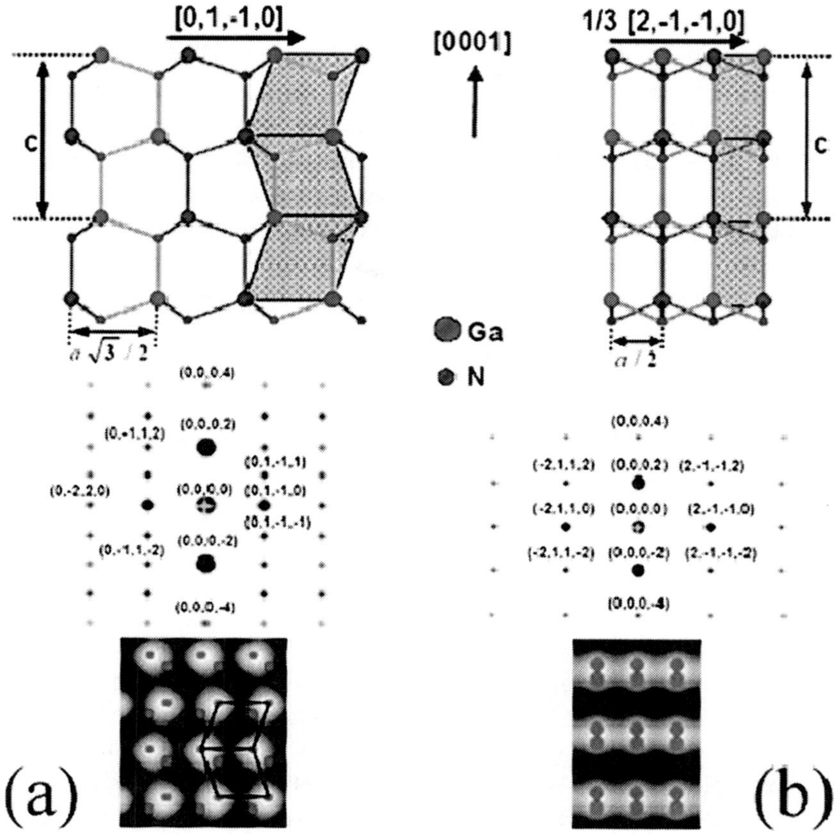

Figure 3. Schematic illustration of GaN (or AlN) wurtzite structure and the corresponding diffraction pattern and typical simulated HRTEM image along a) [2,-1,-1,0] and b) [0,1,-1,0] zone axes. Basically the imaging conditions involve the thickness and the defocus that are around 7 nm and slightly far from Scherzer, respectively. Generally very close ranges of defocus and thickness exist for which both crystals (GaN and AlN) have the maximum contrast.

Figure 3 demonstrates the geometry of the GaN wurtzite structure and the associated diffraction patterns and typical HRTEM simulated images for the two traditional zone axes. A similar scheme can also be considered for AlN. It can be observed that the wurtzite structure is constituted of “(0002)” planes containing Ga and N atoms (Ga-planes alternate with N-planes). In Fig. 3a, each atomic column contains one type of element. The distance between two atoms of the same type is equal to a $\sqrt{3}/2$ along the [0,1,-1,0] direction, and **c**/2 along the [0001]. The stacking along the [0001] axis of the projected structure can be described by consecutive parallelepiped, defined by four atoms of the same type, and which are inclined one toward the right and one toward the left. While in Fig. 3b, [0,1,-1,0] zone axis, the unit cell of the projection is a rectangle whose edge sizes are a/2 and c/2. The stacking along the [0001] axis can be described by consecutive rectangles, which are defined by four atoms of the same type. The distance between two atoms of the same type along the [0001] direction is equal to **c**/2, while along the [2,-1,-1,0] direction this distance becomes **a**/2.

3.5 QUANTITATIVE HRTEM

During the last decades, many efforts have been made to achieve the extraction of quantitative information from HRTEM images.[20] The geometric phase analysis (GPA) has proven to be a powerful tool for measuring and mapping displacements and strain fields from HRTEM images.[21] Basically, for perfect crystals, the phase of any Bragg reflection **g** is constant across the image. However, for a distorted lattice small deformation can be seen as local lateral shifts of the lattice fringes and consequently as small changes in the phase corresponding to **g**. In our case, the interest was focused on the (0002) reflection. For HRTEM images taken along a low-index zone axis namely <5,-1,-4,0>, obtained by tilting the sample in such a manner to keep the {0002} planes parallel to the viewing direction. In a first approximation, observation along this low-index zone axis corresponds to the projection of the (0002) lattice planes. The advantage of such tilted conditions is the enhancement of the image contrast and the improvement

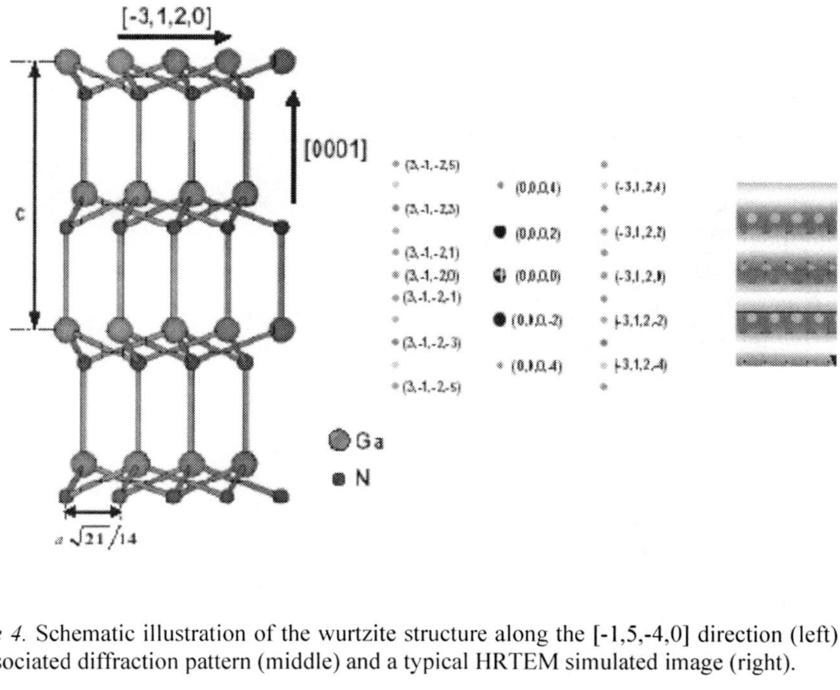

Figure 4. Schematic illustration of the wurtzite structure along the [-1,5,-4,0] direction (left) and the associated diffraction pattern (middle) and a typical HRTEM simulated image (right).

of the signal-to-noise ratio. This is due to the fact that, when being along <5,-1, -4,0> zone axis conditions, a smaller number of beams is excited by the incident beam (mostly (0002)) and the resulting amplitude of these excited beams is greater than that corresponding to those beams obtained along a high-index zone axis, such as <1,1,-2,0> axis for instance.

A schematic illustration of the wurtzite structure along the [-1,5,-4,0] direction together with the associated diffraction pattern and typical HRTEM simulated image can be seen in Fig. 4. Following the [0001] direction the distance between two atoms of the same type is equal to c/2, while along the [-3,1,2,0] direction this distance is equal to $\mathbf{a}\sqrt{21}/14$. Since this interatomic distance cannot be resolved by the electron microscope, the (0002) planes appear therefore as straight lines rather than discontinuous lines of spots. A slight distortion in the crystal lattice can then easily be detected.

4. p-type doping of GaN

GaN is naturally n-type and its doping with substitutional donors is easy and controllable up to a few 10^{19} cm^{-3}. However the development of efficient GaN based optoelectronic devices has long been hampered by the difficulty to dope it p-type. Magnesium (Mg) is today the only reliable p-type dopant in GaN, but its large Hall depth, around 160 meV[22] necessitates large atomic concentrations [Mg], in the 10^{18} cm^{-3} range, leading to free hole densities [*p*] at room temperature saturating around 1–2 x 10^{18} cm^{-3}. A further increase of the Mg concentration, up to typically 10^{19} cm^{-3} leads to a strong decrease of the free hole concentration,[23,24] which is interpreted as auto-compensation due to an increased formation of N vacancies (V_N) and/or MgV_N or Mg_2V_N complexes following the decrease of the crystal's Fermi level with increasing Mg concentration. In fact a complete understanding of the optical and electrical behavior of the resulting compound (GaN:Mg) was subject to vivid discussions. Particularly the origin of the 2.8–2.9 eV blue emission detected in most heavily Mg doped samples.[25] In numerous studies, this origin has been associated: (i) with transitions from the conduction band to a deep acceptor level,[26] (ii) with an Mg complex,[27] or (iii) with a donor-acceptor pair type transition from a deep donor to a shallow Mg acceptor.[28]

In order to shed light on all the above interrogations, a profound knowledge about structure and chemistry of the GaN:Mg compound was needed. To gain the required information, various electron microscopic techniques were applied, such as CBED, HRTEM, energy dispersive X-ray spectroscopy (EDX) and electron energy loss spectroscopy (EELS). One can notice that TEM has an enormous advantage in enabling these approaches to be carried out subsequently for investigating and characterizing the physical, chemical, and crystallographic nature of our GaN:Mg films.

The 1-μm-thick GaN:Mg films studied in this report were grown directly on a nitrited *c*-plane sapphire in a MOVPE vertical reactor using TMGa, NH_3, and $MeCp_2Mg$ as precursors. More details on the growth conditions are described elsewhere.[29] EELS spectra were acquired using an approximate spot size of 1 nm and an energy loss spectral resolution of about 1 eV (as measured from the full width at half maximum of the zero-loss peak). To avoid any beam damage, acquisition times were always kept inferior to 1 s. To gain information on the chemical environment, N-*K*(400 eV), Mg-*K*(1,305 eV), and Ga-$L_{2,3}$(1,115 eV) edges were systematically recorded parallel and perpendicular to the GaN *c*-axis, corresponding to plan-view and cross-section samples respectively. The as-obtained spectra were then processed using a power law for background subtraction in order to elucidate the energy-loss near edge structure (ELNES). This latter is particularly interesting in the range of 50 eV from threshold since it is sensitive to the chemical and crystallographic environment of the excited atom, which also refers to as the partial density of states above the Fermi level. In the case of plan-view samples, specimens were prepared by mechanical polishing and ion milling from the substrate side.

4.1 STRUCTURAL AND CHEMICAL ANALYSES

All our GaN:Mg films exhibited characteristic triangular defects (TDs)[29] having a pyramidal shape with a hexagonal (0001) base and {11-23} facets (Fig. 5). Density measurement of these TDs was found to be close to 10^{18} cm^{-3}. To investigate the thickness variation across TDs in detail, the energy-loss Log-Ratio method was used.[30] Indeed, a series of careful low-loss spectra were recorded on several TDs within very thin areas of the sample, so that the contribution from the TD is fully considered with respect to the bulk. By computing the thickness value obtained from signals acquired inside and in three close regions outside the TD (as indicated in the inset of Fig. 5), no thickness variation was detected suggesting that the observed TDs are not empty volumes.

CBED patterns (performed on larger TDs to avoid thickness ambiguities) obtained from inside and outside regions of the TD (Figs. 6a–c) demonstrate a reversal of polarity between these two regions. A comparison with a simulated CBED pattern for an object thickness of 20 nm (Fig. 6d) leads to the schematic representation of polarity distribution shown in Fig. 6e. In Figs. 6d and e, the white arrows indicate the positive [0001] direction, i.e., pointing to the Ga face. These results clearly show that TDs are inversion domains limited by inversion domains boundaries (IDBs).[31] Similar IDBs were observed when a total reversal

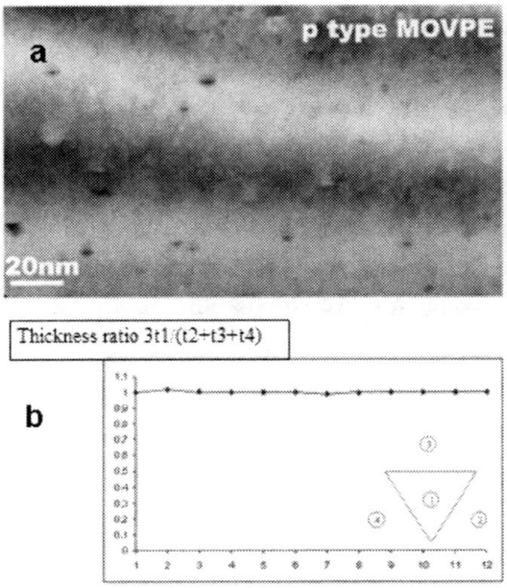

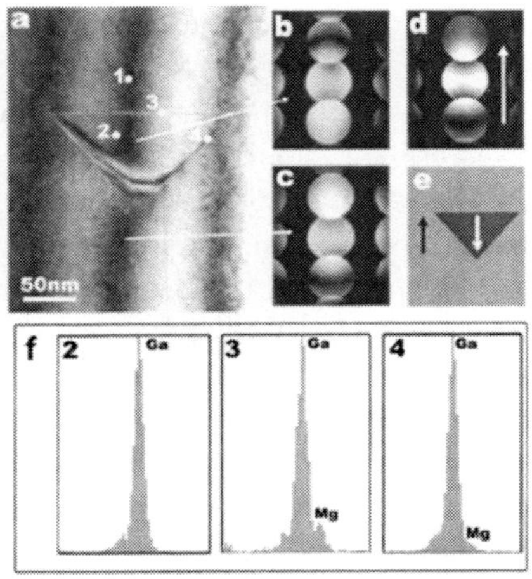

Figure 5. Left (a) Low-magnification TEM image showing the presence of TDs (dark features) within an Mg:GaN film. (b) Thickness ratio measurements within and around 12 TDs.

Figure 6. Right (a) Bright field image of a large TD. The numbers show the different areas where EDX analyses have been performed. (b) Experimental CBED pattern from inside the pyramidal defect. (c) Experimental CBED pattern from the matrix. (d) Simulated CBED pattern for an object thickness of 20 nm. (e) Polarity distribution in the TD. The arrows in Figs. d and e show the [0001] positive direction. (f) EDX spectra from the points shown in (a).

of polarity is induced by Mg doping in MBE and MOVPE-grown GaN films.[32] Let study the distribution of Mg within TDs and along IDBs. Local EDX analyses using a 0.5 nm diameter probe have been performed on four different areas (as depicted in Fig. 6a with white spots). No significant Mg was detected in both the matrix and the center of the TD since Mg concentrations are below the detection limit of our EDX instrument. However, a clear Mg enrichment is visible along the basal IDB, as seen in Fig. 6f, and also along the inclined IDBs but with less intensity as compared to basal IDBs. This Mg-enrichment at IDBs was recently found to cause important distortions which locally change the polarity from Ga to N polar.[33]

4.2 HRTEM AND EELS ANALYSES

Typical HRTEM images of such TDs are shown in Figs. 7a and 8a corresponding respectively to cross-section and top-view observations. In both images, the hexagonal wurtzite atomic arrangement is clearly exhibited. Comparison of Figs. 3a and 7a can be made. In the cross section view, a variation in the contrast is clearly visible along the IDBs. Figures 7b and c show N-*K* edge spectra recorded perpendicularly to the *c*-axis (this latter being in the plane of the image shown in Fig. 7a) outside the TD and close to the basal IDB, respectively. Similar measurements were recorded along the *c*-axis (Figs. 8b and c). As can be observed, all spectra show the typical triplet marked as A, B, and C within a range of 10 eV above threshold where most of the conduction-band states are confined. Analyses of these spectra show, however, a clear change in the relative intensities of the A-B-C features reflecting therefore a change in the electronic properties of the basal IDB as compared to those outside the TD.[34] Indeed, while measurement outside the TD shows a dominant contribution from peak B at about 403 eV, the measured spectrum from the TD shows an increase in the contribution from peaks A (~401 eV) and C (~405 eV). Measurements parallel to the *c*-axis also show a clear difference in the relative intensity of the previously mentioned features, namely A-B-C. Again, a significant increase in the intensity of peak A is noted.

First of all, it is worth mentioning that comparison of spectra in Figs. 7b and 8b (spectra recorded outside the TD) shows a significant change in the ELNES structure, which is basically due to the orientation dependence.[34] Indeed, the double differential scattering cross section for an atom excitation from an initial state $/i>$ to a final state $/f>$ by a fast electron is proportional in the first Born approximation[35] to $|<i/\exp(i\mathbf{qr})/f>|^2$, where $\mathbf{q}(=\mathbf{k}_0 - \mathbf{k})$, $\mathbf{k}_0$ and $\mathbf{k}$ being, respectively, the transfer momentum, the fast electron's wave vectors before and after interaction, and $\mathbf{r}$ the position vector of the electron. Considering the small collection angle used in our experiments (~4 mrad), the dipole-selection rule ($\Delta l=\pm1$, l being the angular-momentum quantum number) is therefore applied giving rise to the approximation $\exp(i\mathbf{qr})\approx1+i\mathbf{qr}$, where only the second term of $|<i/1+i\mathbf{qr}/f>|^2$ becomes nonzero since permitted transitions in the case of nitrogen are from 1*s* initial states ($l=0$) to final states with *p* symmetry ($l=1$). A major advantage of EELS is then the ability of selecting the momentum transfer **q** which results in **q**-resolved ELNES experiments where one single direction (onto which the electron states are projected) is probed. One should thus expect a different ELNES structure when **q** is aligned parallel to the *c*-axis of GaN as compared to that obtained when **q** is aligned normal to the *c*-axis. In other words, anisotropy of the atomic environment, and hence, of the nitrogen conduction states, will result in anisotropy of the obtained spectra when **q** is tilted off the *c*-axis. This anisotropy was also detectable using x-ray absorption spectroscopy depending on the orientation of the polarization vector of the incident photon with respect to the *c* axis.[36]

Let now consider the variation in the ELNES structure observed in Figs. 7c and 8c as compared to that in Figs. 7b and 8b, respectively. It is quite obvious that this variation can only reflect a change in the local chemical and crystallographic environment around nitrogen strongly indicating the presence of an inclusion the electronic properties of which differ from those of the GaN matrix. Indeed, EDX measurements (and quantification of Mg-*K* edges, not

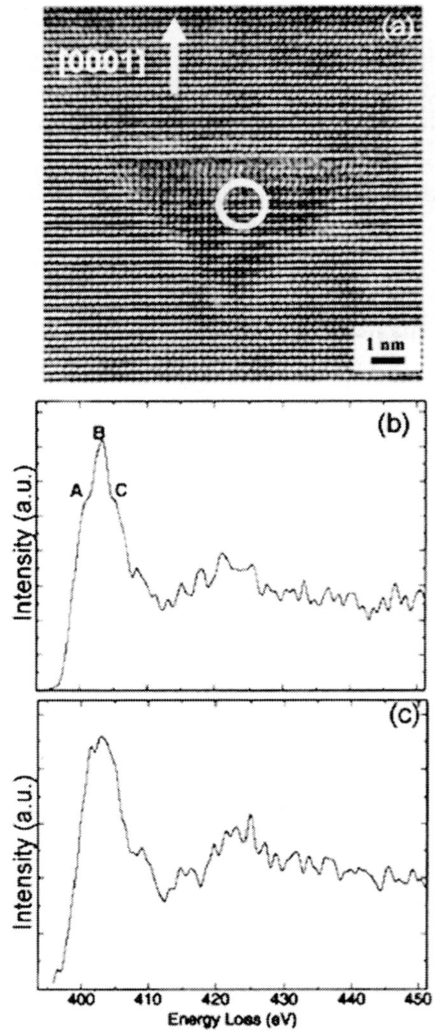

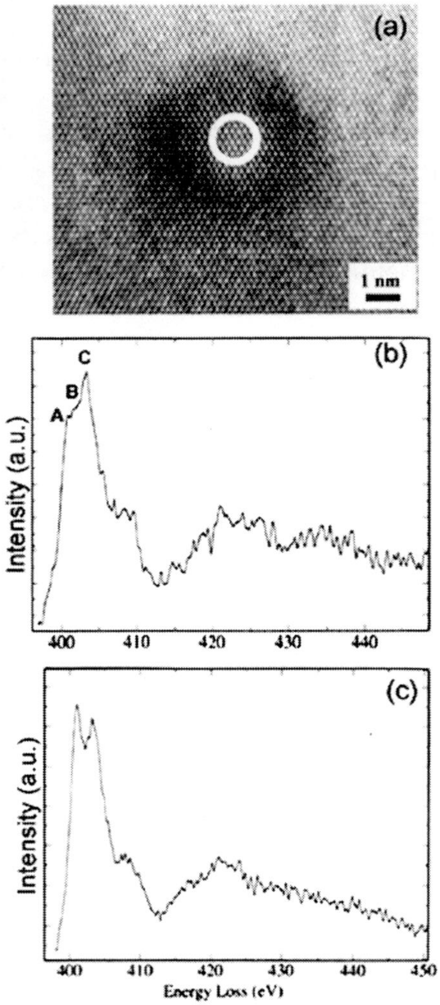

Figure 7. Left HREM of a TD recorded parallel to the c-axis, i.e., along the [0001] crystallographic zone axis. The white circle indicates the location of the probe from which spectrum in (c) was recorded. EELS data in (b) and (c) are N-K edge spectra recorded respectively outside the TD and near the basal IDB. Again, a clear change in the relative intensity of the ELNES structure (400–410 eV range) is observed.

Figure 8. Right (a) HREM image of a typical TD recorded perpendicularly to the c-axis, i.e., along the [11-20] crystallographic zone-axis. The white circle indicates the location of the probe from which spectrum in (c) was recorded. EELS data in (b) and (c) are N-K edge spectra recorded respectively outside the TD and near the basal IDB. Note the clear change in the relative intensity of the ELNES structure (400–410 eV range) in (b) as compared to that in (c).

shown) showed that IDBs are rather richer in magnesium as compared to that of the matrix. This result would suggest that Mg atoms lie preferably parallels to the facets which, to some extent, would probably be at the origin of the inversion of the polarity inside TDs.[31,37] At present, the TD formation mechanism is not fully understood but it is likely linked to the polarity of the matrix, to the amount of Mg incorporated and to the growth temperature. The high temperature growth conditions used in MOVPE (1,080°C) provide a thermodynamic equilibrium which should in principle favor the formation of stable phases based on Mg, such as Mg_3N_2 and/or (GaMg)N as a consequence of the upper limit for the solubility of Mg in GaN. In a recent study, a model of IDBs was proposed using first-principles pseudopotential density functional calculations.[38] The model exhibits GaNMgNGa layers that are stacked in *abcab* registry, with at least 3/4 of the available *c* sites occupied by Mg atoms. From an other HREM study,[31] it was proposed that IDBs consist of Mg_3N_2 building blocks for both the basal and inclined facets of the pyramids. It is clear from all these models that a significant fraction of the Mg present in the film is incorporated in TDs where it is supposedly not active from the electrical point of view. Therefore, TDs formation should be considered as a serious mechanism limiting p-type doping in nitrides, in addition to the fact that their high concentration can reasonably account for the self-compensation phenomenon in *p*-type GaN responsible for the free-hole concentration decrease. To overcome these limitations, promising routes to increase free-hole concentrations were recently attempted by using semipolar GaN layers[39] and Mg insertion in a δ-doping fashion into a GaN matrix.[40]

5. GaN quantum dots

Control and optimization of the radiative and structural properties of single quantum emitters, quantum dots (QDs), is a core issue of present day optical science and technology.[41] Much effort has been devoted to the growth of nitride self-assembled quantum dots. For many applications, a high density of uniform and spatially ordered QDs is desired as, for example, in the realization of QDs based-lasers where the vertical stacking of QDs increases their total density and consequently the gain.[42–44] Such a stacking may also be of great interest for the realization of quantum computation/information processing devices, because of the existence of a strong built-in electric field due to spontaneous and piezoelectric polarization effects.[45] In the case of gallium nitride (GaN) QDs, they have attarcted in the past decade general attention due to their potential in providing highly efficient short-wavelength (ultraviolet) light emitters. This efficiency is attributed to strong exciton localization in QDs.

When grown using an MBE system, structural and optical properties of GaN QDs strongly depend on growth parameters. Indeed, when AlN acts as a spacer layer, it is believed that the formation of GaN QDs proceeds generally according to the Stranski–Krastanow (SK) growth mode that often applies for heteroepitaxy in systems with lattice mismatch superior to 2% (2.5% in GaN/AlN system). Basically, GaN grows in the layer-by-layer mode in Ga-rich conditions (Ga/N >1), while in the opposite case (Ga/N < 1), and at a critical thickness, three-dimensional (3D) QDs develop on top of a remaining wetting layer. In fact, the relief of strain caused by the lattice mismatch between GaN and AlN overcome-pensates the increase in surface area and thus accounts for the formation of self-assembled QDs. The formation of GaN QDs is therefore strictly linked to both kinetics and strain relaxation. However, when ammonia (NH_3) is used as nitrogen source,[17] the realization of good quality GaN/AlN QDs has shown a little deviation from the traditionally known SK mode. In order to increase the surface diffusion length and consequently push the epitaxial growth system closer to the elastic and surface energy equilibrium, a growth interruption was needed.[17] This interruption resulted in an instantaneous 2D–3D transition, giving rise to the fabrication of self assembled GaN QDs, the room-temperature luminescence of which was shown to be successfully tuned from blue to orange depending on the dots size.

Generally, the main drawback of a GaN/AlN system is its high density of structural defects (dislocations) due to the use of foreign substrates which are not adapted in terms of both lattice parameters and chemistry. The density of dislocations in such films is often superior to 10^{10} cm^{-2} and their presence may influence both the nucleation and the physical properties of QDs. It has already been shown that the nucleation of GaN QDs is influenced by the strain field existing around edge-type threading dislocations: QDs are preferentially nucleated on the extensively strained AlN surface regions close to dislocation cores.[46] Even if QDs are efficient to confine carriers, dislocations may play a detrimental role for their physical properties. In fact, in most envisaged devices (light emitting diodes, lasers, etc.), carriers would first be created outside QDs and then would diffuse towards QDs. Threading dislocations would act as trap centers for carriers preventing an efficient electrical injection. Therefore, for the realization of efficient devices it would be of great interest to grow QDs on dislocation-free regions.

5.1 GROWTH CONDITIONS

The growth of GaN/AlN QDs on vicinal Si(111) (5° misorientation) is fully described elsewhere.[47] The misorientation of the Si surface is parallel to <11-2> direction while Si surface steps are aligned along <110> axis. An AlN buffer layer was first deposited at 920°C, which was found to be two dimensional using Reflection High-Energy Electron Diffraction (RHEED). The AlN growth proceeds up to a nominal thickness of 400 nm. This is achieved using a very low growth rate (0.1 μm/h) and a nucleation temperature of 650°C followed by a rapid ramping to the growth temperature. A subsequent 2D GaN layer is then deposited on a fully relaxed AlN buffer-layer at a substrate temperature of 790°C, at a growth rate of 1 μm/h. This GaN 2D strained layer continued up to a thickness of about 9 ML where growth interruption is performed (meaning that both Ga and NH_3 are stopped). At this stage, the RHEED pattern showed an evidence of surface roughness (changes from steaks to spots) reflecting a 2D–3D transition. This gave rise to the formation of the desired GaN QDs. It will be shown later that these QDs leave behind a very thin GaN wetting layer. A second AlN layer (spacer) is deposited afterwards in the same conditions as above up to a thickness of 60 nm. As formerly described, this spacer layer was again followed by a second GaN QDs layer that was leaved on the surface, so that both QD types (buried and surface) can be studied under the same conditions. This type of samples has also the advantage to shed light on the step bunching issue in relationship with the QDs stacking order.

5.2 Si-AlN INTERFACE

First of all, the influence of the Si substrate vicinality on the morphology and the orientation of the AlN layer was studied.

Selected area electron diffraction from the Si/AlN interface in cross-section samples along the Si [1−10] zone axis showed a 2.5° disorientation between the Si [111] and the AlN [0001] axes (Fig. 9a). This finding was also confirmed by HRTEM imaging (Fig. 9b), demonstrating that the AlN growth axis tilt is a

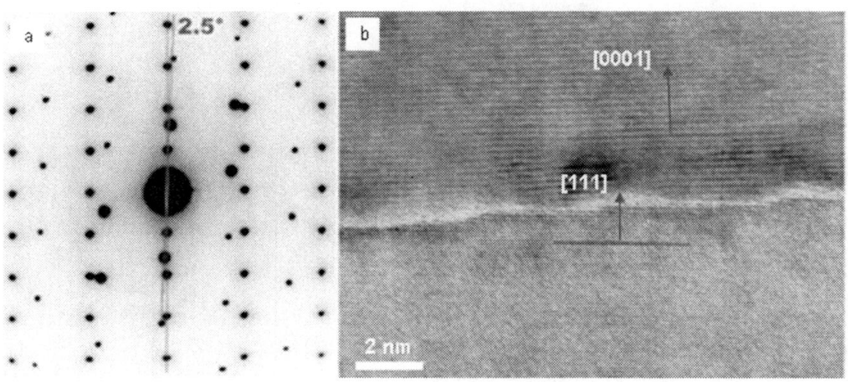

Figure 9. Selected area electron diffraction from the Si/AlN interface in cross-section samples along the Si [1−10] zone axis along with the corresponding HRTEM image.

consequence of the use of a vicinal Si (111) substrate.[48] In fact, initial AlN growth is supposed to start as atomic incorporations at the edge of Si steps, but also by island nucleation on Si terraces. The two processes thus compete for the arriving atoms particularly when terraces are very narrow such as those in our vicinal Si(111) substrate the width of which is around 10 nm. This competition may suppress the growth of large islands (as it is usually the case in MBE samples) giving rise to a slightly tilted growth front due to the AlN lateral accommodation upon the edge of Si steps.

5.3 DISLOCATIONS-FREE SURFACES

Bright-field TEM imaging along a plan-view orientation (Fig. 10) reveals the formation of wide macrosteps (150–250 nm). This phenomenon, also known as step bunching, is particularly pronounced when the miscut angle of the substrate is high (our case), as it was reported elsewhere.[49–52]

In fact, the growth of the first AlN thick layer had led to the formation of higher steps, of about 10 nm height and quite large terraces (see Fig. 11). In other words, our results show that the equilibrium morphology of the vicinal AlN(0001)

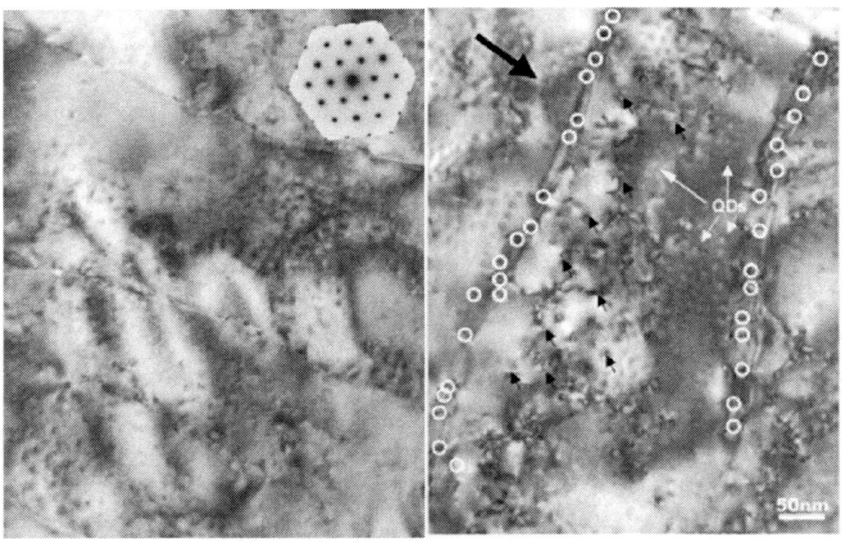

Figure 10. (Left) Bright-field TEM image along a plan-view orientation AlN [0001] axis as confirmed by the diffraction pattern. (Right) Two-beams g = (12-10) image. The dark large arrow in the right image indicates the macrostep displacement direction. All the dislocations emerging on macrostep edges are shown by white open circles. Small black arrows indicate some dislocations emerging on terraces. White arrows indicate some QDs in both dislocation-free and dislocated areas.

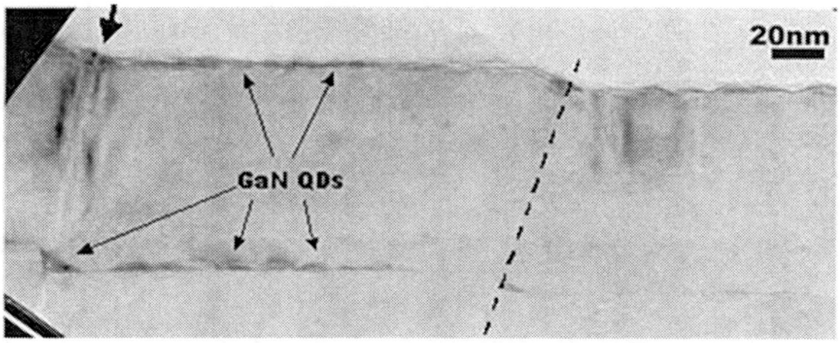

Figure 11. Cross-section TEM image. Surface and buried QD planes are visible. The dotted line shows the displacement of a macrostep edge. The black arrow indicates an inclined threading dislocation.

surface consists of two facets: a large low index facet, AlN(0001), mixed with a high index one in the form of step bunches. It appears therefore that the Si vicinality is partially overcompensated by a slight tilt of the AlN growth direction and by the formation of wide macrosteps.[48]

On the other hand, Fig. 10 illustrates the presence of QDs (dark dots) along edges and onto terraces, with a noticeable contrast corresponding to AlN steps (dark lines), that are aligned parallel to the [11-20] axis, as determined from an electron diffraction pattern (see inset). Another revealed feature is the presence of typical threading dislocations contrasts viewed along their lines. To clearly evidence this contrast, plan-view TEM measurements were carried in a two-beams condition with ***g*** = (12-10) in which all *a*-type and *a*+*c*-type dislocations are in contrast (Fig. 10, right). These measurements elucidated a striking dislocation contrast at the edge of steps, and showed a high dislocation density at these steps as compared to the density measured within terraces. This result suggests that the dislocations were particularly attracted to those parts leaving the other part of the terrace nearly free of dislocations. Further studies using cross-sectional TEM investigations were needed to understand the build-up dislocation network within the AlN film.

On the cross-section TEM image of Fig. 11, macrosteps on the AlN surface are clearly visible. The width of the terraces is confirmed and the macrosteps height is estimated to be 8–12 nm. Thanks to the presence of the buried QD plane, the position and the morphology of the AlN surface at an intermediate stage of the growth is revealed. The macrostep edges are evidenced, thanks to the presence of large GaN QDs. Following the position of macrosteps from the buried plane to the surface, it can be seen that the growth occurs both vertically and laterally with a displacement of the macrostep from left to right on Fig. 11 (the dotted line shows the macrostep displacement). From the measurement of the AlN film thickness (between the buried and the surface QD plans) and macrostep edge lateral displacement, the vertical to latelateral growth rate ratio is estimated to be equal to 4. Figure 11 corresponds to a very thin TEM specimen and only one dislocation is visible (bold arrow on the left). This dislocation is inclined and passes through successive positions of a macrostep edge. A similar behavior of dislocations has already been observed in AlN layers grown on vicinal sapphire-(0001) substrates.[53]

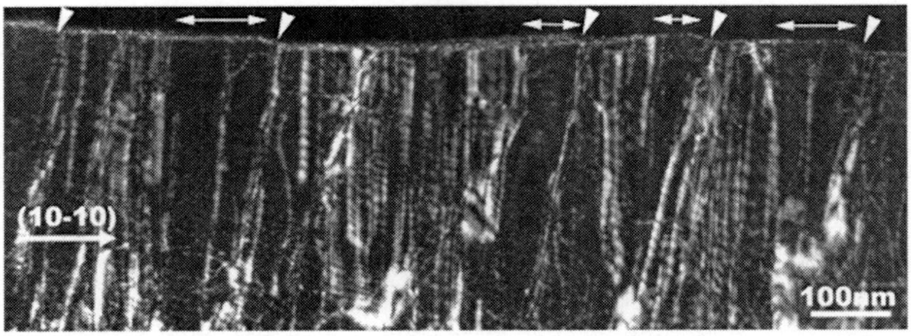

Figure 12. (10-10) dark-field cross-section TEM image. Inclined dislocations emerge on macrostep edges which are indicated by small inclined arrows. The horizontal arrows show the nearly dislocation-free areas above the macrostep edges.

Although analyses in a cross section configuration do not give a true image of the dislocation density located at a given edge, since the edge is only viewed along its profile, a series of dark-field TEM images confirmed that the edge of the steps were found to be attractive to dislocation lines within AlN. Figure 12 is a two-beam (1−100) cross-section dark-field image. This orientation is chosen in order to observe the macrosteps edge on. Two kinds of threading dislocations are observed: most of them have a nearly vertical line but a few of them are clearly inclined. These inclined dislocations emerged on the AlN surface at macrostep edges (white arrows). Due to this dislocation inclination, it is important to note that a part of the terraces above the macrostep edges is nearly free of threading dislocations (horizontal arrows).

The behavior of dislocations can be described as depicted in Fig. 13. We consider a starting situation with an AlN surface formed by macrosteps and with vertical threading dislocations emerging on terraces (Fig. 13a). Due to the lateral displacement of the macrostep edges, they intersect some vertical dislocations (Fig. 13b). At this stage, these dislocations bend and then follow the displacement of macrostep edges, as it is observed in Fig. 12. The area above the bent dislocations is therefore a dislocation-free zone as indicated by arrows in Fig. 13c, where the underneath surface positions are indicated by dotted lines. This situation is what we observed in Fig. 12 in our sample. If the growth proceeds further, we expect that all the vertical threading dislocations intersect a macrostep edge, bend, and therefore emerge in macrostep edges resulting in dislocation-free terraces. This is the case in Fig. 13d, where the macrostep *i*+1 has grown above the first position of the macrostep edge *i*. However, it seems that in the case of our present sample, the mechanism described in Fig. 13 is not complete: the majority of dislocations is still vertical and emerges on the terraces. In fact, macrosteps have only passed through a limited part of the terraces during their lateral displacement. Nevertheless, we already clearly observe dislocation-free areas above macrostep edges where QDs have also nucleated. This result is very promising since one could expect the fabrication of well-suited templates for the growth of QDs. Nonetheless some QDs have still nucleated on surfaces where threading dislocations emerge. White arrows in Fig. 10 indicate some QDs nucleated on both kinds of areas. Obtaining fully dislocation-free terraces under our growth conditions and with the same vicinal angle needs the realization of thicker samples where macrosteps during their lateral displacement should pass through the entire terraces. However, because of the large thermal mismatch between AlN and Si, thick AlN films are drastically cracked. Higher vicinal angles result in the formation of narrower terraces. Another way of resolving this issue is to increase the lateral growth rate. Indeed, the lateral growth rate seems to be

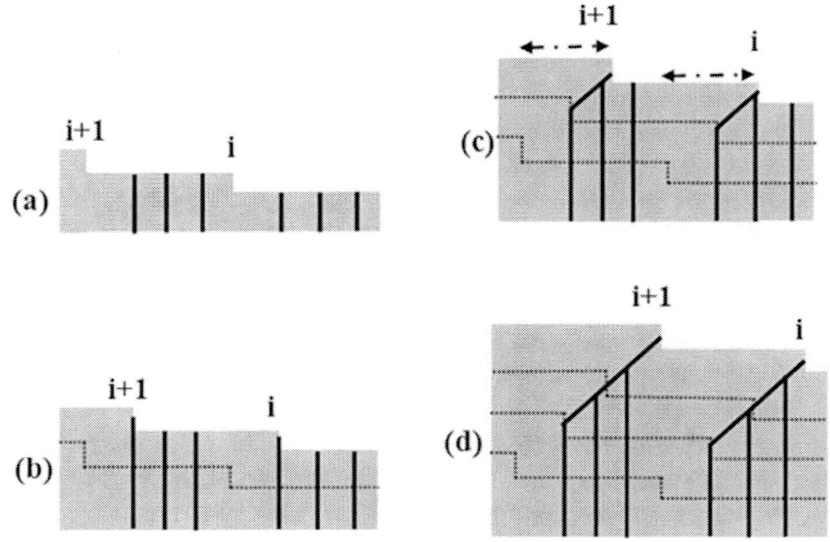

Figure 13. Schematic representation of the dislocation bending phenomena. (a) The starting situation is the following: two macrostep edges (i and i+1) plus some threading dislocations indicated by vertical lines. (b) Intersection of macrostep edges i and i+1 with threading dislocations. The dotted line shows the position of the vicinal surface at the starting point. (c) Intersected dislocations are bent and then they follow the macrostep edges displacement. The horizontal arrows show the dislocation-free areas above the macrostep edges. (d) Macrostep edge i+1 has grown above the starting position of macrostep edge i. All the dislocations are bent and then they emerge on macrostep edges. Terraces are dislocation-free.

four times slower than that of the vertical one, as can easily be deduced from ratio measurements of the step-progress to the spacer-thickness (Fig. 11). This means that in order for a macrostep to intercept all threading dislocations, the AlN layer dimension (thickness) must be four time that of the terrace (width), implying that this dislocation dragging behavior will not concern all threading dislocations, which is in full agreement with dislocations densities measured within terraces and at the edge of the steps (as deduced from our plan-view TEM images). The increase of the lateral growth rate would consequently lead to thin AlN films having large terraces where all threading dislocations are bent and emerge at macrostep edges.

5.4 GaN/AlN QDs

Let's focus now our study on the microstructure of GaN QDs. It is worth mentioning that a preferential nucleation of GaN at the edge of the steps is now clearly evidenced since such a location (either buried or surface) is always occupied by GaN (as shown in Fig. 11).

As far as the QDs shape is concerned, two types were distinguished (those of the surface are not considered in the present review). The cross-sectional HRTEM image of Fig. 14, observed along the [11-20] direction, depicts typical GaN QD embedded in the AlN matrix. The wetting layer aside the QD is also visible. The first type (Fig. 14a) is found to present a truncated pyramidal shape with walls inclined by 30° from the surface plane, corresponding to the usual {101-3} facets, as depicted in Fig. 15. The average height and base diameter are about 3 nm and 10 nm, respectively. While the second type (Fig. 14b), GaN QDs at the edge of AlN steps, shows a rather complex form. It is worth mentioning that our previous atomic force microscopy study[48] showed that the edges of these steps are saturated with GaN QDs in the form of a chain-aligned GaN QDs, or quantum wires, which are simply the result of a QDs coalescence. It can probably be deduced from this observation that a preference for a nucleation site all along the steps occurred up to saturation. A close examination of such a form shows a complex surface section, which in part is a reproduction of the AlN step edge. This complex shape is evidently due to the fact that step bunches do not form a well defined facet, but rather a multifaceted surface, as will clearly be shown in Fig. 16b. Since all these features were also reproduced on the AlN spacer layer surface, it may therefore be deduced that this capacity to reproduce the morphology of AlN layers might lead to the growth of uniform vertically correlated GaN QDs, provided the density of dots is controlled, either, through the optimization of the adatom diffusion length, or, through the reduction of the terraces' width.

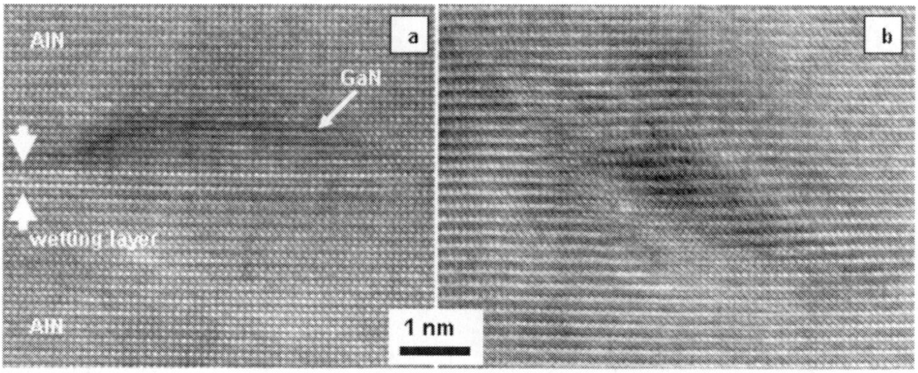

Figure 14. HRTEM images of GaN/AlN QDs. (a) a buried QD showing a truncated shape, and (b) a typical QD located at the edge of an AlN step.

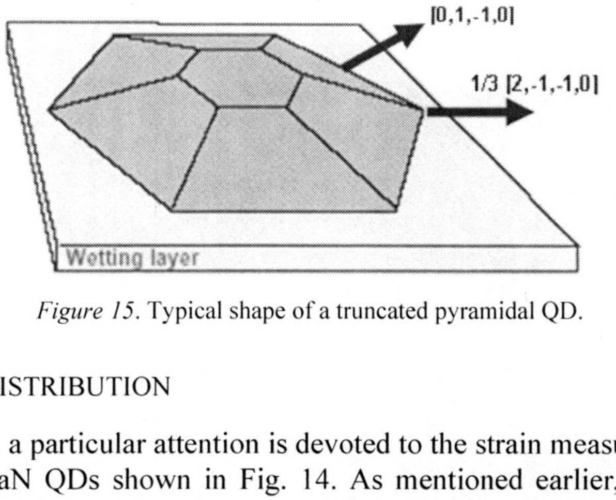

Figure 15. Typical shape of a truncated pyramidal QD.

5.5 STRAIN DISTRIBUTION

In this section, a particular attention is devoted to the strain measurement within and around GaN QDs shown in Fig. 14. As mentioned earlier, GPA method measures the strain **s** with respect to a reference area within a given HRTEM image. Only the $\mathbf{s}_z$ component (perpendicular to (0002) planes) is considered. This strain is given by $\mathbf{s}_z = [\mathbf{g}_{ref} - \mathbf{g}(r)]/\mathbf{g}(r)$ where **g** is the interplanar spacing measured in the image. In our case, $\mathbf{g}_{ref}$ corresponds to the AlN-(0002) planes under the QD (or reference area), supposed to be close to the $\mathbf{g}_{AlN\text{-}bulk}$ since it is considered fully relaxed. While **g**(r) corresponds to the above interatomic distances that include GaN-(0002) planes and AlN-(0002) above the QD.

For such a purpose, QDs shown in Fig. 14 were recorded along the [5,-1, -4,0] zone-axis in order to analyze $\mathbf{s}_z$ strain component (see Figs. 16a and b). As can clearly be observed, only the (0002) planes are visible enhancing therefore the exact location and shape of QDs. The GPA strain (or relative deformation) map of the high resolution image is shown in Figs. 16c and d. It was calculated by placing a Gaussian mask of size approximately **g**/5 around the g = 0002 reflection.

It is clear that on average the $\mathbf{s}_z$ value is almost zero for the reference area. The strain map of Fig. 16c shows positive values within GaN. One can easily deduce that GaN at the heart of the QD is more relaxed as compared to the one within the wetting layer and the upper part of the QD. Same deduction can be made for GaN in Fig. 16d. Although in Fig. 16d a negative value is obtained in the upper AlN layer demonstrating that AlN exhibits a local strain generated by the GaN dot, the upper AlN layer in Fig. 16c indicates however full relaxation (same color as in the bottom AlN layer). In fact, this upper AlN layer should in reality be strained since it is directly in contact with a relatively relaxed GaN, as it was reported in the literature[54] in a GaN/AlN QDs superlattice where the upper AlN layer was ten times thinner than that in our case. On one hand, this difference suggests that the AlN spacer layer thickness played a fundamental role in the strain modulation within such a layer, while on the other, this thickness relaxation effect in the upper AlN layer is less intense when GaN is vertically positioned.

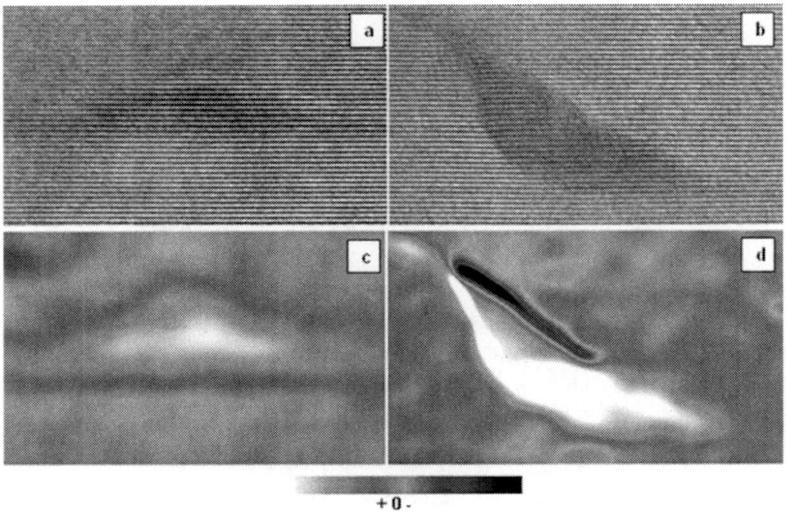

Figure 16. (a) and (b) show HRTEM images viewed along [5,-1,-4,0] zone-axis of two different types of GaN QDs embedded inside an AlN matrix. Corresponding GPA strain map are shown in (c) and (d).

6. Conclusions

The role of microstructure with emphasis on p-type GaN and GaN quantum dots has been illustrated. As has been demonstrated, the techniques for studying the physical and chemical microstructure at the nanometer scale are sufficiently advanced with atomic resolutions which are currently a practical reality using state of the art TEM instrumentation. The current development of luminescence based microscopy, such as cathodoluminescence electron microscopy coupled to suitable optoelectronic modeling based on experimentally observed microstructural parameters should provide additional insight and aid in the understanding of III-nitrides and help the development of new devices with superior properties.

ACKNOWLEDGEMENTS

It is a pleasure to acknowledge the contributions of a number of collaborators involved in this work. In particular, I would like to thank Ph. Vennéguès, B. Beaumont, P. Gibart, F. Semond, and O. Tottereau from CRHEA laboratory where I achieved most of the presented results. This work was performed in the frame of the convention between the Moroccan CNRST and the French CNRS.

References

1. J.I. Pankove, E.A. Miller, and J.E. Berkeyhauser. *RCA Rev.*, **32**, 383 (1971)
2. R. Dingle, K.L. Shaklee, R.F. Leheny, and R.B. Zetterst. *Appl. Phys. Lett.*, **19**(1), 5 (1971)
3. S. Yoshida, S. Misawa, and S. Gonida. *Appl. Phys. Lett.*, **42**(5), 427 (1983)
4. H. Amano, N. Sawaki, I. Akasaki, and Y. Toyoda. *Appl. Phys. Lett.*, **48**(5), 353 (1986)
5. H. Amano, M. Kito, K. Hiramatsu, and I. Akasaki. *Jpn. J. Appl. Phys.*, **28**, 2112 (1989)
6. S. Nakamura and T. Mukai. *Jpn. J. Appl. Phys.*, **31**, 1457 (1992)
7. S. Nakamura, M. Senoh, S. Nagahama, N. Iwasa, T. Matsushita, and T. Mukai. *Appl. Phys. Lett.*, **76**, 22 (2000)
8. S. Nakamura, M. Senoh, S. Nagahama, N. Iwasa, T. Yamada, T. Matsushita, H. Kiyoku, and Y. Sugimoto. *Jpn. J. Appl. Phys.*, **35**, 74 (1996)
9. S. Nakamura, M. Senoh, S. Nagahama, N. Iwasa, T. Yamada, T. Matsushita, H. Kiyoku, and Y. Sugimoto. *Appl. Phys. Lett.*, **69**, 4056 (1996)
10. S. Nakamura and G. Fosol. The Blue Laser Diode. Springer, Berlin (1997)
11. S. Nakamura, M. Senoh, N. Iwasa, and S. Nagahama. *Jpn. J. Appl. Phys.*, **34**, 1332 (1995)
12. H. Morkoç, S. Strite, G.B. Gao, M.E. Lin, B. Sverdlov, and M. Burns. *J. Appl. Phys.*, **76**, 1363 (1994)
13. J.L. Rouviere, J.L. Weyher, M. Seelmann-Eggebert, and S. Porowski. *Appl. Phys. Lett.*, **73**(5), 668 (1998)
14. S. Porowski. *J. Cryst. Growth*, **189–190**, 153 (1998)
15. A.N. Efimov, A.O. Lebedev, and A.M. Tsaregorodtsev. *J. Appl. Cryst.*, **31**, 461 (1998)
16. B. Beaumont, P. Gibart, N. Grandjean, and J. Massies. *Comptes Rendus de l'Académie des Sciences - Series IV - Physics*, **1**, 35 (2000)
17. B. Damilano, N. Grandjean, F. Semond, J. Massies, and M. Leroux. *Appl. Phys. Lett.*, **75**, 962 (1999)
18. J.P. Hirth and J. Lothe. Theory of Dislocations. MacGraw-Hill, New York (1968)
19. P.A. Stadelmann. *Ultramicroscopy*, **21**, 131 (1987)

20. J.L. Rouviere, P. Bayle-Guillemaud, G. Radtke, S. Groh, and O. Briot. *Inst. Phys. Conf. Ser.*, **169**, 17 (2001)
21. M.J. Hÿtch, E. Snoeck, and R. Kilaas. *Ultramicroscopy*, **74**, 131 (1998)
22. W. Götz, N.M. Johnson, J. Walker, D.P. Bour, and R.A. Street. *Appl. Phys. Lett.* **68**, 667 (1996)
23. U. Kaufmann, P. Schlotter, H. Obloh, K. Köhler, and M. Maier. *Phys. Rev. B* **62**, 10 867 (2000)
24. L.T. Romano, M. Kneissi, J.E. Northrup, C.G. Van de Walle, and D.W. Treat. *Appl. Phys. Lett.* **79**, 2734 (2001)
25. M. Leroux, N. Grandjean, B. Beaumont, G. Nataf, F. Semond, J. Massies, and P. Gibart. *J. Appl. Phys.* **86**, 3721 (1999)
26. E. Oh, H. Park, and Y. Park. *Appl. Phys. Lett.* **72**, 70 (1998)
27. M. Smith, G.D. Chen, J.Y. Lin, H.X. Jiang, A. Salvador, B.N. Sverdlov, A. Botchkarev, H. Morkoc, and B. Goldenberg. *Appl. Phys. Lett.* **68**, 1883 (1996)
28. U. Kaufmann, M. Kunzer, M. Maier, H. Obloh, A. Ramakrishnan, B. Santic, and P. Schlotter, *Appl. Phys. Lett.* **72**, 1326 (1998)
29. P. Vennéguès, M. Benaissa, B. Beaumont, E. Feltin, P. De Mierry, S. Dalmasso, M. Leroux, and P. Gibart. *Appl. Phys. Lett.* **77**, 880 (2000)
30. R.F. Egerton. Electron Energy Loss Spectroscopy in the Electron Microscope, Plenum, New York, 1989
31. P. Vennéguès, M. Leroux, S. Dalmasso, M. Benaissa, P. De Mierry, P. Lorenzini, B. Damilano, B. Beaumont, J. Massies, and P. Gibart. *Phys. Rev. B* **68**, 235214 (2003)
32. N. Grandjean, A. Dussaigne, S. Pezzagna, and P. Vennéguès. *J. Cryst. Growth* **251**, 460 (2003)
33. Q. Sun, A. Selloni, T.H. Myers, and W. Alan Doolittle. *Phys. Rev. B* **73**, 155337 (2006)
34. M. Benaissa, P. Vennéguès, B. Beaumont, and P. Gibart, W. Saikaly and A. Charai. *Appl. Phys. Lett.*, **77**, 2115 (2000)
35. M. Inokuti. *Rev. Mod. Phys.* **43**, 297 (1971)
36. W.R.L. Lambrecht, S.N. Rashkeev, B. Segall, K. Lawniczak-Jablonska, T. Suski, E.M. Gullikson, J.H. Underwood, R.C.C. Perera, J.C. Rife, I. Grzegory, S. Porowski, and D.K. Wickenden. *Phys. Rev. B* **55**, 2612 (1997)
37. M. Leroux, P. Vennéguès, S. Dalmasso, M. Benaissa, E. Feltin, P. de Mierry, B. Beaumont, B. Damilano, N. Grandjean, and P. Gibart. *Phys. Stat. Sol. (a)* **192**, 394 (2002)
38. J.E. Northrup. *Appl. Phys. Lett.* **82**, 2278 (2003)
39. J.F. Kaeding, H. Asamizu, H. Sato, M. Iza, T.E. Mates, S.P. DenBaars, J.S. Speck, and S. Nakamura. *Appl. Phys. Lett.* **89**, 202104 (2006)
40. C. Simbrunner, M. Wegscheider, M. Quast, T. Li, A. Navarro-Quezada, H. Sitter, A. Bonanni, and R. Jakiela. *Appl. Phys. Lett.* **90**, 142108 (2007)
41. P. Michler, A. Kiraz, C. Becher, W.V. Schoenfeld, P.M. Petroff, L. Zhang, E. Hu, and A. Imamoglu. *Science* **290**, 2282 (2000); J.N. Farahani, D.W. Pohl, H.-J. Eisler, and B. Hecht. *Phys. Rev. Lett.* **95**, 017402 (2005)
42. B. Daudin, F. Widmann, G. Feuillet, Y. Samson, M. Arlery, and J.L. Rouvière. *Phys. Rev. B* **56**, R7069 (1997)
43. K. Tachibana, T. Someya, and Y. Arakawa. *Appl. Phys. Lett.* **75**, 2605 (1999)
44. K. Hoshino, S. Kako, and Y. Arakawa. *Appl. Phys. Lett.* **85**, 1262 (2004)
45. S. De Rinaldis, I. D'Amico, E. Biolatti, R. Rinaldi, R. Cingolani, and F. Rossi. *Phys. Rev. B* **65**, 081309(R) (2002)
46. J.L. Rouvière, J. Simon, N. Pelekanos, B. Daudin, and G. Feuillet. *Appl. Phys. Lett.* **75**, 2632 (1999)

47. F. Semond, Y. Cordier, N. Grandjean, F. Natali, B. Damilano, S. Vézian, and J. Massies. *Phys. Stat. Sol. (a)* **188**, 501 (2001)
48. M. Benaissa, P. Vennégués, O. Tottereau, L. Nguyen, and F. Semond. *Appl. Phys. Lett.* **89**, 231903 (2006)
49. J. Tersoff, Y.H. Phang, Z. Zhang, and M.G. Lagally. *Phys. Rev. Lett.* **75**, 2730 (1995); F. Liu, J. Tersoff, and M.G. Lagally. *ibid.* **80**, 1268 (1998)
50. J.Y. Tsao. Materials Fundamentals of Molecular Beam Epitaxy. Academic, Boston, MA (1993), Chap. 6.
51. M.V. Ramana Murty, P. Fini, G.B. Stephenson, C. Thompson, J.A. Eastman, A. Munkholm, O. Auciello, R. Jothilingam, S.P. DenBaars, and J.S. Speck. *Phys. Rev. B* **62**, R10661 (2000)
52. J. Brault, S. Tanaka, E. Sarigiannidou, J.-L. Rouvière, B. Daudin, G. Feuillet, and H. Nakagawa. *J. Appl. Phys.* **93**, 3108 (2003)
53. X.Q. Shen, H. Okumura, and H. Matsuhata. *Appl. Phys. Lett.* **87**, 101910 (2005)
54. E. Sarigiannidou, E. Monroy, B. Daudin, J.L. Rouvière, and A.D. Andreev. *Appl. Phys. Lett.* **87**, 203112 (2005)

PROSPECTIVE TERAHERTZ APPLICATIONS OF CARBON NANOTUBES

M. E. PORTNOI[*, 1], O. V. KIBIS[2] AND M. ROSENAU DA COSTA[3]
[1]*School of Physics, University of Exeter, Stocker Road, Exeter EX4 4QL, United Kingdom*
[2]*Department of Applied and Theoretical Physics, Novosibirsk State Technical University, Novosibirsk 630092, Russia*
[3]*International Center for Condensed Matter Physics, University of Brasilia, 70904-970 Brasilia DF, Brazil*

Abstract. We formulate and justify several proposals utilizing the unique electronic properties of different types of carbon nanotubes for a broad range of applications to THz optoelectronics, including THz generation by hot electrons in quasi-metallic nanotubes, frequency multiplication in chiral-nanotube-based superlattices controlled by a transverse electric field, and THz radiation detection and emission by armchair nanotubes in a strong magnetic field.

Keywords: Carbon nanotubes, terahertz radiation.

1. Introduction

Creating a compact reliable source of terahertz (THz) radiation is one of the most formidable tasks of contemporary applied physics.[1] One of the latest trends in THz technology[2] is to use carbon nanotubes – cylindrical molecules with nanometer diameter and micrometer length[3–6] as building blocks of novel high-frequency devices. There are several promising proposals of using carbon nanotubes for THz applications including a nanoklystron utilizing extremely efficient high-field electron emission from nanotubes,[2, 7, 8] devices based on negative differential conductivity in large-diameter semiconducting nanotubes,[9, 10] high-frequency resonant-tunneling diodes[11] and Schottky diodes,[12, 13, 14, 15] as

[*] To whom Correspondence should be sent: M. E. Portnoi Email: m.e.portoi@exeter.uk

well as electric-field-controlled carbon nanotube superlattices,[16] frequency multipliers,[17, 18] THz amplifiers,[19] switches[20] and antennas[21].

In this paper we formulate and discuss several novel schemes to utilize physical properties of single-wall carbon nanotubes (SWNTs) for generation and detec-tion of THz radiation.

2. Quasi-metallic carbon nanotubes as terahertz emitters

This first scheme[22] is based on the electric-field induced heating of electron gas in a nanotube resulting in the inversion of population of optically active states with the energy difference within the THz spectrum range. It is well-known that the elastic backscattering processes in metallic SWNTs are strongly suppressed,[23] and in a high enough electric field charge carriers can be accelerated up to the energy allowing emission of optical/zone-boundary phonons. At this energy, corresponding to the frequency of about 40 THz, the major scattering mechanism switches on abruptly resulting in current saturation.[24–28] In what follows we show that for certain types of carbon nanotubes the heating of electrons to the energies below the phonon-emission threshold results in the spontaneous THz emission with the peak frequency controlled by an applied voltage.

The electron energy spectrum of a metallic SWNT, $\varepsilon(k)$, linearly depends on the electron wave vector k close to the Fermi energy and has the form $\varepsilon(k) = \pm\hbar\upsilon_F\,|k - k_0|$, where $\upsilon_F \approx 9.8 \times 10^5$ m/s is the Fermi velocity of graphene, which corresponds to the commonly used tight-binding matrix element γ_0 = 3.033 eV.[4, 6] Here and in what follows the zero of energy is defined as the Fermi energy position in the absence of an external field. When the voltage, V, is applied between the SWNT ends, the electron distribution is shifted in the way shown by the heavy lines in Fig. 1a corresponding to the filled electron states. This shift results in inversion of population and, correspondingly, in optical transitions between filled states in the conduction band and empty states in the valence band. The spectrum of optical transitions is determined by the distribution function for hot carriers, which in turn depends on the applied voltage and scattering processes in the SWNT. It is well-known that the major scattering mechanism in SWNTs is due to electron-phonon interaction.[24–28] Since the scattering processes erode the inversion of electron population, an optimal condition for observing the discussed optical transitions takes place when the length of the SWNT $L < l_{ac}$, where the electron mean-free path for acoustic phonon scattering is $l_{ac} \approx 2$ μm.[26] Below we consider only such short SWNTs with ideal Ohmic contacts[25] and in the ballistic transport regime, when the energy acquired by the electron on the whole length of the tube, $\Delta\varepsilon = eV$, does not exceed the value of $\hbar\Omega = 0.16$ eV at which the fast emission of high-energy phonons begin.[26] In this so-called low-bias regime,[24–26] in which the current in

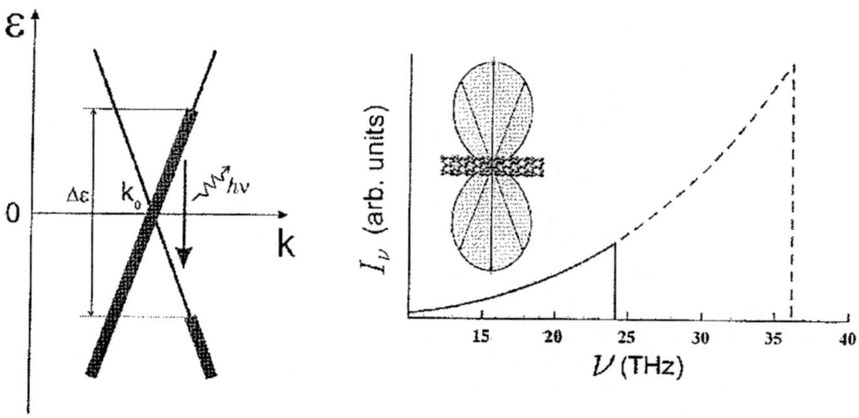

Figure 1. (a) The scheme of THz photon generation by hot carriers in quasi-metallic SWNTs. (b) The spectral density of spontaneous emission as a function of frequency for two values of applied voltage: solid line for V = 0.1 V; dashed line for V = 0.15 V. The inset shows the directional radiation pattern of the THz emission with respect to the nanotube axis.

the nanotube is given by the Büttiker-Landauer-type formula, $I \approx (4e^2 / h)V$, the distribution function of hot electrons is

$$f_e(k) = \begin{cases} 1, & 0 < k - k_0 < \Delta\varepsilon / 2\hbar\upsilon_F \\ 0, & k - k_0 > \Delta\varepsilon / 2\hbar\upsilon_F \end{cases}. \tag{1}$$

The distribution function for hot holes, $f_h(k)$, has the same form as $f_e(k)$.

Let us select a SWNT with the crystal structure most suitable for observation of the discussed effect. First, the required nanotube should have metallic conductivity and, second, the optical transitions between the lowest conduction subband and the top valence subband should be allowed. The crystal structure of a SWNT is described by two integers (n, m), which completely define its physical properties.[3, 4, 6] SWNTs with true metallic energy band structure, for which the energy gap is absent for any SWNT radius, are armchair (n, n) SWNTs only.[6, 29–32] However, for armchair SWNTs the optical transitions between the first conduction and valence subbands are forbidden.[33, 34] So we propose to use for the observation of THz generation the so-called quasi-metallic (n, m) SWNTs with $n - m = 3p$, where p is a non-zero integer. These nanotubes, which are gapless within the frame of a simple zone-folding model of the π-electron graphene spectrum.[4] are in fact narrow-gap semiconductors due to curvature effects. Their bandgap is given by $\varepsilon_g = \hbar\upsilon_F a_{C-C} \cos 3\theta/(8R^2)$[29, 32] where a_{C-C} = 1.42 Å is the nearest-neighbor distance between two carbon atoms, R is the nanotube radius, and $\theta = \arctan[\sqrt{3}m/(2n+m)]$ is the chiral angle.[4] It can be seen from the expression for ε_g that the gap is decreasing rapidly with increasing the nanotube radius. For large values of R this gap can be neglected even in the case of moderate applied voltages due to Zener tunneling of electrons across the gap. It is easy to show in the fashion similar to the original Zener's work[35] that the tunneling probability in quasi-metallic SWNTs is given by $\exp(-\alpha\varepsilon_g^2/eE\hbar\upsilon_F)$, where α is a numerical factor close to unity.[36] For example, for a zigzag (30, 0) SWNT the gap is $\varepsilon_g \approx 6$ meV and the Zener breakdown takes place for the electric field $E \sim 10^{-1}$ V/μm. Since almost the whole voltage drop in the ballistic regime occurs within the few-nanometer regions near the contacts[37] a typical bias voltage of 0.1 V corresponds to an electric field, which is more than sufficient to achieve a complete breakdown. In what follows all our calculations are performed for a zigzag (3p, 0) SWNT of large enough radius R and for applied voltages exceeding the Zener breakdown, so that the finite-gap effects can be neglected. The obtained results can be easily generalized for any quasi-metallic large-radius SWNT.

Optical transitions in SWNTs have been a subject of extensive research (see, e.g., Refs.[33, 34, 38–42]). We treat these transitions using the results of the nearest-neighbor orthogonal π-electron tight binding model.[4] Despite its apparent simplicity and well-known limitations, this model has been extremely successful in describing low-energy optical spectra and electronic properties of single-walled SWNTs (see, e.g., Ref.[43] for one of the most recent manifestations of this model success). Our goal is to calculate the spectral density of spontaneous emission, I_ν, which is the probability of optical transitions per unit time for the photon frequencies in the interval (ν, $\nu + d\nu$) divided by $d\nu$. In the dipole approximation[44] this spectral density is given by

$$I_\nu = \frac{8\pi e^2 \nu}{3c^3} \sum_{i,f} f_e(k_i) f_h(k_f) |\langle \Psi_f | \hat{\upsilon}_z | \Psi_i \rangle|^2 \delta(\varepsilon_i - \varepsilon_f - h\nu). \quad (2)$$

Equation (2) contains the matrix element of the electron velocity operator. In the frame of the tight binding model, this matrix element for optical transitions between the lowest conduction and the highest valence subbands of the (3p, 0) zigzag SWNT can be written as (cf. Ref.[34, 38])

$$\langle \Psi_f | \hat{\upsilon}_z | \Psi_i \rangle = \frac{a_{\text{C-C}}\,\omega_{if}}{8} \delta_{k_f, k_i} \quad (3)$$

where $\hbar\omega_{if} = \varepsilon_i - \varepsilon_f$ is the energy difference between the initial (i) and the final (f) states. These transitions are associated with the light polarized along the nanotube axis z, in agreement with the general selection rules for SWNTs.[33] Substituting Equation (3) in Equation (2) and performing necessary summation, we get

$$I_\nu = L f_e(\pi\nu / \upsilon_F) f_h(\pi\nu / \upsilon_F) \frac{\pi^2 e^2 a_{\text{C-C}}^2 \nu^3}{6c^3 \hbar \upsilon F}. \quad (4)$$

Equation (4) has broader applicability limits than the considered case of $L < l_{ac}$ and $eV < \hbar\Omega$, in which the distribution functions for electrons and holes are given by Equation (1). In the general case there is a strong dependence of I_ν on the distribution functions, which have to be calculated taking into account all the relevant scattering mechanisms.[24–28] In the discussed ballistic regime the spectral density has a universal dependence on the applied voltage and photon frequency for all quasi-metallic SWNTs. In Fig. 1b the spectral density is shown for two values of the voltage. It is clearly seen that the maximum of the spectral density of emission has strong voltage dependence and lies in the THz frequency range for experimentally attainable voltages. The directional radiation pattern, shown in the inset of Fig. 1b, reflects the fact that emission of light polarized normally to the nanotube axis is forbidden by the selection rules for the optical transitions between the lowest conduction subband and the top valence subband.

For some device applications it might be desirable to emit photons propagating along the nanotube axis, which is possible in optical transitions between the SWNT subbands characterized by angular momenta differing by one.[6, 33] To achieve the emission of these photons by the electron heating, it is necessary to have an intersection of such subbands within the energy range accessible to electrons accelerated by attainable voltages. From our analysis of different types of SWNTs, it follows that the intersection is possible, e.g., for the lowest conduction subbands in several semiconducting zigzag nanotubes and in all armchair nanotubes. However, for an effective THz emission from these nanotubes it is necessary to move the Fermi level very close to the subband intersection point.[45] Therefore, obtaining the THz emission propagating along the nanotube axis is a more difficult technological problem than generating emission shown in Fig. 1b.

3. Chiral carbon nanotubes as frequency multipliers

Another proposal for using SWNTs for THz applications is based on chiral nanotubes, which represent natural superlattices. For example, a (10, 9) single-wall nanotube has a radius which differs from the radius of the most commonly studied (10, 10) nanotube by less than 5%, whereas a translational period T along the axis of the (10, 9) SWNT is almost 30 times larger than the period of the (10, 10) nanotube. Correspondingly, the first Brillouin zone of the (10, 9) nanotube is 30 times smaller than the first zone for the (10, 10) tube. However such a Brillouin zone reduction cannot influence electronic transport unless there is a gap opening between the energy subbands resulting from the folding of graphene spectrum. In our research we show how an electric field normal to he nanotube axis opens noticeable gaps at the edge of the reduced Brillouin zone, thus turning a long-period nanotube of certain chirality into a 'real' superlattice. The field-induced gaps are most pronounced in (n, 1) SWNTs.[16, 46] Figure 2a shows the opening of electric-field induced gap near the edge of the Brillouin zone of a (6, 1) SWNT. This gap opening results in the appearance of a negative effective-mass region in the nanotube energy spectrum. The typical electron energy in this part of the spectrum of 15 meV is well below the optical phonon energy $\hbar\Omega \approx 160$ meV, so that it can be easily accessed in moderate heating electric fields. The negative effective mass results in the negative differential conductivity (NDC), as can be seen from Fig. 2b. The effect of the negative effective mass also leads to an efficient frequency multiplication in the THz range. The results of our calculations of the electron velocity in the presence of the time dependent longitudinal electric field are presented in Fig. 3. One of the advantages of a frequency multiplier based on chiral SWNTs, in comparison with the conventional superlattices,[47] is that the dispersion relation in our system can be controlled by the transverse electric field $E_\perp$.

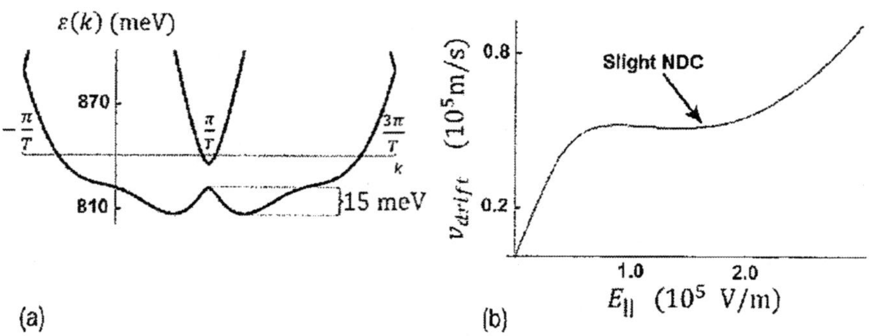

Figure 2. (a) Energy spectrum of the (6, 1) SWNT in a transverse electric field, $E_\perp = 4$ V/nm. (b) The electron drift velocity in the lowest conduction subband of a (6, 1) SWNT as a function of the longitudinal electric field, in the presence of acoustic-phonon scattering.

4. Armchair nanotubes in a magnetic field as tunable THz detectors and emitters

The problem of detecting THz radiation is known to be at least as challenging as creating reliable THz sources. Our proposal of a novel detector is based on several features of truly gapless (armchair) SWNTs. The main property to be utilized is opening of a band gap in these SWNTs in a magnetic field along the nanotube axis.[4, 6] For a (10, 10) SWNT this gap corresponds to approximately 1.6 THz in the field of 10 T. For attainable magnetic fields, the gap grows linearly with increasing both magnetic field and the nanotube radius. It can be shown[46] that the same magnetic field also allows dipole optical transitions between the top valence subband and the lowest conduction subband, which are strictly forbidden in armchair SWNTs without the field.[33]

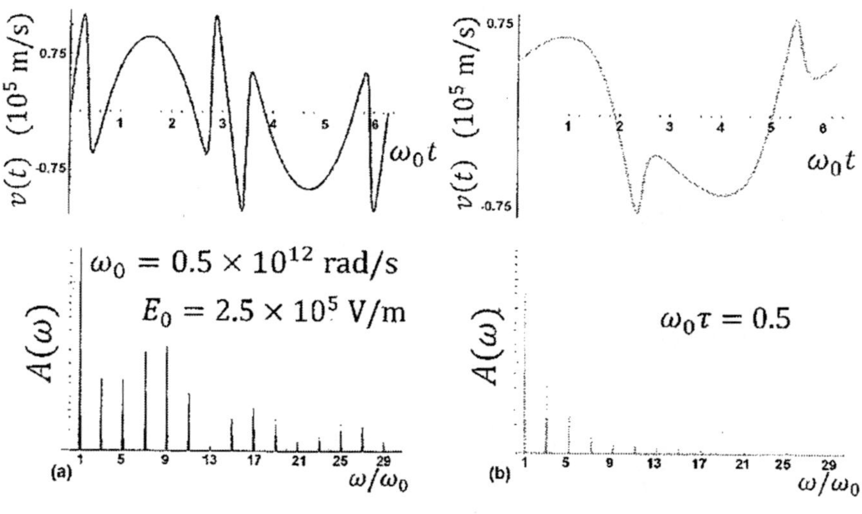

Figure 3. Time dependence of the electron velocity in the lowest conduction subband of a (6, 1) SWNT under the influence of a pump harmonic longitudinal electric field, $E_{\parallel}(t) = E_0 \sin(\omega_0 t)$, and its correspondent spectral distribution A (ω): (a) in the ballistic transport regime; (b) in the presence of scattering with the relaxation time $t = 10^{-12}$ s.

In Fig. 4 we show how the energy spectrum and matrix elements of the dipole optical transitions polarized along the nanotube axis are modified in the presence of a longitudinal magnetic field. In the frame of the nearest-neighbor tight binding model, one can show that for a (*n*, *n*) armchair nanotube the squared matrix element of the velocity operator between the states at the edge of the gap opened by the magnetic field is given by a simple analytic expression:

$$\left|\left\langle \Psi_n^{\upsilon} \middle| \hat{\upsilon}_z \middle| \Psi_i^{c} \right\rangle\right|^2 = \frac{4}{3}\left[1 - \frac{1}{4}\cos^2\left(\frac{f}{n}\pi\right)\right]\upsilon_F^2, \tag{5}$$

where $f = eBR^2/(2\hbar)$. For experimentally attainable magnetic fields, when the magnetic flux through the SWNT is much smaller than the flux quantum, the absolute value of the velocity operator is close to υ_F. Equation (5) is relevant to the transitions between the highest valence subband and the the lowest conduction subband only for $f \leq 1/2$, since for the higher values of f the order of the nanotube subbands is changed. Notably, the same equation allows to obtain the maximum value of the velocity operator in any armchair SWNT for the transitions polarized along its axis: this value cannot exceed $2\upsilon_F/\sqrt{3}$ (see panel (c) in Fig. 4).

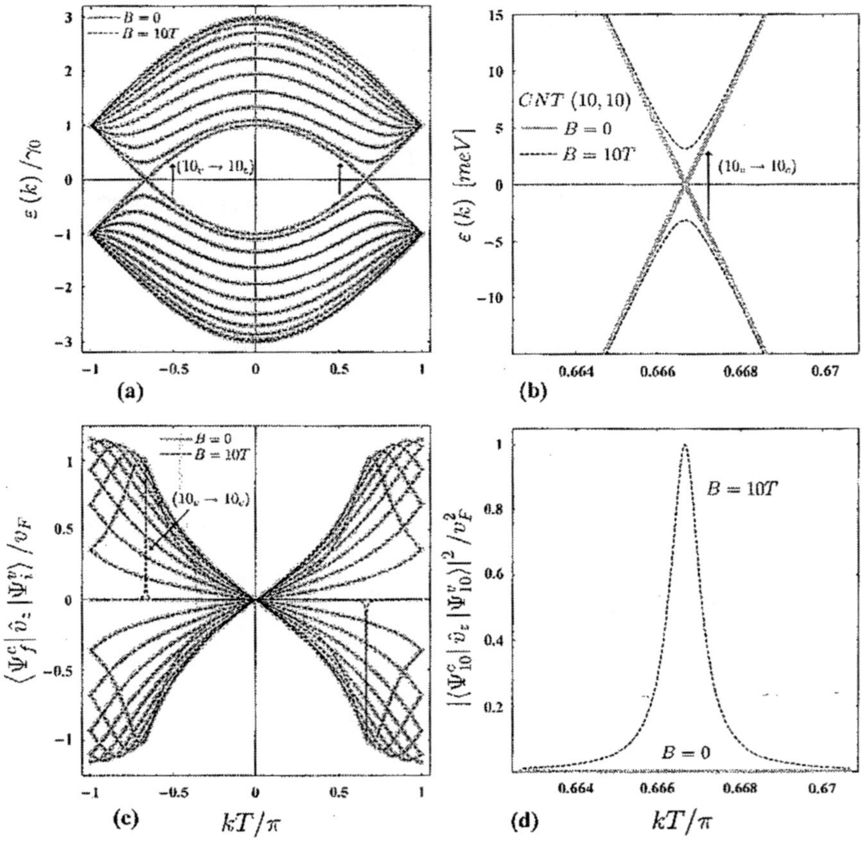

Figure 4. (a) Band structure of a (10, 10) nanotube, with and without an external magnetic field along the nanotube axis. (b) Detailed view of the gap, which is opened between the top valence subband and the lowest conduction subband in an external field B = 10 T. (c) The change in the dipole optical transitions matrix elements, for the light polarized along the SWNT axis, due to the introduction of the external magnetic field. The only appreciable change is in the appearance of a high narrow peak associated with the transition ($10_\upsilon \rightarrow 10_c$), which is not allowed in the absence of the magnetic field. Here and in what follows the energy subbands are numbered in the same way as in Ref.[4] (d) Dependence of the squared dipole matrix element for the transition ($10_\upsilon \rightarrow 10_c$) on the 1D wave vector k, with and without an external magnetic field.

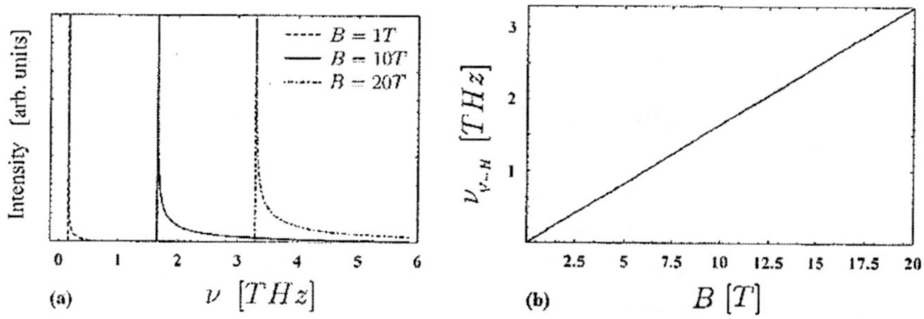

Figure 5. (a) Calculated photon absorption spectra for a (10, 10) SWNT, for three different magnetic field values. The absorption intensity is proportional to the product of $\left|\langle \Psi^{\upsilon}_{10}|\hat{\upsilon}_z|\Psi^{c}_{10}\rangle\right|^2$ and the joint density of states. (b) Dependence of the position of the peak in the absorption intensity, associated with the Van Hove singularity, on the magnetic field.

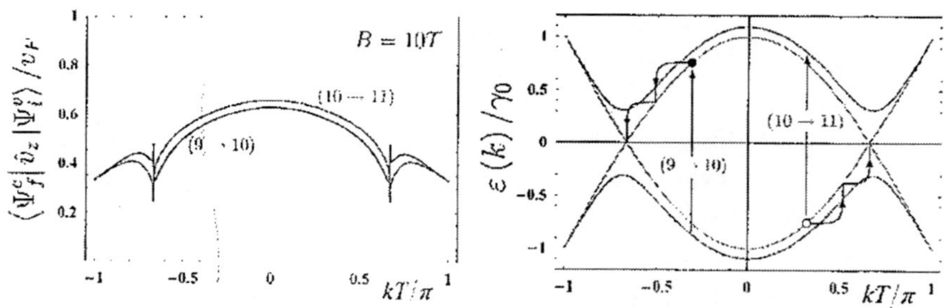

Figure 6. A scheme for creating a population inversion between the lowest conduction subband and the top valence subband of an armchair SWNT in a magnetic field. The left plot shows the calculated matrix elements of the relevant dipole optical transitions polarized normally to the axis of a (10, 10) SWNT. The right plot shows several energy subbands closest to the Fermi level and illustrates the creation of photoexcited carriers and their non-radiative thermalization.

The electron (hole) energy spectrum near the bottom (top) of the band gap produced by the magnetic field is parabolic as a function of a carrier momentum along the nanotube axis. This dispersion results in a Van Hove singularity in the reduced density of states, which in turn leads to a very sharp absorption maximum near the band edge and, correspondingly, to a very high sensitivity of the photocurrent to the photon frequency, see Fig. 5.

Notably, the same effect can be used for the generation of a very narrow emission line having the peak frequency tunable by the applied magnetic field. A population inversion can be achieved, for example, by optical pumping with the light polarized normally to the nanotube axis, as shown in Fig. 6.

5. Conclusions

We have demonstrated that a quasi-metallic carbon nanotube can emit the THz radiation when the potential difference is applied to its ends. The typical required voltages and nanotube parameters are similar to those available in the state-of-the-art transport experiments. The maximum of the spectral density of emission is shown to have the strong voltage dependence, which is universal for all quasi-metallic carbon nanotubes in the ballistic regime. Therefore, the discussed effect can be used for creating a THz source with frequency controlled by applied voltage. Appropriately arranged arrays of nanotubes should be considered as promising candidates for active elements of amplifiers and generators of coherent THz radiation.

We have also shown that an electric field, which is applied normally to the axis of long-period chiral nanotubes, significantly modifies their band structure near the edge of the Brillouin zone. This results in the negative effective mass region at the energy scale below the high-energy phonon emission threshold. This effect can be used for an efficient frequency multiplication in the THz range.

Finally, we have discussed the feasibility of using the effect of the magnetic field, which opens energy gaps and allows optical transitions in armchair nanotubes, for creating tunable THz detectors and emitters.

ACKNOWLEDGEMENTS

This research was supported by the EU Foundation INTAS (Grants 03-50-4409 and 05-1000008-7801), the Russian Foundation for Basic Research (Grants 06-02-16005 and 06-02-81012), the Russian Ministry for Education and Science (Grant RNP.2.1.1.1604), the Royal Society (UK), MCT and FINEP (Brazil).

References

1. B. Ferguson and X.C. Zhang, "Materials for terahertz science and technology", *Nat. Mater.* **1**, 26–33, 2002.
2. D. Dragoman and M. Dragoman, "Terahertz fields and applications", *Prog. Quant. Electron.* **28**, 1–66, 2004.
3. S. Iijima, "Helical microtubules of graphitic carbon", *Nature* **354**, 56–58, 1991.

4. R. Saito, G. Dresselhaus, and M.S. Dresselhaus, *Physical Properties of Carbon Nanotubes*, Imperial College Press, London, 1998.
5. *Carbon Nanotubes: Synthesis, Structure, Properties, and Applications*, edited by M.S. Dresselhaus, G. Dresselhaus, and Ph. Avouris, Springer, Berlin, 2001.
6. S. Reich, C. Thomsen, and J. Maultzsch, *Carbon Nanotubes: Basic Concepts and Physical Properties*, Wiley, Berlin, 2004.
7. H.M. Manohara, M.J. Bronikowski, M. Hoenk, B.D. Hunt, and P.H. Siegel, "High-current-density field emitters based on arrays of carbon nanotube bundles", *J. Vac. Sci. Technol. B* **23**, 157–161, 2005.
8. A. Di Carlo, A. Pecchia, E. Petrolati, and C. Paoloni, "Modeling of carbon nanotube-based devices: from nanoFETs to THz emitters", *Proc. SPIE* **6328**, p. 632808, 2006.
9. A.S. Maksimenko and G. Ya. Slepyan, "Negative differential conductivity in carbon canotubes", *Phys. Rev. Lett.* **84**, 362–365, 2000.
10. G. Pennington and N. Goldsman, "Semiclassical transport and phonon scattering of electrons in semiconducting carbon nanotubes", *Phys. Rev. B* **68**, p. 045426, 2003.
11. D. Dragoman and M. Dragoman, "Terahertz oscillations in semiconducting carbon nanotube resonant-tunneling diodes", *Physica E* **24**, 282–289, 2004.
12. A.A. Odintsov, "Schottky barriers in carbon nanotube heterojunctions", *Phys. Rev. Lett.* **85**, 150–153, 2000.
13. F. Léonard and J. Tersoff, "Negative differential resistance in nanotube devices", *Phys. Rev. Lett.* **85**, 4767–4770, 2000.
14. M.H. Yang, K.B.K. Teo, W.I. Milne, and D.G. Hasko, "Carbon nanotube Schottky diode and directionally dependent field-effect transistor using asymmetrical contacts", *Appl. Phys. Lett.* **87**, p. 253116, 2005.
15. C. Lu, L. An, Q. Fu, J. Liu, H. Zhang, and J. Murduck, "Schottky diodes from asymmetric metal-nanotube contacts", *Appl. Phys. Lett.* **88**, p. 33501, 2006.
16. O.V. Kibis, D.G.W. Parfitt, and M.E. Portnoi, "Superlattice properties of carbon nanotubes in a transverse electric field", *Phys. Rev. B* **71**, 035411, 2005.
17. G. Ya. Slepyan, S.A. Maksimenko, V.P. Kalosha, J. Herrmann, E.E.B. Campbell, and I.V. Hertel, "Highly efficient high-order harmonic generation by metallic carbon nanotubes", *Phys. Rev. A* **60**, 777–780, 1999.
18. G. Ya. Slepyan, S.A. Maksimenko, V.P. Kalosha, A.V. Gusakov, and J. Herrmann, *Phys. Rev. A* **63**, 053808, 2001.
19. D. Dragoman and M. Dragoman, "Terahertz continuous wave amplification in semiconductor carbon nanotubes", *Physica E* **25**, pp. 492–496, 2005.
20. M. Dragoman, A. Cismaru, H. Hartnagel, and R. Plana, "Reversible metal-semiconductor transitions for microwave switching applications", *Appl. Phys. Lett.* **88**, 073503, 2006.
21. G. Ya. Slepyan, M.V. Shuba, S.A. Maksimenko, and A. Lakhtakia, "Theory of optical scattering by achiral carbon nanotubes and their potential as optical nanoantennas", *Phys. Rev. B* **73**, 195416, 2006.
22. O.V. Kibis, M. Rosenau da Costa, and M.E. Portnoi, *Nano Lett.* **7**, 3414, 2007.
23. T. Ando, T. Nakanishi, and R. Saito, "Impurity scattering in carbon nanotubes — absence of back scattering", *J. Phys. Soc. Jpn.* **67**, 1704–1713, 1997.
24. Z. Yao, C.L. Kane, and C. Dekker, "High-field electrical transport in single-wall carbon nanotubes", *Phys. Rev. Lett.* **84**, 2941–2944, 2000.
25. A. Javey, J. Guo, M. Paulsson, Q. Wang, D. Mann, M. Lundstrom, and H. Dai, "High-field quasiballistic transport in short carbon nanotubes", *Phys. Rev. Lett.* **92**, 106804, 2004.

26. J.-Y. Park, S. Resenblatt, Yu. Yaish, V. Sazonova, H. Üstünel, S. Braig, T.A. Arias, P.W. Brouwer, and P.L. McEuen, "Electron-phonon scattering in metallic single-walled carbon nanotubes", *Nano Lett.* **4**, 517–520, 2004.
27. M. Freitag, V. Pereibenos, J. Chen, A. Stein, J.c. Tsang, J.A. Misewich, R. Martel, and Ph. Avouris, "Hot carrier electroluminescence from a single carbon nanotube", *Nano Lett.* **4**, 1063–1066, 2004.
28. V. Perebeinos, J. Tersoff, and P. Avouris, "Electron-phonon interaction and transport in semiconducting carbon nanotubes", *Phys. Rev. Lett.* **94**, 086802, 2005.
29. C.L. Kane and E.J. Mele, "Size, shape, and low energy electronic structure of carbon nanotubes", *Phys. Rev. Lett.* **78**, 1932–1935, 1997.
30. M. Ouyang, J.-L. Huang, C.L. Cheung, and C.M. Lieber, "Energy gaps in "metallic" single-walled carbon nanotubes", *Science* **292**, 702–705, 2001.
31. Y. Li, U. Ravaioli, and S.V. Rotkin, "Metal-semiconductor transition and Fermi velocity renormalization in metallic carbon nanotubes", *Phys. Rev. B* **73**, 035415, 2006.
32. D. Gunlycke, C.J. Lambert, S.W.D. Bailey, D.G. Pettifor, G.A.D. Briggs, and J.H. Jefferson, "Bandgap modulation of narrow-gap carbon nanotubes in a transverse electric field", *Europhys. Lett.* **73**, 759–764, 2006.
33. I. Milošević, T. Vuković, S. Dmitrović, and M. Damnjanović, "Polarized optical absorption in carbon nanotubes: a symmetry-based approach", *Phys. Rev. B* **67**, 165418, 2003.
34. J. Jiang, R. Saito, A. Grüneis, G. Dresselhaus, and M.S. Dresselhaus, "Optical absorption matrix elements in single-wall carbon nanotubes", *Carbon* **42**, 3169–3176, 2004.
34. C. Zener, "A theory of the electrical breakdown of solid dielectrics", *Proc. Royal. Soc. (London)* **145**, 523, 1934.
36. For the energy spectrum near the band edge given by $\varepsilon = \pm\left[\varepsilon_g^2/4 + \hbar^2 \upsilon_F^2 \left(k - k_0\right)^2\right]^{1/2}$, it can be shown that $\alpha = \pi/4$.
37. A. Svizhenko and M.P. Anantram, *Phys. Rev. B* **72**, 085430, 2005.
38. A. Grüneis, R. Saito, G.G. Samsonidze, T. Kimura, M.A. Pimenta, A. Joria, A.G. Souza Filho, G. Dresselhaus, and M.S. Dresselhaus, "Inhomogeneous optical absorption around the K point in graphite and carbon nanotubes", *Phys. Rev. B* **67**, 165402, 2003.
39. V.N. Popov and L. Henrard, "Comparative study of the optical properties of single-walled carbon nanotubes within orthogonal and nonorthogonal tight-binding models", *Phys. Rev. B* **70**, 115407, 2004.
40. R. Saito, A. Grüneis, G.G. Samsonidze, G. Dresselhaus, M.S. Dresselhaus, A. Jorio, L.G. Cançado, M.A. Pimenta, and A.G. Souza Filho, "Optical absorption of graphite and single-wall carbon nanotubes", *Appl. Phys. A* **78**, 1099–1105, 2004.
41. S.V. Goupalov, "Optical transitions in carbon nanotubes", *Phys. Rev. B* **72**, 195403, 2005.
42. Y. Oyama, R. Saito, K. Sato, J. Jiang, G.G. Samsonidze, A. Grüneis, Y. Miyauchi, S. Maruyama, A. Jorio, G. Dresselhaus, and M.S. Dresselhaus, "Photoluminescence intensity of single-wall carbon nanotubes", *Carbon* **44**, 873–879, 2006.
43. M.Y. Sfeir, T. Beetz, F. Wang, L. Huang, X.M.H. Huang, M. Huang, J. Hone, S. O'Brien, J.A. Misewich, T.F. Heinz, L. Wu, Y. Zhu, and L.E. Brus, "Optical spectroscopy of individual single-walled carbon nanotubes of defined chiral structure", *Science* **312**, 554–556, 2006.
44. V.B. Berestetskii, E.M. Lifshitz, and L.P. Pitaevskii, *Quantum Electrodynamics*, Butterworth-Heinemann, Oxford, 1997.

45. O.V. Kibis and M.E. Portnoi, "Carbon nanotubes: a new type of emitter in the terahertz range", *Technical Phys. Lett.* **31**, 671–672, 2005.
46. M. Rosenau da Costa, O.V. Kibis, and M.E. Portnoi, *to be published.*
47. K.N. Alekseev, M.V. Gorkunov, N.V. Demarina, T. Hyart, N.V. Alexeeva, and A.V. Shorokhov, "Suppressed absolute negative conductance and generation of high-frequency radiation in semiconductor superlattices", *Europhys. Lett.* **73**, 934–940, 2006.

Part II

ENERGY RELATED MATERIALS

SMART MATERIALS AND CONCEPTS FOR PHOTOVOLTAICS: DYE SENSITIZED SOLAR CELLS

A. DI CARLO*, A. REALE, T. M. BROWN, M. CECCHETTI, F. GIORDANO, G. ROMA, M. LIBERATORE, V. MIRUZZO AND V. CONTE
CHOSE – Centre for Hybrid and Organic Solar Energy, Department of Electronics Engineering University of Rome "Tor Vergata", via del Politecnico 1, 00133 Rome (Italy)

Abstract. Recently the issue of clean energy generation for powering an ever growing and developing civilization has come to the fore in international communities. A range of strategic actions have been identified in order to reach the renewable energy goals of 2020. The photovoltaic will play an important role with annual growth rate of the order of 30%. Particular emphasis has been given to organic solar cells and in particular to dye solar cells (DSC). These have the potential to greatly reduce materials and fabriccation costs. In this chapter we will be giving a brief survey of dye solar cell science and technology: both the material aspects, highlighting the contribution to the photovoltaic process played by the electrolyte and counter electrode materials, and the engineering issues involved in creating a module of interconnected cells, using different architectures, on a common substrate. It is the possibility of using solution processed techniques with many of the active materials of the cell that contributes to make organic semiconductor and hybrid DSC technology so promising.

Keywords: Photovoltaics, solar cells, dye sensitized solar cells, electrolyte, PEDOT.

1. Introduction

On the 10th January 2007 the European Community defined the strategic guidelines for the development of renewable energy sources (Renewable Energy

*To whom correspondence should be sent: Aldo Di Carlo, Department of Electronics Engineering University of Rome "Tor Vergata", via del Politecnico 1, 00133 Rome (Italy): Email: aldo. dicarlo@uniroma2.it

Road Map).[1] The document indicates the goal of rising by 20% renewable energy sources by 2020. In 2020 the contribution from renewable sources should reach 33.8% of the total. Among the different sources, photovoltaic energy will see the highest annual growth rate, estimated about 30%, which will set electric power production by 9% from renewable sources. This goal will be reached only to the condition that the research development will introduce innovative solutions. The European Photovoltaic Industry Association (EPIA) has defined a range of strategic actions in order to reach the goals of 2020.[2] Besides the increase of efficiency of the Si modules (that should rise to 22% for crystalline Si and 20% for polysilicon), EPIA indicates that the development of new materials and new concepts is of fundamental importance in achieving these goals. Among the new concepts especial emphasis is given to organic cells and in particular to dye sensitized solar cells (DSC). EPIA identifies this technology as the most mature among the organic solutions and hopes a clear-cut research and development plan that will take in short time to the first production lines. The use of organic techonologies should solve one of the most critical problems of traditional photovoltaic industry, namely Si harvest. Silicon material affects more than 50% of a photovoltaic module. Recent estimates[3] show that silicon costs 2 euros/Wp, whereas the remaining costs are only 0.5 euros/Wp. It is obvious that reducing the material cost and using cheaper fabrication methods it will be possible to reach 0.5 euros/Wp, as set by the IMEC roadmap for organic solar cells.

In this paper we will discuss important aspect of DSC science and technology and the importance of developing new material for improving cell performance in terms of efficiency and stability.

2. The dye sensitized solar cell

Dye sensitized solar cells are interesting both from the academic and the commercial point of view, thanks to their ability to convert light into electricity at a low cost, with a relatively high efficiency[4,5] and using relatively simple fabrication methods, similar to those used in printing. The kernel of the system is represented by the mesoporous material (typically TiO_2) made of nanoparticles, on which a layer of sensitizing dye is adsorbed.[6] The function of the sensitizer (Fig. 1) is to convert the light energy into electrons, which are transferred to the semiconducting oxide: light excites an electron from an occupied to an empty molecular orbital and the electron is transferred from the excited state to the conduction band of the semiconductor. The traditional systems are based on polypiridinic Ru-complexes. The lowest excited state is of charge-transfer nature and involves the direct transfer of an electron mainly localized on the Ru atom to a molecular orbital of the carboxy-bipiridine ligand.[7] Therefore, the absorption bands are from the metal to the ligand (MLCT) and a careful choice of the ligands can give the desired energy levels and proper redox properties. Indeed, the possibility of influencing the relative position of the orbitals involved allows to engineering the complex properties; for instance the introduction of electron-donors ligand groups enhances the charge on the Ru atom, with the consequence of moving towards the blue the MLCT transition. Also the non-chromoforic ligands (e.g., NCS-) play an important role in changing the relative energy levels and consequently influence the solvatochromatic properties of the complex.

With this type of sensitizers it has been possible to reach efficiencies of 11%[8] and noticeable long-term stabilities.[9] Recently the interest on fully organic dye sensitizers is growing, particularly driven by their lower cost with respect to Ru-complexes. Moreover, the organic dyes show (i) very high extinction coefficients, (ii) simple synthesis and purification processes, (iii) considerable potentialities in terms of molecular engineering. To date the best photovoltaic conversion efficiencies obtained with organic dyes in DSCs have reached a promising 8%,[10,11] but long-term performance needs to be improved.

Another component of the cell is the electronic mediator, which should carry the electric charge from the anode to the cathode of the solar cell. The universally used component is the I^-/I_3^- couple which, if on the one hand has almost ideal kinetic properties, on the other has a considerable draw backs[12]: (1) it is corrosive against the vast majority of metals, limiting or making very difficult to introduce metallic grids that would increase the efficiency of gathering and transport of the electric charges. The resistivity of the conducting glass limits considerably the efficiency of the solar cells when their dimensions exceed few square centimeters; (2) it is quite volatile; (3) cathodic reduction of I_3^- requires the presence of Pt catalysts which introduce problems of costs and material availability. Some authors report catalyst instabilities in the presence of the I^-/I_3^- couple.[13] It is possible to use non corrosive Co (II) or Cu (I) complexes[14]; these species are inert with respect to metals and plastic materials and show interesting electrochemical properties, among which a noticeable electrochemical inertia on conductive oxides surfaces.[15] However they exhibit worst performances with respect to I^-/I_3^- couples, mainly because of a lower regeneration rates of the oxidized dye.

The DSC fabrication typically employs printing technology. This makes the fabrication process low cost and easy to implement. Among the different techniques we find (a) serigraphy (screen printing), (b) spray coating, (c) blade coating, and (d) ink-jet printing. The latter has been developed for LED applications and polymer transistors in new high-tech industries like CDT, Plastic Logic and EPSON. A great advantage of screen printing and spray coating resides in the fact that these are additive processes, which allow to deposit only the needed amount of material and up to 90% savings with respect to the other methods, also reducing impact on the environment. Furthermore, these methods are compatible with glass substrates and plastic roll-to-roll production methods, with further cost reduction with respect to silicon which is a rigid and fragile material. Scanning techniques are easily extensible to large area panel production, leading to lower production costs, integration and installation for square meter. As opposed to silicon, the material processes for device realization do not need high temperatures and are compatible with plastic substrates like PEN or PET, with further advantages in terms of costs, integration over existing surfaces and assembly.

In the DSC fabrication process, particular attention has been devoted to the sealing procedure of the cell, since the I^-/I_3^- spilling may cause damage to the metallic contacts and grids. Encapsulation of the glass substrates is obtained using thermo sealing component like Suryln® and Bynel® by Dupont. In order to pass the aging tests at high temperature it is necessary that the materials do not degrade. The thermo sealing resins are not sufficient to guarantee good hermetic properties since their fusion point is around 80°C. In order to circumvent this problem it is possible to encapsulate the DSSC using high fusion temperature glass powders (glass frit) that are used today for plasma screen and LCD sealing.[16–19] The advantage of glass-frit is to be economic and it can be applied using screen printing techniques since it forms more or less viscous powders. It is also quite thermally and chemically stable. The main disadvantage of this technique is that requires high deposition temperatures (about 400°C), and both dye and electrolyte needs to be injected afterwards.

Once the encapsulation and hermetic sealing of the cell, it is possible to study the intrinsic mechanisms of the cell itself that degrades performances. A number of studies have assessed that DSC do not significantly degrade under normal illumination.[20,21] A factor that limits the lifetime of the cells are the exposition to UV light.[22] The degradation mechanism due to UV light has been studied in detail[23] and considerable improvements have been obtained by adding CaI_2 or MgI_2 to the electrolyte. These compounds have the ability to increase the cell lifetimes up to 3300 hours.[20] In order to go beyond this figure it is possible to integrate a polymeric film in the cell structure[22] that filters the UV radiation. In this way it is possible to easily reach 20 years stability over UV light. Another factor that limits lifetime is the oxygen or water presence. Accelerated lifetime tests have demonstrated a substantial invariance of performance for 1400 hours at 45°C. This leads to the conclusion that DSC have lifetimes of 22 years under typical sun conditions of central Europe and 13 years in south Europe.[24] These data lets envisage a commercial use of this technology.

Several industrial realities, like Konarka Technologies, Dyesol, Aisin Seki, Hitachi and Sharp are investing large resources in developing this technology with the intention of a future commercialization. Sharp has the efficiency record of 10.4% on a cell of 1 cm^2. On lower area cells of 0.219 cm^2 an efficiency of

11.2% has been demonstrated. Toyota has tried a different scheme in order to improve efficiency of large area cells, by inserting metal grids with the function of gathering effectively the photogenerated charges. In this way it should be possible to decrease the inter-electrode distance and decrease the series resistance of the TCO glass. Toyota also reported[25] that ambient light illumination effectively increases efficiency by 10–20% with respect to indoor simulations. Dyesol is an Australian company that has been growing considerably in the last years and is now one of the largest DSSC manufacturer. Some year ago it has realized a demonstration cover of a building using DSSC over an area of 200 m^2. Dyesol mainly realize cells based on glass substrates and the basic module comprises six cells connected in series. For commercial applications many modules can be connected in series or in parallel to form a solar panel of required dimensions and power supply.

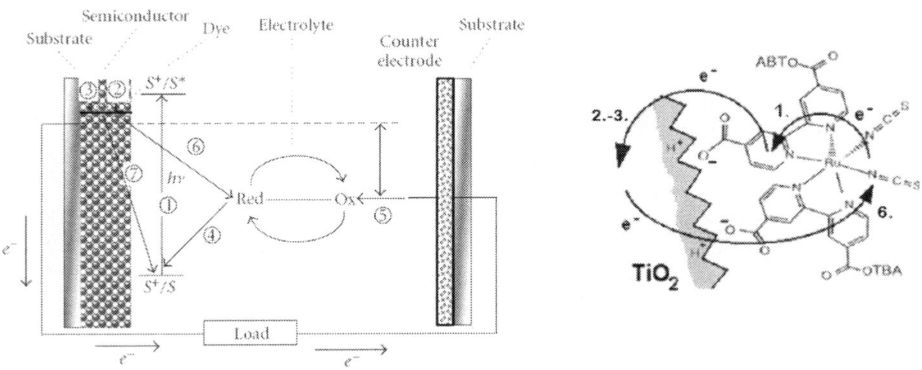

Figure 1. Standard DSC structure and working principle (1) exitation of the dye; (2–3) injection of an e- into the TiO_2 band; (4) reduction of the dye oxidated state; (5) regeneration of the electrolyte ions by the catalyzer; (6, 7) recombinations.[26–28]

3. Electrolytes for DSC

The electrolyte is one of key components for dye solar cells and its properties have much effect on the conversion efficiency and stability of the solar cells. It is the medium for:

- Charge regeneration
- Charge transport in the form of charged ions of the redox couple; equilibrating redox concentration
- Gradients between TiO_2 and counter electrode (CE) surfaces

The requirements for the electrolyte include:

- Good charge transfer characteristics to the dye after electron injection:

$$Dye^+ + 3/2\ I^- \rightarrow Dye + \frac{1}{2}\ I_3^-$$

- Good charge transfer characteristics at the CE (low charge transfer resistance):

$$\frac{1}{2}I_3^- + e^- \rightarrow 3/2\ I^-$$

There are also some "loss process" in DSC as shown in Fig. 2. An excited dye molecule may directly relax into its ground state, without injection of an electron into the TiO_2. This process is negligible, as injection is about 1,000 times faster. Also, electrons from the conduction band of the TiO_2 may recombine with the oxidized dye molecule, before the dye is reduced by the electrolyte. However, reduction by the electrolyte is about 100 times faster. The most significant loss mechanism in the DSC is the recombination of TiO_2 conduction band electrons with the holes in the electrolyte, i.e. I_3^-. The electron transport by diffusion in the TiO_2, and their recombination with the electrolyte are the two competing processes in the DSC.[26]

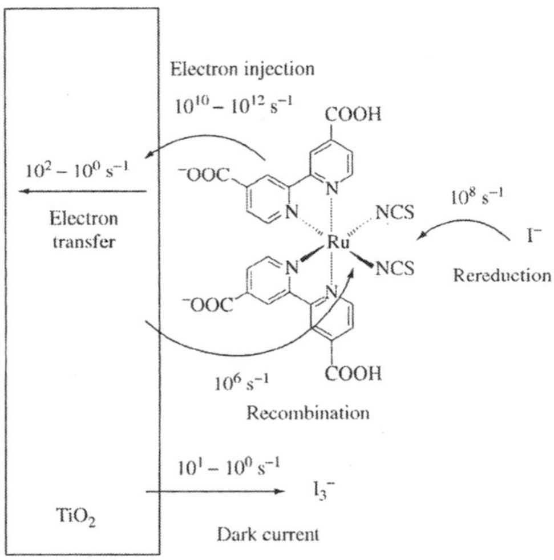

Figure 2. Kinetics in a DSC.[29]

The I^-/I_3^- redox couple is very efficient in these cells because of a fortunate confluence of the right kinetics for at least four different hetereogeneous electron-transfer reactions:

- The photoexcited dye must inject an e^- faster than it reacts with the mediator.
- The oxidized dye must be reduced by the mediator more rapidly than it recombines with the photoinjected electron.
- The oxidized mediator must, itself, react slowly with electrons in both the TiO_2 and the fluorine-doped tin oxide (SnO_2:F) contact.
- Finally, the reduction of the oxidized mediator at the cathode (CE) must be rapid.

The electrolyte used in DSC usually contains organic solvent, redox couple, and additive. It is divided into three types: liquid electrolyte, quasi-solid state electrolyte, and solid electrolyte. Liquid electrolyte could be divided into organic solvent electrolyte and ionic liquid electrolyte according to the solvent used. Organic solvent electrolytes were widely used and investigated in dye-sensitized solar cells for their low viscosity, fast ion diffusion, high efficiency, easy to be designed, and high pervasion into nanocrystalline film electrode.

Orgnic solvent used in organic liquid electrolyte include nitrile such as acetonitrile, valeronitrile, 3-methoxypropionitrile, and esters such as ethylene carbonate (EC), propylene carbonate (PC), γ-butyrolactone. A good solvent should have high boiling point, high dielectric constant and permittivity to separate the salts. Moreover it need low viscosity to get higher mobility because the diffusion of the redox couple is inversely related to the viscosity of the electrolyte.

A number of additives have been discovered, which are able to facilitate self-assembly of the dye molecules on to the TiO_2 electrode surface, making it more impermeable which result in a reduced dark current of the cell. Example are carboxylic like hexadecylmalonic acid) and phosphonic acid derivates (like decylphosphonic acid). The additives reduce recombination and maintain dye integrity. Secondary cations were also used to screen electronic charge of electrons injected into TiO_2 nanoparticles to enable higher photocurrents. The commonly used additive in the electrolytes for dye-sensitized solar cells contained 4-*tert*-butylpyridine (TBP) and N-methylbenzimidazole (NMBI). The addition of these additives suppresses the dark current and improves the photoelectric conversion efficiency. TBP reduces the recombination of electrons in the conduction band of the semiconductor and the electron acceptor in the electrolyte through the coordination between N atom and the Ti ion in incomplete coordination state on the surface of TiO_2 film. This results in a large increase of the photovoltage fill factor and the conversion.[27,30]

The influence of various cations added to the electrolyte upon the performance of DSCs has been well investigated (see Figs. 3 and 4). Studies have revealed that, due to the small size of the Lithium ions (Li+, added as LiI), it intercalates or strongly adsorbs at the TiO_2 surface (see Fig. 5). The surface adsorption results in a positive shift of the conduction band and consequently causes a dramatic drop in the open-circuit voltage of the cell.

Lithium ions show the best characteristics in terms of accelerating the rate of oxidized dye reduction by I^-, slowing the rate of direct combination of electrons with the dye and slowing the rate of recombination of electrons with I_3^-.

In addition, several cationic species have been investigated. Among those, guanidinium and imidazolium ions have shown the most promising results in terms of efficiency improvement (Fig. 6). Replacing, for instance, LiI by guanidinium

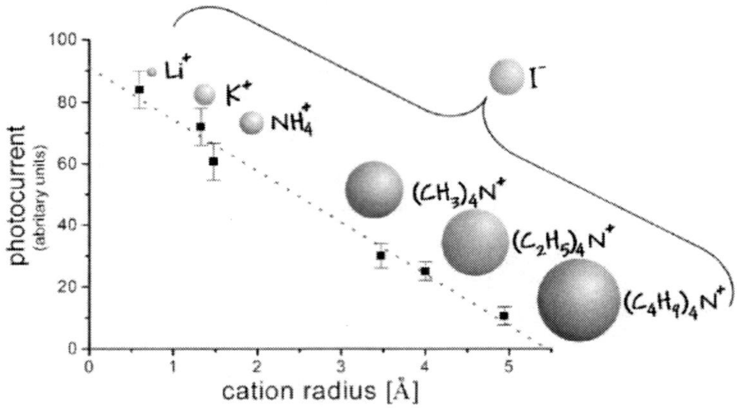

Figure 3. Photocurrent as a function of cation size in iodide salt. Iodide concentration is 50 mM; solvent is 95% CH_3CN: 5% H_2O v = 540 nm.[31–33]

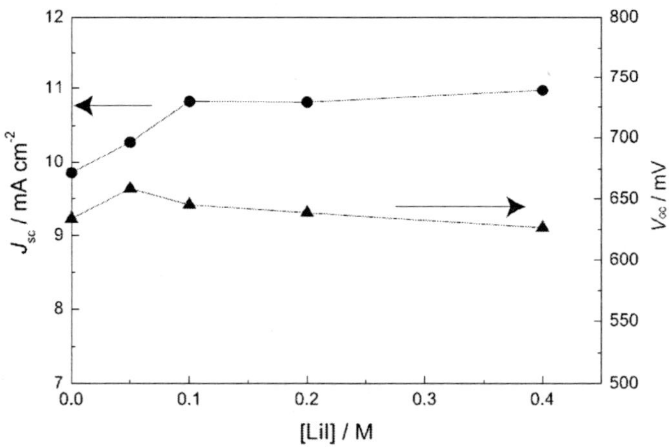

Figure 4. Influence of Lithium iodide concentration in the electrolyte.[34,35]

thiocyanate (GuSCN, Fig. 7) in a standard electrolyte composition reduces the dark reduction current at the mesopourous TiO_2 electrode, resulting in a increased cell voltage (>800 mV) and an increased overall efficiency of the device (5–10% relative).

The recently discovered voltage enhancing additive GuSCN seems also to act as a voltage stabilizer, showing most stable DSC in aging measures.[36]

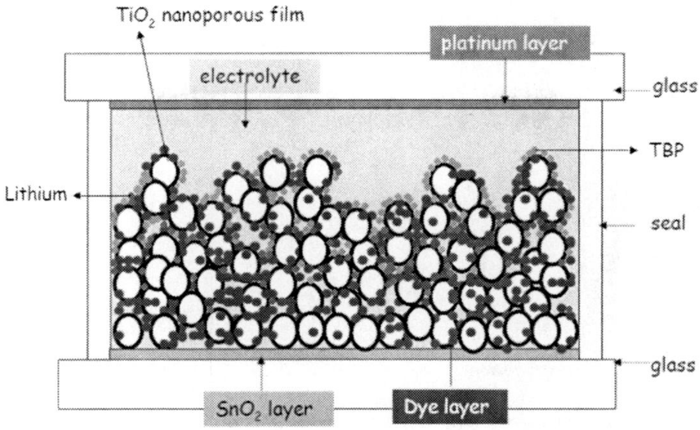

Figure 5. I DSC's structure: Adsorbtion of TBP molecules (grey) and intercanalation of Lithium ions (red) on TiO_2 surface.

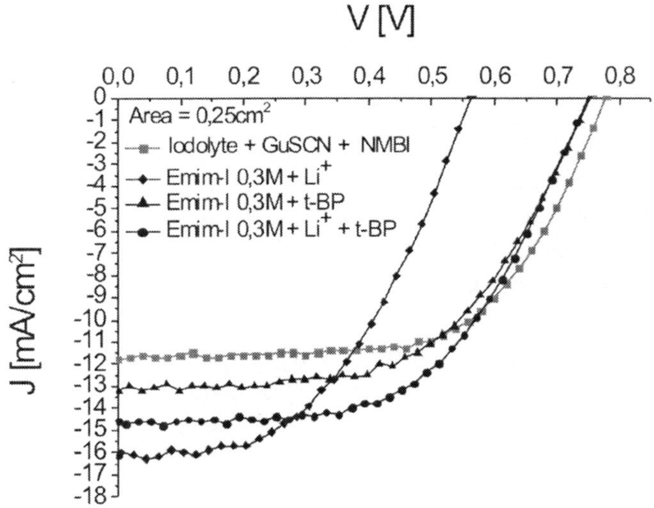

Figure 6. I-V curves of 0.25 cm^2 active area DSC with different electrolyte developed at CHOSE laboratory. Comparison of a standard commercial iodolyte electrolyte with GuSCN and NMBI in MPN and a ionic liquid 1-Ethyl-3Methylimidazoliumiodide based electrolyte with different additives (4-tert-butylpyridine and Lithium ions) in MPN. Measure under sun simulator with A.M. 1.5.

Triiodide/iodide is the common choice of redox couple for obtaining high efficiency in liquid electrolytes. Despite its good performance, the triiodide/ iodide couple has its own disadvantages: the triiodide ion absorbs a significant part of the visible light when employed in high concentrations, its low redox potential limits the open-circuit voltage available, and its aggressiveness towards

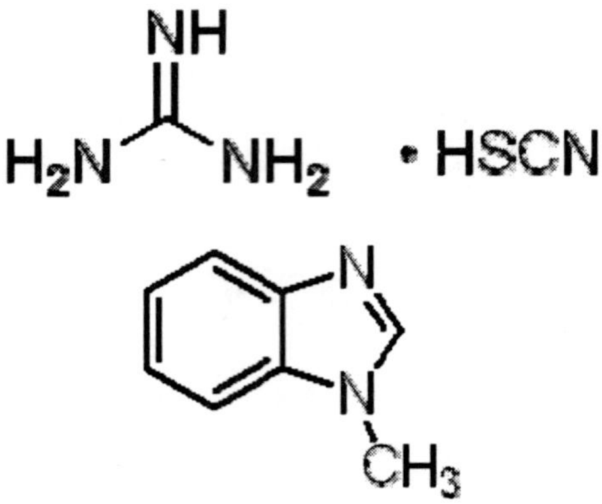

Figure 7. Chemical structure of Guanidyne Thiocyanate (GuSCN), 1-Methylbenzimidazole (NMBI).

silver prevents the use of this metal as current collector in large cells.[37] Several other redox couples have been considered such Br^-/Br_2 $SCN^-/(SCN)_2$ and $SeCN^-/(SeCN)_2$. Sapp[38] reported the substituted bipyridyl cobalt($^{III}/^{II}$) couple as redox couple in DSC. However, the performance of this couple can hardly match that of I_3^-/I^- couple. Moreover, it requires different catalyser on counterelect-rode. Gold and carbon cathodes gives higher J_{sc} than platinum since, for efficient mediators, the reduction of Co(III) is much slower on platinum. Beside the redu-ced efficiency, cobalt based electrolytes are important for the limited corrosive action, allowing a best engineering of the DSC module.

The efficiency record of DSC is typically obtained in solar cells based on organic solvent electrolyte, especially when highly volatile organic solvent elect-rolyte is used (due to the efficient infiltration of organic electrolyte in nanocry-stalline films). Typical organic solvent electrolyte is a solution of LiI/I_2 (0.5/ 0.05 M) in Acetonitrile (ACN) or in Methoxypropionitrile (MPN).

When current passes through the cell (Fig. 8), the steady state concentrations of the mediator redox couple (iodide, I^- and triiodide, I_3^-) are space dependent and different from the initial or open-circuit concentrations. The corresponding concentration potential loss, the concentration overpotential, is then obtained. When the photon flux irradiating the cell becomes sufficiently large, and the electric load appropriate, the maximum current density becomes limited by the mass transport of the species.[39,40] Mass transport manifests itself as a deviation from the linear dependence of the current density with light intensity. When the mass transport is limiting, it is preferable to provide irradiation from the countere-lectrode side.

Solar cells based on organic electrolyte have the disadvantages such as less long-term stability, difficulty in robust sealing and leakage of electrolyte due to the volatility of organic solvent. Room-temperature Ionic liquid (RTILs) were

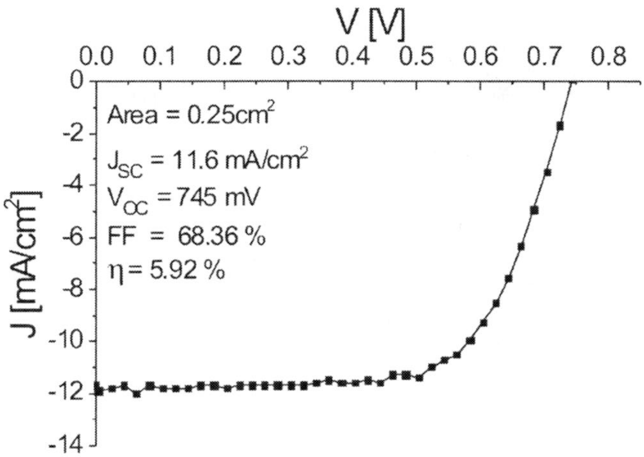

Figure 8. I-V curve of a 0.25 cm^2 active area DSC with standard electrolyte LiI/I_2 with TBP in ACN, measured under sun simulator with A.M. 1.5.

developed in recent year in view of the disadvantage of organic solvent electrolyte. Compared with normal organic solvent electrolyte, ionic liquid have a lot of advantage, such as good chemical and thermal stability, negligible vapor pressure, non flammability, high ionic conductivity and high solubility for organic or inorganic materials, and a wide electrochemical window, which has been intensively pursued as alternative electrolytes for DSC and also for other electrochemical devices. When incorporated into DSCs, they can be both the source of iodide and the solvent themselves.

In recent years, ionic liquid electrolyte was developed rapidly. Was estimate that the viscosity of the molten salts increases with increasing alkyl chain length because of van der Waals forces? The conductivity of the molten salts decreases with increasing viscosity since the diffusion of ions in a liquid depends on its viscosity. The electrolyte with 1-hexyl-3-methylimidazolium iodide gave the highest photoelectric conversion efficiency. Among these ionic liquids, alkyl imidazolium-based ionic liquids are both iodide sources and solvents of electrolytes in solar cells. The properties of ionic liquids can be tuned by controlling the structures of cations and anions. The counterions in the alkyl imidazolium-based ionic liquid included I^-, $N(CN)_2^-$, $B(CN)_4^-$, $(CF3COO)_2N^-$, $BF4^-$, $PF6^-$, NCS^-, and so forth 1-Alkyl-3-methylimidazolium iodides are viscous liquids, whose viscosity is much higher than that of organic-solvent-based liquid electrolyte. Then the transport I_3^- in the electrolyte is very slow and the mass transfer process occupies the leading position. To improve the mobility of redox couple in the electrolyte and the photovoltaic performance, various ionic liquid with low viscosity were developed. Gratzel et al. reported the solar cells based on low-viscosity ionic liquid and MPII mixture.

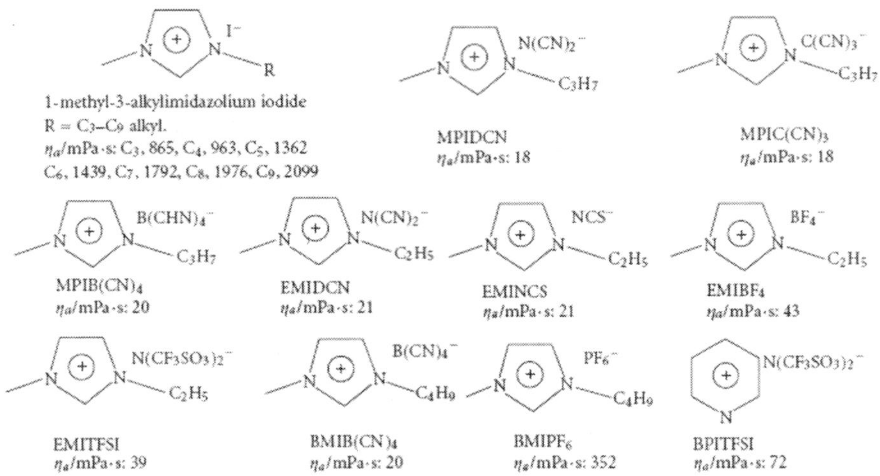

Figure 9. Structure and the viscosity of several ionic liquids.[27]

Figure 9 summarize the structure and viscosity of some ionic liquid and the photovoltaic performance of DSC based on ionic liquid electrolyte of different composition.

The evaporation of the liquid electrolyte often caused some practical limitations of sealing and long-term operation. Recently some attempts were made to improve the long-term stability by using a p-type semiconductor or hole transporting organic materials to replace a liquid electrolyte. However, their conversion efficiencies are not comparable with those of the liquid solar cells. Thus, it can be seen that quasi-solid-state ionic liquid electrolytes might be a better choice to increase stability. Owing to their unique physicochemical properties such as high thermal stability, negligible vapor pressure, relatively high ionic conductivity, and good stability, Ionic liquids were widely used in dye-sensitized solar cells.

The recent progress in ionic liquid at the CHOSE laboratory include the synthesis of different ionic liquids like EMIM-I, EMIM-DCN, PMI-I.

For the EMIM-I (Fig. 10): The salt is obtained trough a quaternization reaction between an alkyl-substituted imidazole and an alkyl iodide. The reaction was performed without solvent, since in these conditions a product of better quality

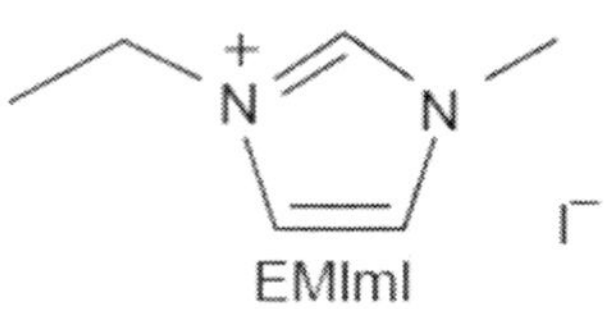

Figure 10. Chemical structure of 1-Ethyl-3-Methylimidazoliumiodide.

and with higher yields is obtained. In a typical experiment, 1-iodoethane (20 g, 0.128 mol) was slowly added – in a 3 h period – to an equimolar amount of 1-methylimidazole (10 g), stirred and refrigerated with an ice bath. Then reaction was stirred 1.2 h, allowing it to reach room temperature slowly. The product was kept under vacuum (oil pump) with a gentle heating, to remove possible unreacted starting materials. The yellowish solid was repeatedly crystallized from acetonitrile, until white crystals were obtained. The solid was filtered, washed with ethyl acetate and diethyl ether, and kept 6 h under reduced pressure (1 mbar). Another method to obtain emimI implies metathesis reaction between emimBr (synthesized with identical procedure described above) and KI. Reaction is carried out in acetone: being the side product KBr insoluble in this solvent, it is easily separated by filtration. The emimI is recovered evaporating acetone and, finally, dried under high vacuum. Purity was checked by 1H NMR. Yield of reactions: 80%.[41–43]

Working with this ionic liquid electrolyte with different concentration from 0.3–1.5 M the best efficiency achieved is about 7.19% using red dye as shown in Fig. 11.

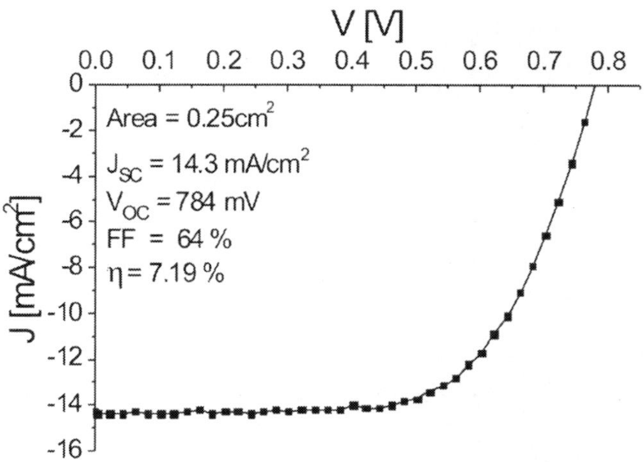

Figure 11. I-V curves of 0.25 cm^2 active area DSC of ionic liquid 1-Ethyl-3-Methylimidazoliu-miodide with Lithium ions and TBP in MPN solvent; Measure under sun simulator with A.M. 1.5.

4. Materials for the counter electrode

In recent years, many investigations have been focused on different DSC issues as synthesis of new dye molecules, analysis of charge-transport processes, development of quasi-solid state or solid-state electrolyte. However, less attention has been paid and only a few studies have been performed on the counter electrode of the DSC.

The mostly used counter electrode is a conducting glass substrates coated with Pt films, where the platinum acts as a catalyst. As this type of counter electrode has the better performance, it is expensive and in addition, the glass substrate has shape limitations and is fragile. Therefore, in order to reduce the cost of the DSCs and broaden their applications, such as a flexible film cell, it is necessary to develop novel counter electrodes based on other substrates and other materials.[44]

In order to try to reduce the Platinum amount in DSC fabrication and correspondingly the counter electrode costs, maintaining the high performance due to the Pt characteristics, some studies have been also performed on the thickness of the Pt layer. These works revealed that the layer thickness, although leads to a higher conductivity of the electrode, has no significant influence on the performance of the DSC and a quite high conversion efficiency can be obtained by a few tens nm thickness (Table 1).[45]

TABLE 1. Performance characteristics of DSCs based on Pt films of different thickness.[45]

Pt film thickness (nm)	V_{oc} (mV)	J_{sc} (mA/cm^2)	Fill factor (%)	η (%)
0	120	1.90	17	0.38
10	705	11.35	61	4.92
25	702	11.60	59	4.99
50	690	12.10	62	5.17
100	701	11.25	62	4.89
200	698	11.40	64	5.08
300	685	11.60	63	5.03
415	694	12.50	60	5.18

However, to suit the possible DSCs application as better as possible, different kind of CEs materials and substrate were investigated.

The high interest in the flexible cells, due to their versatility, for example, leads to the introduction and the study of new substrates. Metal sheets or foils, have excellent electrical and thermal conductivity and are inexpensive. On other hand, plastic materials have many advantages in term of flexibility, weight and handling.

One of the studies performed on such a characteristics, report all the chemical and physical properties of different new substrates (Tables 2 and 3) and their performances when used as counter electrode substrate in DSCs, using Pt layer (Tables 4 and 5). The DSCs performances were also studied depending on their different active areas.[44]

Furthermore, mainly for flexible application using plastic substrate, alternative methods of platinization are required. In one of this alternative method to manufacture Pt counter electrodes at low temperatures called POLYOL, a platinum salt (i.e., $PtCl_4$ or K_2PtCl_6) is diluted in a mixture of organic solvents and

TABLE 2. Properties of different metal substrate used as counter electrode.[44]

Substrates	Thickness of substrates (μm)	10^{-3} R_s of substrates (Ω/sq)	10^{-3} R_s of counter electrodes (Ω/sq)	Stabilities in electrolyte
Stainless steel	100	3.8	3.7	Stable
Nickel	100	0.35	0.46	Stable
Copper	100	0.09	0.15	Dissolves
Aluminum	100	0.19	0.27	Dissolves

TABLE 3. Properties of different plastic substrate used as counter electrode.[44]

Substrate	Thickness of substrate (μm)	R_{sh} of substrate (Ω/sq)	R_{sh} of Pt sputtered CE (Ω/sq)	Stability in electrolyte
ITO-PEN110	125	10	16	Stable
ITO-PEN120	188	20	27	Stable
ITO-PEN150	188	50	62	Stable
Polyester film	100	Insulation	61	Stable
PET	1000	Insulation	48	Unstable
Polyethylene	1000	Insulation	82	Unstable
Polystyrene	1506	Insulation	81	Unstable

TABLE 4. Performance characteristics of DSCs based on different metal substrates and different platinization method.[44]

Substrate	Platinizing methods	V_{oc} (mV)	j_{sc} (mA cm^{-2})	FF (%)	η (%)
FTO glass	Sputtering	695	13.10	65	5.94
Stainless steel	Sputtering	703	12.40	60	5.24
	Plating	613	5.20	61	1.95
	Thermal deposition	576	12.40	46	3.26
Ni	Sputtering	685	12.30	61	5.13
	Electro-deposition	650	13.30	46	3.96

TABLE 5. Performance characteristics of DSCs based on different plastic substrates and different platinization method.[44]

Substrate	Platinizing method	V_{oc} (mV)	j_{sc} (mA cm^{-2})	FF (%)	η (%)
ITO-PEN110	Sputtering	693	12.90	60	5.39
	Thermal deposition	624	16.32	46	4.65
ITO-PEN120	Sputtering	675	12.30	58	4.81
ITO-PEN150	Sputtering	671	11.90	53	4.24
Polyester	Sputtering	647	12.50	49	3.99
Polystyrene	Sputtering	712	7.56	59	3.19
Polyethylene	Sputtering	632	6.90	43	1.90

by a reduction process carried out at temperatures ranging from 90°C to 160°C. In this way, Pt is precipitated directly on the supporting material forming a fine, highly nanoparticulate sized layer of Pt on the substrate at a relatively low temperature. It has been shown that a reaction temperature of 90°C could be enough to prepare counter electrodes with a sufficiently low RCT (around 0.5 Ω cm^2) to be applied in high efficiency DSCs.[46]

However, for DSCs commercial standard applications, which based on glass substrates CE, is desirable to develop alternative low cost material to use as catalyst, for the reduction of the triiodide ion, and corrosion-resistant current collector.

Any of the most widely investigated material for such a purpose are the carbon ones. Many studies have pointed out the physical property of carbon material as graphite, used to increase the lateral conductivity of the CE, and carbon black, used because, having a very high surface area, is be able to act as a catalyst for the redox couple.

Due to their large scale using, for example as printing toners, carbon blacks (CBs) are already set for cheap industrial mass production method and seems to be one of the promising alternative to the Pt layers. Different studies point out the physical-chemical properties of this material, depending on its thickness, leading to a base for future developments.[47]

The comparison between Pt and Carbon material catalyst effect, was pointed out using an impedance spectroscopy measurement (EIS) of a dummy cell consisting of a platinized FTO-glass and a carbon coated FTO-glass placed face-to face in a sandwich configuration. Its equivalent circuit is shown in Fig. 12, where R_{CT} value gives a measure of the catalytic activity for reducing the tri-iodide ion, W is the Warburg impedance for triiodide diffusion emerges and R_S is the DSC serial resistance.

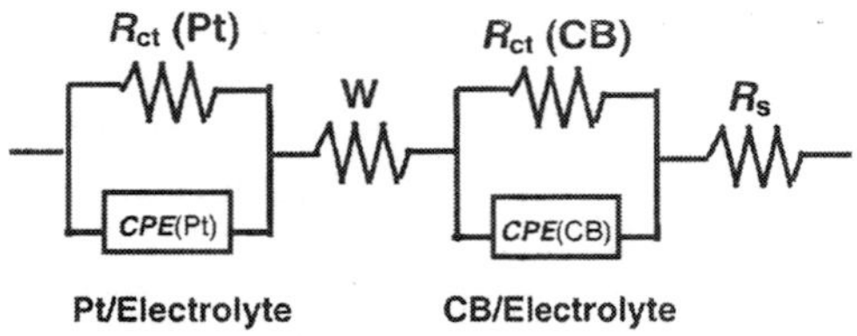

Figure 12. Equivalent circuit diagram to fit the EIS measurement observed spectra. The cell was measured by applying no voltage (0V) and using a solution of 0.60 M 1-methyl-3-butyl-imidazolium iodide, 0.03 M I_2, 0.10 M guanidinium thiocyanate, 0.50 M 4-tert.butylpiridine in a mixed solvent of acetonitrile and valeronitrile (volume ratio 85:15).[47]

Furthermore always new carbon material, with always high performances and specific behaviour are developed. In recent works new Hard Carbon Sphe-rules (HCS) and Nanocarbon powder were investigated to produce new real alternative to the expensive Pt layer.[48,49]

Other promising alternatives to the Pt CE, were founded in poly (3,4-ethylene dioxythiophene) (PEDOT) compounds. This material shows catalytic action for the triiodine ion and different physical properties, as highest conductibility, in relation to its dopant and solvent. PEDOT doped with poly(4-styrene sulfonate) (PSS) has attracted much attention, however, the films are characterized by a low conductivity, that badly affects some of its applications (Fig. 13).

Recently, it was discovered that the conductivity of a PEDOT:PSS film can be enhanced by using a liquid or solid organic compound, as methyl-sulfoxide (DMSO).

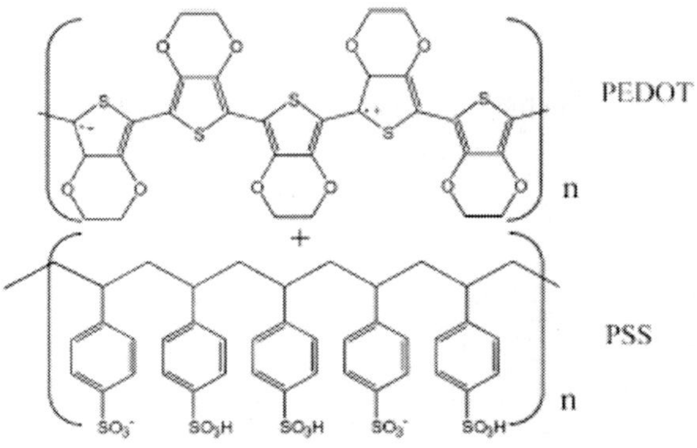

Figure 13. Chemical structure of PEDOT:PSS.[50]

The effect of the organic solvent on PEDOT:PSS film conductivity, the improving of its catalytic action for the electrolyte redox couple, the layer surface roughness and DSCs performances were investigated and compared to the standard Pt film.[50]

The use of the DMSO solvent, pointed out an improving of the physical-chemical properties of the doped PEDOT, in relation with the bare PEDOT:PSS, mainly adding a little amount of carbon black.

In fact, the highest conductivity, enhanced electrochemical activity, roughness surface and consequently the reduced charge transfer resistance due to the more reaction sites available for reducing the triiodide ion in the electrolyte, and cell performances have been reached by using a PEDOT:PSS (C: 0.1 wt.%) compound.

Our studies have focused on different carbon compound and doped PEDOT:PSS. The results we would like to consider are based on different solution processed materials, including Pt (Solaronix, Transparent Pt Catalyst Paint), a graphite based paste (Acheson 423SS), pencil (8B hardness) and two different formulations (low and high conductivity) of poly (3,4-ethylene dioxythiophene):poly(styrene sulphonate), PEDOT:PSS (Figs. 14 and 15).

As we discuss above, the graphite paste (deposited via blade coating) and the PEDOT formulations (spin-coated) are interesting as they can be processed at temperatures compatible with plastic substrates such as PET or PEN. We have seen that the high conductivity PEDOT:PSS has comparable short circuit currents (ISC) and open circuit voltages to the Pt counter electrode (taken as standard) although it has a ~35% (in relative terms) lower fill factor.

The low conductivity of the PEDOT:PSS film leads to a significantly poorer device output.

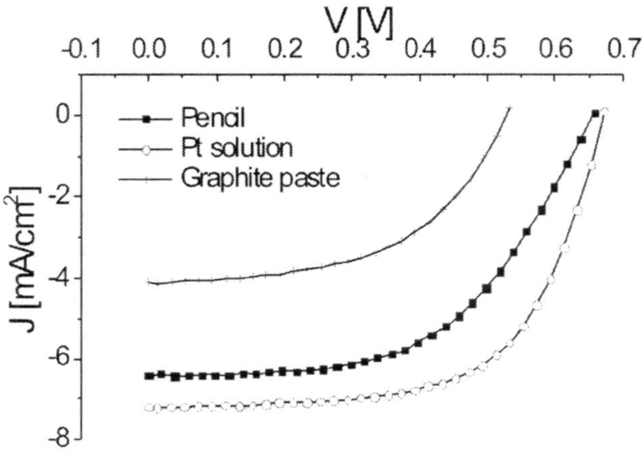

Figure 14. I-V characteristics of DSCs based on different carbon counter electrode. Performances are compared with the standard cell using Pt CE.
Pencil: J_{SC} = 6.44 mA/cm^2, V_{OC} = 658 mV, FF = 53%, η = 2.30%; Pt solution: J_{SC} = 7.21 mA/cm^2, V_{OC} = 673 mV, FF = 62.4%, η = 3%; Graphite paste: J_{SC} = 4.09 mA/cm^2, V_{OC} = 534 mV, FF = 50%, η = 1.15%.

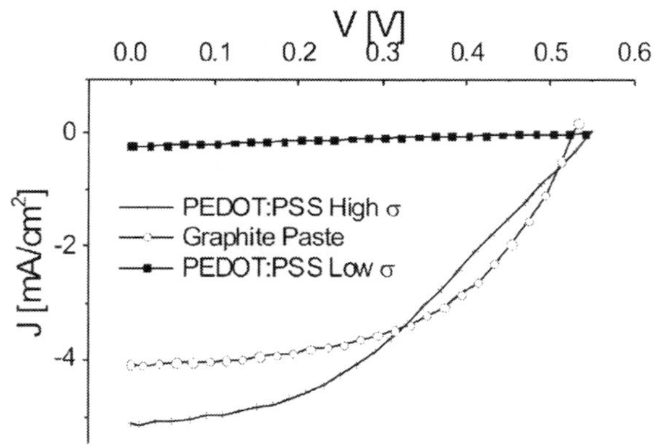

Figure 15. I-V characteristics of DSCs based on different counter electrode. DSCs performances are investigated for PEDOT:PSS with different conductivity and graphite paste CE.
PEDOT:PSS High conductivity: J_{SC} = 5.14 mA/cm^2, V_{OC} = 550 mV, FF = 38.4%, η = 1.12%; Graphite Paste: J_{SC} = 7.21 mA/cm^2, V_{OC} = 673 mV, FF = 50.54%, η = 1.15%; PEDOT:PSS Low conductivity: J_{SC} = 0.236 mA/cm^2, V_{OC} = 543 mV, FF = 19.69%, η = 0.03%.

The graphite paste has an ISC which is ~20% lower than those of both Pt and the high conductivity PEDOT:PSS (Fig. 16).

In order to understand the underlying physical-chemical properties of the photovoltaic cells and correlate these to the I-V characteristics of the cells under

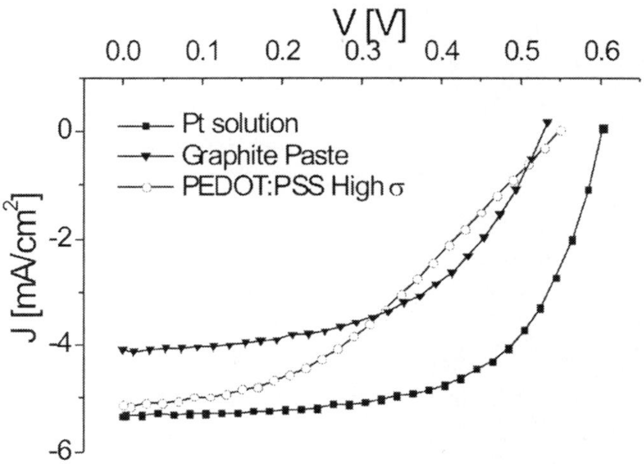

Figure 16. I-V characteristics of DSCs based on different counter electrode. DSCs performances are investigated for PEDOT:PSS with different conductivity and graphite paste CE.
Pt solution: $J_{SC} = 5.35$ mA/cm^2, $V_{OC} = 604$ mV, FF = 61.31%, $\eta = 2\%$; Graphite Paste: $J_{SC} = 7.21$ mA/cm^2, $V_{OC} = 673$ mV, FF = 50.54%, $\eta = 1.15\%$; PEDOT:PSS High conductivity: $J_{SC} = 5.14$ mA/cm^2, $V_{OC} = 550$ mV, FF = 38.4%, $\eta = 1.12\%$.

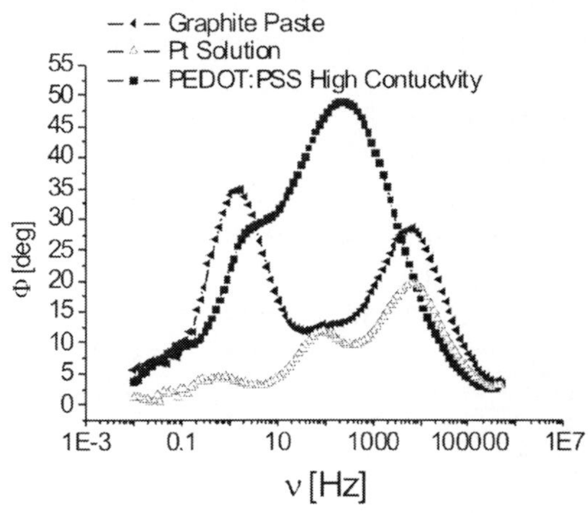

Figure 17. EIS spectra. The Bode diagrams for the investigated materials are reported.

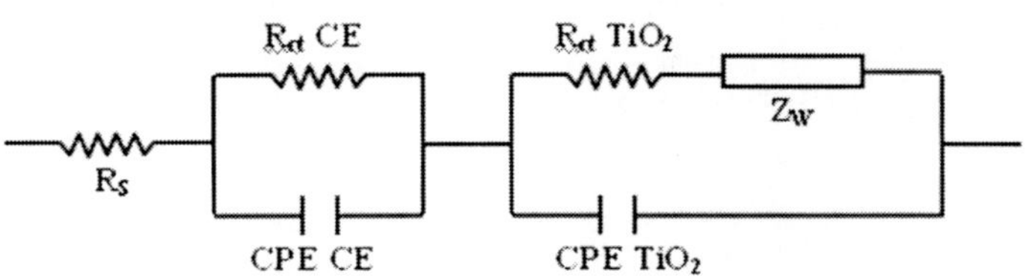

Figure 18. Equivalent circuit to fit impedance spectra. Each resistance act as a single part of the whole DSC.

AM1.5 solar simulator we have performed electrochemical impedance spectroscopy experiments (Fig. 17). In this respect we separate the elements of the model equivalent circuit (Fig. 18) especially regarding the series and charge transfer resistances (R_{ct}(Pt) < R_{ct}(Graphite) < R_{ct}(PEDOT)), which we see can vary significantly amongst the materials employed (including the low and high conductivity formulation of PEDOT:PSS).

This information is useful to assist us in the choice of suitable additives with which to load the inks and pastes in order to increase the performance further, besides giving us a correct interpretation of the whole DSC system including the electrolyte.

Indeed, the EIS measurement can show how good could be the interaction between the CE film and the electrolyte system and consequentially the right choice for all the DCS components.

5. Photovoltaic module engineering

When one moves from the lab scale cell (less than 1 cm^2) you to large area cell or modules, efficiencies tend to halve themselves. This reduction of efficiency, present in any photovoltaic technology, is due primarily to the increase of series resistance as consequence of the increase of cell size. To solve this problem, part of the total area of the module is dedicated to conductive fingers whose aim is to reduce this resistance. These areas occupy space and these impacts on the module efficiency. These considerations lead to a clear trade off linked to optimization of geometries but also of materials. The presence of a liquid electrolyte is an aspect that complicates the choice of materials with regard to metal contacts and materials for encapsulation.

Increasing the active area is one of the obstacles the DSCs have to overcome before these can be made in form of a commercial product.[51]

From a module design point of view a DSC is characterized by its electronic circuit (see Fig. 19).

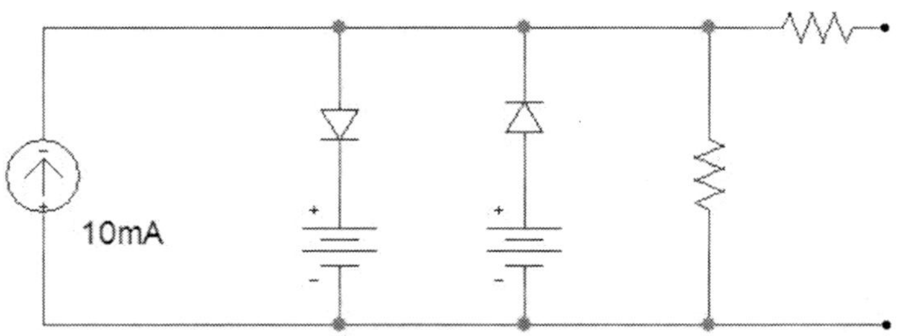

Figure 19. Equivalent circuit of a DSC.

The circuit in Fig. 19 is a PSPICE model used at CHOSE laboratory, developed from measurements and modelling. Voltage generators are used to translate diode threshold potential, V_T, while the second diode is necessary to characterize the reverse bias behavior.[52] Reverse bias is an essential aspect in module design but this topic will not be treated here. We found a good match between our measured curve and the curve of simulated circuit (Figs. 20 and 21).

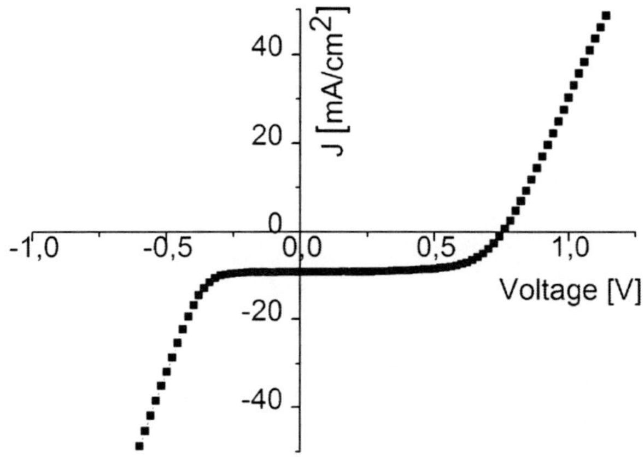

Figure 20. I-V curve of a 0.25 cm^2 active area DSC @1000 W/m^2.

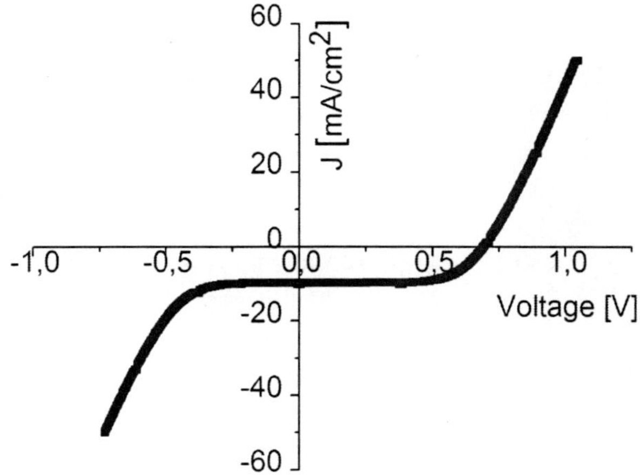

Figure 21. I-V curve obtained from PSPICE DC simulation of circuit presented in Fig. 19.

The series resistance R_s depends mainly from the sheet resistance of transparent conductive oxide (TCO); in our case Fluorine doped Tin Oxide (FTO), the thickness of the different layers and the quality of contacts, while the parallel resistance R_p depends on pathways that allow charge recombination before they can be collected from external electrodes. Figure 22 shows the influence of the R_s and the decrease of R_p on I-V curves.

So the effect of the increment of cell size is to drop the maximum power point to lower values.

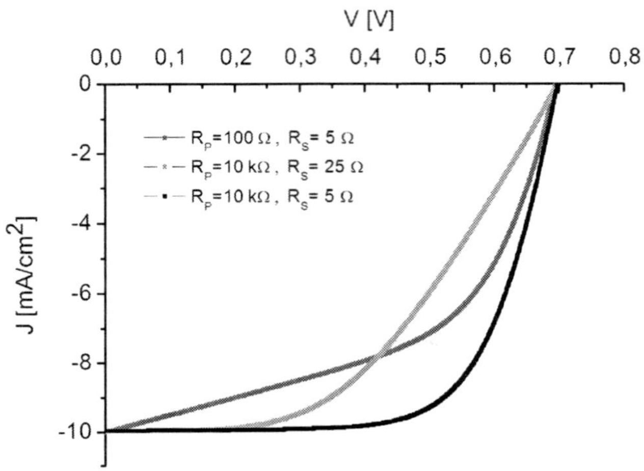

Figure 22. Effect of variation of resistance on equivalent circuit in Figure 1.

5.1 MODULE TECHNOLOGY

As we have seen, photovoltaic technology allows the direct conversion of solar energy into electricity. The conversion takes place through photovoltaic cells that must be electrically connected with each other (modules) to achieve the required voltage and current levels. The photovoltaic modules can be used individually (for example 36 silicon cells can load approximately a battery from 12 V) or connected in series or in parallel to form strings and fields.

When interconnecting single cells together to form a module, a number of new issues arise from this system. These aspects have partly already been studied in other solar cell technologies, but DSC technology introduces additional problems. In fact the wide experience gained with previous technologies can be exploited by DSC, however, we must not forget that the DSC are electrochemical devices and in particular the presence and management of a liquid electrolyte is a specific topic of this technology.

The connection of two or more cells (or modules) can be performed in the following ways:

- Parallel: obtaining high current and low voltage
- Series: obtaining low current and high voltage

Connections can be external (this requires a dedicated post-processing of cells to built them in a module) or integrated. In integrated connection the cells in a module are built simultaneously, and are contacted with some integrated connection strategy. Integrated connection is generally preferable (when it is technically feasible) as it involves lower production costs.

5.2 DSC CONNECTIONS

DSC Cells allow the integrated connection into different ways. The individual cell not only should not come into contact with each other electrically, but we must also ensure a full isolation between cell electrolyte chambers. Moreover electrical connections (for example a silver contact) must be protected against corrosive effect of Redox couple of' the electrolyte.

5.2.1 Parallel connection

The structure is shown in Fig. 23. Current flows perpendicular to the module section shown in figure.

Cells are built interspersed with conductive finger, for example. The encapsulation is necessary to avoid contact between silver and electrolyte, but is also important as it avoids possible short circuits.

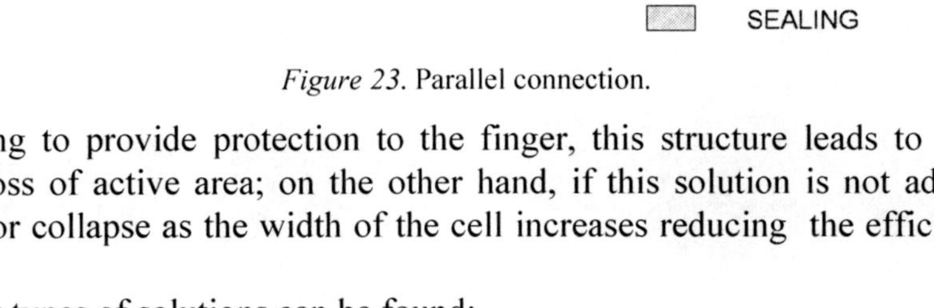

Figure 23. Parallel connection.

Having to provide protection to the finger, this structure leads to a significant loss of active area; on the other hand, if this solution is not adopt, the fill factor collapse as the width of the cell increases reducing the efficiency of the cell.

Other types of solutions can be found:

- Use of conductors that are more resistant to corrosive effect of' electrolyte (Nickel for example[53]).
- Protect fingers with a thick layer of TiO_2.[54]
- Protect fingers with the TCO layer.[53]

- Focus on the development of highly conductive transparent conductive oxides.[55]

5.2.2 W connection

The W structure is shown in Fig. 24. The single cell is electrically isolated removing TCO by Laser scribing.

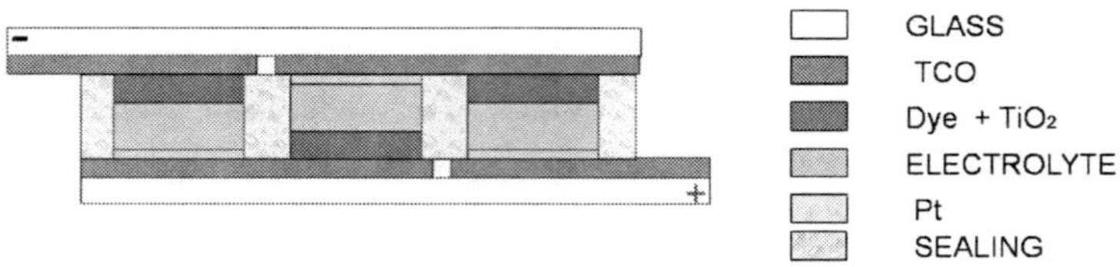

Figure 24. W connection.

The removal of conductive oxide occurs every two cells and, on the same glass, a counter electrode is deposed alternately against a photo electrode. The series connection is simply made by conductive oxide that is common between adjacent counter electrode and photo electrode (Fig. 25). The best advantage of this connection is that the introduction of external elements of connection are not necessary. These external elements are in fact difficult to manage in relation to their possible contact with electrolyte and their thermal stability. On the other hand, in this configuration the counter electrode is dipped into dye (process that degrades the functioning of cells[56]). Moreover, half of the cells are exposed on the counter electrode side. These cells have a performance disadvantage since part of the light is absorbed by electrolyte and platinum and this leads to a permanent electrical mismatch in module[52] (Fig. 26).

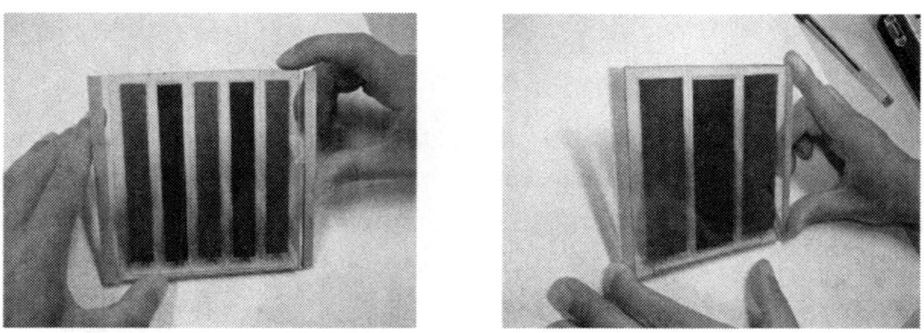

Figure 25. Example of W connection modules.

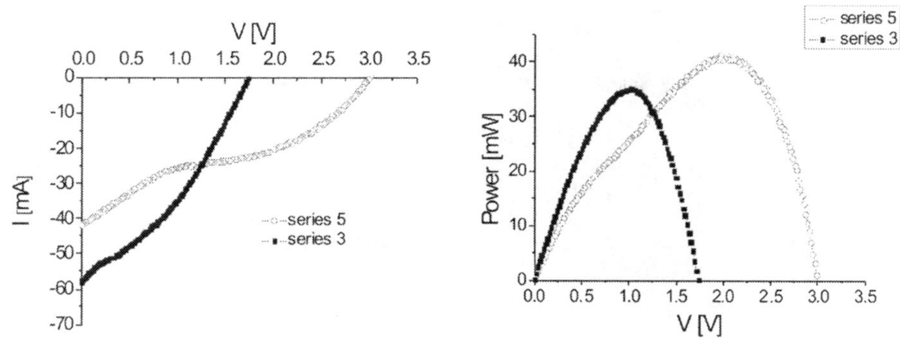

Figure 26. I-V and power curves for the W connected modules with three (black curves) and five (grey curves) cells.

5.2.3 Z connection

The Z connection is another series connection (see Fig. 27) but it includes a vertical connection into it (that is a very critical step). Another disadvantage of this structure in contrast with W connection is a lower ratio of the active area on total area.

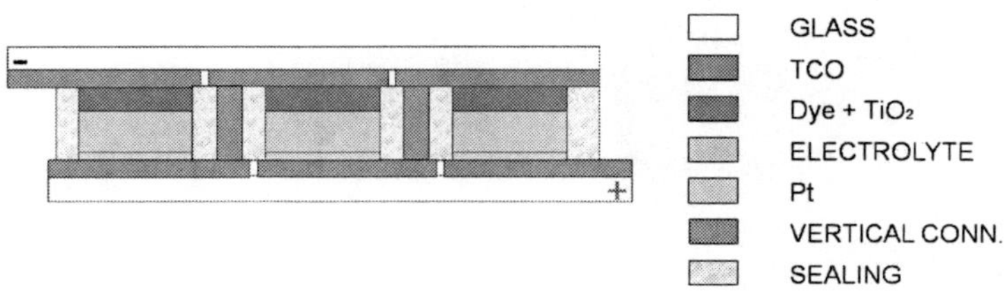

Figure 27. Z connection.

However, this structure allows placing all counter electrodes on the same glass as well as the photo electrodes (all cells have maximum performance).

In Fig. 28 is shown an example of Z modules build at the CHOSE laboratories, while Fig. 29 shows the electrical characteristics of the module.

Figure 28. Example of W connection modules.

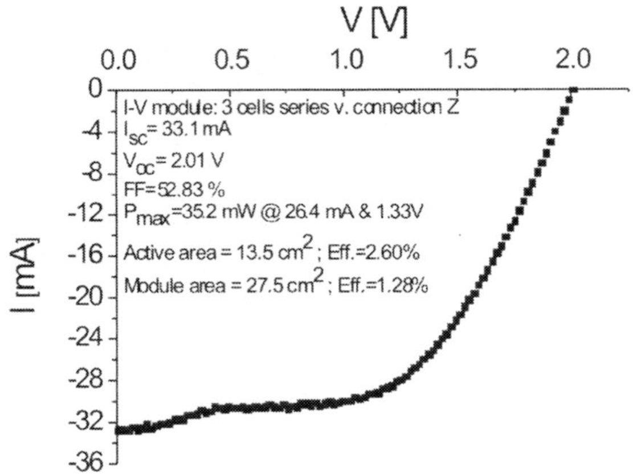

Figure 29. I-V curves for the Z connected modules.

6. Conclusions

Dye solar cells (DSCs) have been earmarked by the European Photovoltaic Industry Association as a leading technology that can help achieve the targets of renewable energy generation set by the European Union for the future. In this chapter we have given a brief survey of DSC science and technology. We have highlighted some of the technology aspects where the development of smart materials and concepts are required: the roles played by the electrolyte and counter electrode materials to the photovoltaic process and strategies and issues encountered in creating a solar cell module as a series of cells integrated in a common sandwich structure between two conductive glass plates.

Electrolytes in their liquid phase are currently still the favoured choice in terms of yielding the highest solar cell performance. The most efficient electrolytes contain the I^-/I_3^- redox couple as it is found to quickly regenerate the oxidized dyes and has a redox potential which leads to a relatively high V_{OC}. Solvent viscosity plays also a part in facilitating ion transport through the cell. High boiling points solvents are particularly favourable for applications as these are less volatile in outdoor conditions. Additives are also added to the electrolyte solution to increase efficiencies. One of these is 4-tert-butylpyridine (TBP) which shifts the band edge of the TiO_2 thereby increasing the open circuit voltage.

A number of coadsorbents have been included in electrolyte solutions to increase the Voc and suppress the aggregation of dye molecules on TiO_2 surface. Recently NMBI (N-Methylbenzimidazole) and GuSCN (Guanidine Thiocyanate) were found to be respectively useful to increase long term stabi-lity and to suppress recombination between electrons in the conduction band of TiO_2 and the electron acceptor in the electrolyte. The use of GuSCN results in an increase in V_{OC} of around 100 mV whilst NMBI is interesting since it was found to lead to more stable devices in ageing tests.

Ionic liquids have also been investigated as they are non-flammable, non-volatile, and have high ionic conductivity and are a typical ingredient of "Green Chemistry". The ionic liquid can be both source of the redox couple and be used "solventless". However adding solvents decreases their viscosity resulting in higher efficiencies which we show can be comparable to those of a typical standard liquid electrolyte.

Catalysing the redox reaction at the counter electrode is usually a thin film of platinum which can be either deposited via sputtering or via precursors in liquid form. Replacing Pt with alternatives which are less costly and can be annealed at lower temperatures is an important issue for commercial modules and flexible cells respectively. We have shown that carbon based electrodes, mainly deposited in liquid form, such as graphite and carbon black pastes and PEDOT: PSS inks, can lead to photovoltaic cells with relatively good performances and thus are leading candidates for alternative counter electrode materials.

Finally, this chapter has outlined two main approaches to extend the development of the technology from the individual cell to large area modules: one where the series connections are carried out by vertical interconnects, and one that does not require these but that necessitates an alternation of vertically inverted cell structures. Issues which need to be addressed when considering large areas are the relatively high sheet resistances of transparent conductors and the increase of the inactive area when integrating many cells on a single substrate. Furthermore, we show that matching of the cell performance for each cell of the module is important to avoid losses and device degradation due to some of the cells operating in reverse bias in some operating conditions. An equivalent circuit developed at chose laboratories is presented that is used to study, together with the experimental IV characteristics extracted from our modules, the aspects of module physics related to fabricating an efficient module that considers electrical and lighting condition mismatch of the cells composing the module.

ACKNOWLEDGEMENTS

This work was supported by Lazio Region project "Polo Solare Organico".

References

1. Renewable Energy Road Map, Renewable energies in the 21st century: building a more sustainable future. Communication from the commission to the council and the European Parliament, 10 January 2007.
2. EPIA Roadmap, June 2004, http://www.epia.org/fileadmin/EPIA_docs/publications/epia/EPI-Aroadmap.PDF
3. Fonte EMSS, comunicazione private. Vedi anche J. Kazmerski JESRP 2005.
4. M. Graetzel, Nature 2001, 414, 338.
5. M. K. Nazeeruddin in Special issue: Michael Graetzel Festschrift, A tribute for this 60th birthday: dye sensitized solar cells., Vol. 248 (Ed. M. K. Nazeeruddin), Elsevier, Amsterdam, 2004, pp. 1161–1511.
6. M. K. Nazeeruddin, M. Graetzel in semiconductor photochemistry and photophysics, Vol. 10 (Eds. V. Ramamurthy and K. S. Schanze), Marcel Dekker, New York, 2003, pp. 301–344.
7. K. Kalyanasundaram, M. Graetzel, Coord. Chem. Rev. 1998, 77, 347–414.
8. M. K. Nazeeruddin, F. De Angelis, S. Fantacci, A. Selloni, G. Viscardi, P. Liska, S. Ito, B. Takeru, M. Graetzel, J. Am. Chem. Soc. 2005, 127, 16835–16847.
9. D. Kuang, S. Ito, B. Wenger, C. Klein, J.-E. Moser, R. Humphry-Baker, S. M. Zakeeruddin, M. Graetzel, J. Am. Chem. Soc. 2006, 128, 4146–4154.
10. T. Horiuchi, H. Miura, K. Sumioka, S. Uchida, J. Am. Chem. Soc. 2004, 126, 12218–12219.
11. S. Kim, J. K. Lee, S. O. Kang, J. Ko, J. H. Yum, S. Fantacci, F. De Angelis, D. Di Censo, M. K. Nazeeruddin, M. Graetzel, J. Am. Chem. Soc. 2006, 128, 16701–16707.
12. B. Gregg, F. Pichot, S. Ferrere, C. R. Fileds, J. Phys. Chem. B 2001, 105, 10461.
13. K. Okada, H. Matsui, T. Kawashima, T. Ezure, N. Tanabe, J. Photochem. Photobiol. A 2004, 164, 193–198.
14. S. A. Sapp, C. M. Elliott, C. Contado, S. Caramori, C. A. Bignozzi, J. Am. Chem. Soc. 2002, 124, 11215–11222; M. Brugnati, S. Caramori, S. Cazzanti, L. Marchini, R. Argazzi, C. A. Bignozzi, Int. J. Photoenergy 2007; submitted.
15. C. M. Elliott, S. Caramori, C. A. Bignozzi, Langmuir 2005, 21, 3022–3027.
16. R. Sastrawan, J. Beier, U. Belledin et al., New interdigital design for large area dye solar modules using a lead-free glass frit sealing, Prog. Photovoltaics Res. Appl. 2006, 14(8), 697–709.
17. R. Sastrawan, Photovoltaic modules of dye solar cells Dissertation zur Erlangung des Doktorgrades der Fakultät für Mathematik und Physik der Albert-Ludwigs-Universität Freiburg im Breisgau, 2006.
18. R. Knechtel, Glass frit bonding: an universal technology for wafer level encapsulation and packaging, Springer Berlin/Heidelberg ISSN 0946-7076, 2005.

19. W. J. Lee, E. Ramasamy, D. Y. Lee, J. S. Song, Glass frit overcoated silver grid lines for nano-crystalline dye-sensitized solar cells. Photochem. Photobiol. A 2006.
20. A. Hinsch, J. M. Kroon, R. Kern, I. Uhlendorf, Longterm stability of dye-sensitised solar cells, Progress in Photovoltaics: Research and Applications 2001, 9(6), 425–438.
21. R. Sastrawan, Photovoltaic modules of dye solar cells Dissertation zur Erlangung des Doktorgrades der Fakultät für Mathematik und Physik der Albert-Ludwigs-Universität Freiburg im Breisgau, 2006.
22. H. Pettersson, T. Gruszecki, Long-term stability of low-power dyesensitised solar cells prepared by industrial methods, 2001.
23. R. Kern, Untersuchungen zur Langzeitstabilität von Farbstosolarzellen mittels optischer und elektrischer Impedanzspektroskopie Fakultät für Physik, Albert-Ludwigs Universität, Doctoral dissertation, 2001.
24. H. DeSilvestro, in II DSC Industrialization conference, 12–14 September, 2007, St. Gallen, Switzerland.
25. T. Toyoda, T. Sano, J. Nakajima, S. Doi, S. Fukumoto, Outdoor performance of large scale DSC modules, J. Photochem. Photobiol. A 2004.
26. R. Sastrawan, Photovoltaic modules of dye solar cells - Freiburg im Breisgau Juni, 2006.
27. F.-T. Kong, S.-Y. Dai, K.-J. Wang, Review of recent progress in dye-sensitized solar cells — Hindawi publishing Corporation Advances in Optoelectroics Volume 2007, Article ID 75384.
28. J. Halme, Dye-sensitized nanostructured and organic photovoltaic cells: technical review and preliminary tests, Master's thesis submitted in partial fulfillment of the requirements for the degree of Master of Science in Technology, Espoo, February 12, 2002.
29. A. Luque, Handbook of photovoltaic science and engineering, 2003, John, p. 674, ISBN 0471491969.
30. D. Kuang, C. Klein, H. J. Snaith, J.-E. Moser, R. Humphry-Baker, P. Comte, S. M. Zakeeruddin, M. Gratzel, Ion coordinating sensitizer for high efficiency mesoscopic dye-sensitized solar cells: influence of Lithium ions on the photovoltaic performance of liquid and solid-state cells, Laboratory for Photonics and Interfaces, Institute of Chemical Sciences and Engineering, Ecole Polytechnique Fédérale de Lausanne, 1015 Lausanne, Switzerland, February 20, 2006.
31. G. Wolfbauer, A. M. Bond, J. C. Eklund, D. R. MacFarlane, A channel flow cell system specifcally designed to test the efficiency of redox shuttles in dyesensitized solar cells, Solar Energy Materials & Solar Cells 2001, 70, 85–101.
32. L. C. Pauling, The nature of the chemical bond and the structure of molecules and crystals, an introduction to modern structural chemistry, 3rd Edition, Cornell University Press, Ithaca, NY, 1960.
33. R. A. Robinson, R. H. Stokes, Electrolyte solutions: the measurement and interpretation of conductance, chemical potential and diffusion in solutions of simple electrolytes, Butterworths Scientific Publications, London, 1959.
34. R. Kawano, H. Matsui, C. Matsuyamaa, A. Sato, Md. A. B. H. Susan, N. Tanabe, M. Watanabe, High performance dye-sensitized solar cells using ionic liquids as their electrolytes, J. Photochem. Photobiol. A 2004, 164, 87–92.
35. Y. Saito, T. Kitamura, Y. Wada, S. Yanagida, Synth. Met. 2002, 131, 185.
36. J. M. Kroon, N. J. Bakker, H. J. P. Smit, P. Liska, K. R. Thampi, P. Wang, S. M. Zakeeruddin, M. Gratzel, A. Hinsch, S. Hore, U. Wurfel, R. Sastrawan, J. Durrant, E. Palomares, H. Petterson, T. Gruszecki, J. Walter, K. Skupien, G. E. Tulloch, Nanocrystalline dye-sensitized solar cells having maximum performance, Prog. Photovolta: Res. Appl. 2007, 15, 1–18.
37. H. Nusbaumer, S. M. Zakeeruddin, J.-E. Moser, M. Gratzel, An alternative efficient redox couple for the dye-sensitized solar cell system, Chem. Eur. J. 2003, 9, 3756–3763.

38. S. A. Sapp, C. M. Elliott, C. Contado, S. Caramori, C. A. Bignozzi, Substituted polypyridine complexes of cobalt (II/III) as efficient electron-transfer mediators in dye-sensitized solar cells, J. Am. Chem. Soc. 2002, 124(37), 11215–11222.
39. N. Papageorgiou, M. Gratzel, P. P. Infelta, On the relevance of mass transport in thin layer nanocrystalline photoelectrochemical solar cells, Sol. Energy Mater. Sol. Cells 1996, 44, 405–438.
40. G. P. Kalaignan, Y. S. Kang, A review on mass-transport in dye-sensitized nanocrystalline solar cells, J. Photochem. Photobiol. C: Photochem. Rev. 2006, 7, 17–22.
41. G. S. Owens, M. M. Abu-Omar, J. Mol. Catal. A: Chemical 2002, 187, 215.
42. V. Farmer, T. Welton, Green Chem. 2002, 4, 97.
43. P. Bonhôte, A. P. Dias, N. Papageorgiou, K. Kalyanasundaram, M. Grätzel, Inorg. Chem. 1996, 35, 1168.
44. T. Ma et al., J. Electroanal. Chem. 2004, 574, 77–83.
45. X. Fang et al., J. Photochem. Photobiol. A 2004, 164, 179–182.
46. J. M. Kroon et al., Prog. Photovolt: Res. Appl. 2007, 15, 1–18.
47. T. N. Murakami, M. Gratzel, Inorg. Chim. Acta 2008, 361, 572–580.
48. Ramasamy et al., Appl. Phys. Lett. 2007, 90, 173103.
49. Huang et al., Electrochem. Commun. 2007, 9, 596–598.
50. J. G. Chen et al., Sol. Energy Mater. Sol. Cells 2007, 91, 1472–1477.
51. S. Dai, Dye-sensitized solar cells, from cell to module, Sol. Energy Mater. Sol. Cells 2004, 84, 125–133.
52. Sastrawan, Interconnecting dye solar cells in modules I–V characteristics under reverse bias, J. Photochem. Photobiol. 2006, 178, 33–40.
53. K. Okada, 100 mm × 100 mm large-sized dye sensitized solar cells, J. Photochem. Photobiol. 2004, 164, 193–198.
54. K. C. Mandal, Progress in producing large area flexible dye sensitized solar cells, Mater. Res. Soc. Symp. Proc. 2005, 836, L121–L127.
55. K. Goto, Heat-resisting TCO films for PV cells, Sol. Energy Mater. Sol. Cells 2006, 90, 3251–3260.
56. G. E. Tulloch, Light and energy—dye solar cells for the 21st century, J. Photochem. Photobiol. 2004, 164, 209–219.

MAGNETOTRANSPORT PROPERTIES OF COPPER TERNARIES: NEW SOLAR CELLS MATERIALS

LAHCEN ESSALEH[1*] AND SYED M. WASIM[2]
[1]*Laboratoire de Matière Condensée et Nanostructures (L.M.C.N.)*
Université Cadi-Ayyad, Faculté des Sciences et Techniques, Département de Physique, Guéliz, B.P. 549, Marrakech, Maroc
[2]*Centro de Estudios de Semiconductores, Facultad de Ciencias Universidad de Los Andes, Merida 5101, Venezuela*

Abstract. Ever since the first solar cell was developed by Bell Laboratories,[1] new materials are being investigated that could yield solar energy conversion efficiency comparable in device production cost to the standard electric power generators. Continuing with this effort, in recent years solar cells with efficiency close to 20%, prepared with alloys of $CuIn_{1-x}Ga_xSe_2$ has been reported. On the other hand, it is suggested that n-$CuIn_3Se_5$ combined with p-$CuInSe_2$ could play an important role in the optimization of solar cells based on Cu-ternaries. In addition to their technological importance, these compounds are also of academic interest. Because of the presence of shallow donor and acceptor levels due to the presence of cation-cation disorders in $Cu(In,Ga)Se_2$ and ordered defects in $Cu(In,Ga)_3$ $(Se,Te)_5$, the impurity band in these compounds starts to form between liquid helium and nitrogen temperatures. This permits to study the variable range hopping (VRH) conduction in the impurity band over a much wider temperature range. In the case of Mott type VRH, the electrical resistivity follows the relation $\rho = \rho_o \exp(T_o / T)^{1/4}$, where ρ_o is the pre-exponential factor and T_o is the characteristic temperature. In this paper we report on a comparative study of the variation of electrical resistivity over a wide temperature range at different magnetic field values in single crystals of $CuInSe_2$, $CuInTe_2$, $CuGaTe_2$, $CuIn_3Se_5$, $CuIn_3Te_5$ and $CuGa_3Te_5$. Results related to the variable range hopping and metallic conduc-tion mechanisms are discussed. The magnetoresistance data of n–$CuInSe_2$ in the variable range and p–$CuGaTe_2$ in the metallic regimes, measured in pulsed magnetic field up to 25 T between 2 and 300 K are analyzed and compared with the existing theoretical models.

*To whom correspondence should be addressed: Lahcen Essaleh, L.M.C.N, Université Cadi-Ayyad, Faculté des Sciences et Techniques, Département de Physique, Guéliz, B.P. 549, Marrakech, Maroc: Email: lessaleh@ yahoo.com/lessaleh@fstg-marrakech.ac.ma

I. A. Luk'yanchuk and D. Mezzane (eds.), *Smart Materials for Energy, Communications and Security.* 127

Keywords: Photovoltaic devices, solar cells, semiconductors, ternary compounds, ordered defect compounds, variable range hopping conduction, magnetoresistance.

1. Introduction

1.1 PHOTOVOLTAIC MATERIALS

Photovoltaic is a high-technology approach to converting sunlight directly into electrical energy. In its simplest form, conceptually, a photovoltaic device is a solar–powered battery whose only consumable is the solar energy. The first practical solar cell was developed[1] nearly half a century ago at Bell Laboratories. However, the disruption of oil supplies in the early seventies to the industrialized world led to serious research activities in the development of renewable energy resources, in particular in photovoltaic electric generators as terrestrial power sources.[2]

Because of their adequate band gap in the visible region of the solar spectrum and relatively high absorption coefficient α, crystalline Si, thin films of CdTe and hydrogenated α-Si have been investigated extensively.[3,4] The highest thin films cell efficiency has been confirmed around 16% for CdTe and Si and 12% for α-Si. Although crystalline Si has many valuable properties as a solar cells material, because of its lower absorption coefficient as compared to other materials (to be discussed later), a greater thickness is required in a sample homojunction solar cells. In the case of CdTe, that uses CdS as its n-type window layer, concern is raised about its toxicity, whereas α-Si devices are found to degrade due to the effect of solar radiation.[4]

Another material that has turned out to be very promising in the fabrication of thin film photovoltaic devices is the ternary compound $CuInSe_2$ (CIS). This is because, in the main part of the solar spectrum, in the visible range, it has very high absorption coefficient α above 10^5 cm^{-1}.[3,4] It is radiation hard and shows chemical stability. Already, in the early nineties, thin film solar cells based on this material with addition of Ga had shown efficiency around 16%, comparable to that of CdTe.

In the recent years, solar cells with efficiency close to 20%, prepared with alloys of $CuIn_{1-x}Ga_xSe_2$ (CIGS),has been reported.[5,6] Another important development has been a three–step CIGS co-evaporation on polymer substrates at temperatures below 450°C.[7]

An ordered defect ternary compound that is formed on the In-rich side of the pseudo-binary phase diagram of Cu_2Se-In_2Se_3 is $CuIn_3Se_5$. It segregates as a secondary phase on the surface of In-rich $CuInSe_2$ thin films. Since high efficiency solar cells can be prepared by p-n junction between p-type $CuInSe_2$ and n-type $CuIn_3Se_5$, it is expected that $CuIn_3Se_5$ could play an important role in the optimization of solar cells based on $CuInSe_2$.[8]

During the last 2 decades several review articles[2–4,9] on the growth of single crystals and preparation of thin films by different techniques of $CuInSe_2$, their physical characterizations and its possible use in photovoltaic devices, have been reported. Since these compounds, in addition to their technological importance, are also of great academic interest, in the present paper we report on the electrical conduction mechanisms between 2 and 300 K in both low and high magnetic field up to 25 T in single crystal samples of several ternary compounds of the I-III-VI_2 family such as $Cu(In,Ga)(Se,Te)_2$ and also the ordered defect compounds $Cu\,(In,Ga)_3(Se,Te)_5$.

1.2 COPPER TERNARIES OF THE I-III-VI_2 AND I-III_3-ϒ-VI_5 FAMILIES

In lightly doped semiconductors and at low temperatures, the electrical conduction occurs in the impurity band. The main contribution to the electrical conductivity in the case of n–type semiconductors comes from electrons hopping directly between impurity states without any excursion to the conduction band. Electrons jump from occupied donors to empty donor states, and therefore the presence of empty states is a necessary condition. The electrical conduction is attributed to hopping mechanism when the electrons jump are associated with a weak overlap of wave function tails from neighboring donors. This leads to very low electrical mobility. The variable range hopping (VRH) conduction mechanism occurs at low temperatures when the hopping length R_{hop} is higher than the localization length ξ. The energy necessary for a hop is small and the electron prefers to make a hop to a long distance to find a site with lowest activation energy. With increasing doping, the metal–insulator transition can occur at certain impurity concentration.

Measurement of the magnetoresistance, in doped materials, is a common tool in semiconductor physics. When the conduction is due to carriers in the conduction band, the resistivity has a power-law dependence with the magnetic field. The theory of this phenomenon is well developed. In heavily doped semiconductors on the metallic side of the metal-insulator transition (MIT), the observed negative magnetoresistance is explained by considering quantum corrections due to the weak localization and electron-electron interactions.[10,11]

Other experimental works confirmed that at low temperature and high magnetic fields quantum oscillations occur in all electronic properties in degenerate semiconductors. With increasing temperature the oscillations gradually vanish. In some cases the oscillations are superimposed on the monotonic magnetoresistance. The term 'Shubnikov-de Haas effect' means quantum oscillations of the electronic transport coefficients. In the narrower sense, it is the oscillations in the magnetoresistance.[12,13]

In materials, located on the insulator side of the MIT, a characteristic feature of the hopping mechanism of conduction is the presence of a giant positive magnetoresistance which varies exponentially with the strength of the magnetic field. Such magnetoresistance has been observed in many semiconductors, namely, InSb, InP, Ge and GaAs.[14] As it is well known, in a strong magnetic field the wave functions of impurity electrons are squeezed in the transverse direction. In the simplest case, spherically symmetric wave functions become cigar-shaped with the application of magnetic field. This leads to a sharp decrease in the overlap of the wave function tails for an average pair of neighboring impurities, and hence to an exponentially increased resistivity. Shklovskii and Efros[14] have developed a positive magnetoresistance theory based on the percolation method. In many experimental works, it is observed that in the VRH regime the magnetoresistance is negative at low magnetic field. New theoretical models, based on localizations effects, are proposed to explain this behavior.[15,16] The theory of quantum corrections, used to explain the negative magnetoresistance (NMR) in heavily doped materials, is not valid for semiconductors situated on the insulating side of the MIT. On the insulating side, near the MIT, the existing models are base essentially on the manifestation of the quantum interference among random paths in the hopping process.[16] This interference increases the hopping probability and decreases the resistivity.

In the present work, we report on the analysis of the temperature and magnetic field dependence of the magnetoresistance in samples of Cu–ternaries like $CuInSe_2$, $CuGaTe_2$, $CuIn_3Se_5$, $CuIn_3Te_5$ and $CuGa_3Te_5$. We will present, essentially, some important results related to the variable range hopping and metallic conduction mechanisms in these materials. $CuIn_3Se_5$ with n = 5 is one member of the ordered defect compounds which can be obtained from the formula $Cu_{n-3}In_{n+1}(Se,Te)_{2n}$, where n = 4, 5,…9.[17] Other members of this family with n = 5 are $CuGa_3Se_5$, $CuIn_3Te_5$ and $CuGa_3Te_5$. They can also be derived from a repeat of one unit of the interacting donor–acceptor $\left\{(In,Ga)_{Cu}^{+2}, 2V_{Cu}^{-1}\right\}$ defect pair (DADP) in each *n* units of Cu (In,Ga)$(Se,Te)_2$.[18] In addition to their technological importance, these compounds are also of academic interest. Electrically active intrinsic defects caused by vacancies and interstitials play an important role in the electrical conduction, especially at low temperatures.[8,19,20] Because of the presence of the cation–cation defects in $Cu(In,Ga)(Se,Te)_2$ and ordered vacancy defects in $Cu(In,Ga)_3(Se,Te)_5$, the impurity band in these materials, as compared to the elemental and binary compounds, starts to form at much higher temperatures. The growth of single crystals of these compounds has been reported earlier.[8]

2. Brief theoretical considerations

2.1 VARIABLE RANGE HOPPING CONDUCTIVITY

The idea of impurity band conduction in a doped and compensated semiconductor was first introduced by Gudden and Schottky,[21] who mentioned the possibility of a process in which a carrier is transferred from an occupied to an equivalent empty donor state of higher energy. This is the hopping process which can only take place in a compensated material. The conductivity shows a small activation energy ε_3 which depends on concentration and compensation, and a small pre-exponential factor that varies as $exp(-2R/a_H)$, where R is the mean distance between centers and a_H the hydrogen radius. Miller and Abrahams[22] propose a treatment for the hopping to nearest neighbours. The process of variable range hopping, expected at low temperatures, was developed by Mott.[23] The model is of tight binding lower Hubbard band, not fully occupied, in which Anderson localization is produced (a) by the random fields of the charged acceptors and, (b) the random positions of the donors. As the concentration is increased, the bandwidth B increases and eventually states in mi-band become delocalized. A mobility edge (E_C) will then exist between localized and delocalized states. As the concentration increases, E_C will eventually move to the Fermi energy, the activation energy $\varepsilon_2 = /E_C - E_F/$ will disappear and the transition to metallic conduction will take place.

Near the metal-insulator transition when the Fermi level is very close to the mobility edge, the localization length and the dielectric constant have critical behaviors. These have a strong effect on the conduction regimes. On the insulating side of the metal–insulator transition (MIT) and in a three dimensional system, the electrical conduction corresponding to the hopping mechanism takes place at temperatures which are so low that typical resistances between neighboring impurities become larger than those connecting some remote impurities whose energy levels happen to be very close to the Fermi level. In this case, the characteristic hopping length (R_{hop}) increases with lowering temperature. The necessary energy for a hop is small and the electron prefers to make a hop to a long distance to find a site with lowest activation energy. This site is not necessarily the nearest one. This mechanism is known as variable range hopping. If the density of states at the Fermi level is estimated to be constant, one obtains the celebrated Mott's law[23]:

$$\sigma(T) = \sigma_o \exp\left[-\left(\frac{T_o}{T}\right)^{1/4}\right] \quad \text{and} \quad T_o = \frac{\beta}{k_B N(E_F)\xi^3} \tag{1}$$

where $N(E_F)$ is the density of states at the Fermi level, ξ the localization length and β is a numerical coefficient whose value as 21.2 is given by Shklovskii and Efros.[14]

The theory of the presence of a gap in the density of states at the Fermi level (Coulomb gap) was first described by Efros and Shklovskii.[24] According to them, because of Coulombian correlations, the density of states vanishes at the Fermi level. This has an important effect on the temperature dependence of hopping conduction, especially in the variable range hopping. At low impurity concentrations, the quantum effects associated with the overlap of neighboring states can be regarded as small, and the states are strictly localized. The expression for the conductivity in this case is

$$\sigma(T) = \sigma'_o \exp\left[-\left(\frac{T'_o}{T}\right)^{1/2}\right] \quad \text{and} \quad T'_o = \frac{\beta_1 e^2}{\kappa \xi}, \tag{2}$$

where κ is the dielectric constant and β_l is a numerical coefficient whose the value is given by Shklovskii and Efros[14]; $\beta_1 = 2.8$.

2.2 MAGNETORESISTANCE IN THE VARIABLE RANGE HOPPING REGIME

2.2.1 The positive magnetoresistance

The effect of the shrinkage due to magnetic field of the wave function becomes stronger with increasing distance from the impurity center. Therefore in the variable range hopping (VRH) conduction regime at sufficiently low temperatures, one should expect a large positive magnetoresistance. We will first consider the Mott VRH regime with a nonvanishing density of states at the Fermi level. In the case of weak magnetic field, the magnetoresistance is obtained from the expression

$$Ln \frac{\rho(B)}{\rho(0)} = t_1 \frac{e^2 a^4}{c^2 \hbar^2} B^2 \left[\frac{T_o}{T}\right]^{3/4}; \qquad t_1 = 5/2016 \tag{3}$$

and for strong magnetic field, Shklovskii and Efros[14] showed that,

$$\rho(B) = cte.\exp\left\{\left[\frac{T_o(B)}{T}\right]^{1/3}\right\}, \quad T_o(B) = \frac{2.1\, e}{N(E_F)c\hbar\, a_B\, k_B} B \tag{4}$$

In these expressions a and a_B represent the state radius in the absence and presence of a magnetic field B, respectively.

The situation of the Coulomb gap is of special interest. The density of states is proportional to $(E - E_F)^2$ and vanishes at the Fermi level. The expressions for the magnetoresistance are

$$Ln\,\frac{\rho(B)}{\rho(0)} = t_2\,\frac{a^4}{\lambda^4}\left[\frac{T_o^{'}}{T}\right]^{3/2}; \qquad t_2 = 0.0015 \text{ in the case of weak field} \tag{5}$$

and for the case of strong magnetic field

$$\rho(B) = cte.\exp\left\{\left[\frac{T_o^{'}(B)}{T}\right]^{3/5}\right\}, \quad T_o^{'}(B) = \frac{\beta_2 e^2}{\kappa a^*}, \quad a^* = (\lambda^2 a_B)^{1/3} \tag{6}$$

where $\beta_2 = 3.17$. In these expressions, $\lambda = \sqrt{\frac{c\hbar}{eB}}$ is the magnetic length which physically corresponds to the amplitude of the zero-point motion of a free electron in the lowest Landau level which is the smallest length of electron localization due to magnetic field.

2.2.2 The negative magnetoresistance

In many experimental works, it is observed that in the VRH regime the magnetoresistance is negative at low magnetic field. New theoretical models are proposed to explain this behavior. These are by Nguyen (1985),[15] Entin-Wohlman (1989),[25] Schirmacher (1990),[26] and Raikh (1990).[27] In heavily doped semiconductors on the metallic side of the metal-insulator transition (MIT), the negative magnetoresistance is explained by considering a quantum corrections due to the weak localization and electron-electron interactions[10,11] which is not valid for semiconductors situated on the insulating side of the MIT. In the insulator side near the MIT, the existing models are base essentially on:

a) *The dependence of the activation energy of the hop on magnetic field. It is shown*[27] *that the shrinkage of the wave functions of the localized electronic states in magnetic field that reduces the repulsion of the energy levels of the neighboring sites, can decrease on the average the activation energy of the hop and therefore increases its probability.*
b) *The manifestation of the quantum interference among random paths in the hopping process.*[15] *These interference influences the hopping probability.*[16]

To evaluate the negative magnetoresistance in the VRH regime, these authors have applied the percolation theory to the system formed by elementary resistances. They have calculated the average of the conductivity and found that the negative magnetoresistance varies as B^2 in the weak field regime and takes the form

$$\left(\frac{\Delta\rho}{\rho}\right)_{-} \propto T^{-3/4} B^2 \tag{7}$$

Another model is proposed by Schirmacher.[26] He considered the interference between the amplitude for a direct transition and that for an indirect one involving a single third site. This site is situated below the Fermi level. With the effective medium approximation developed by Movaghar (1980),[28] Shirmacher showed that the magnetoconductivity can be expressed as:

$$\frac{\Delta\sigma}{\sigma} = a(T)B, \quad a(T) \propto T^{-7/8} \tag{8}$$

2.3 MAGNETORESISTANCE IN THE METALLIC REGIME

The explanation of the observed negative magnetoresistance (NMR) on the metallic side of the metal-insulator transition (MIT) at low magnetic field is based generally on the weak localization phenomenon. The weak localization regime corresponds to the fact that the localization effects are considered as small corrections to the classical Boltzmann conduction. Bergman (1983)[29] described qualitatively the localization as a quantum interference between the diffused waves by the impurity centers in the system.

In the presence of an external magnetic field B, the partial waves describing the buckles in the inverse sense are dephased. This occurs when the electron return to its original point 0, by $\delta\Phi = e\Phi/\hbar$ where Φ is the flux of magnetic field through the buckles. Then the magnetic field destroys the phase coherence which is in the origin of constructive interference leading to localization. The application of the magnetic field then leads to a delocalization and to negative magnetoresistance. This implies that the existence of the NMR does not depend only on the presence of a magnetic field, but also on the position of the studied system towards the metal-insulator transition (MIT) and on the temperature. We know that to have localization, the phase coherence must be maintained through the diffusion paths. All collisions must be elastic. If the elastic diffusion time τ_e is much greater than the inelastic diffusion time τ_i, the inelastic energy appears randomly in the expression of the phase. Thus, higher is the τ_i, higher is the chance to observe NMR. Isawa (1984)[30] noted that τ_i is proportional to T^{-1}. This relation shows that the probability to observe NMR increases with lowering temperature. In the regime of weak localization, the corrections to the classical Boltzmann theory can be calculated by a perturbation method. The results of these theories lead to interpret the experimental results, and hence to determine the values of the electronic relaxation times. In the case of a tridimensional system, three contributions are considered, namely, the localization effects, the inter-electronic interactions and the superconducting fluctuations.[31, 32]

3. Experimental Results

3.1 ELECTRICAL CONDUCTION MECHANISMS AT ZERO MAGNETIC FIELD

The variation of the electrical resistivity ρ and Hall coefficient R_H as a function of temperature up to 300 K is given in Fig. 1 for one representative

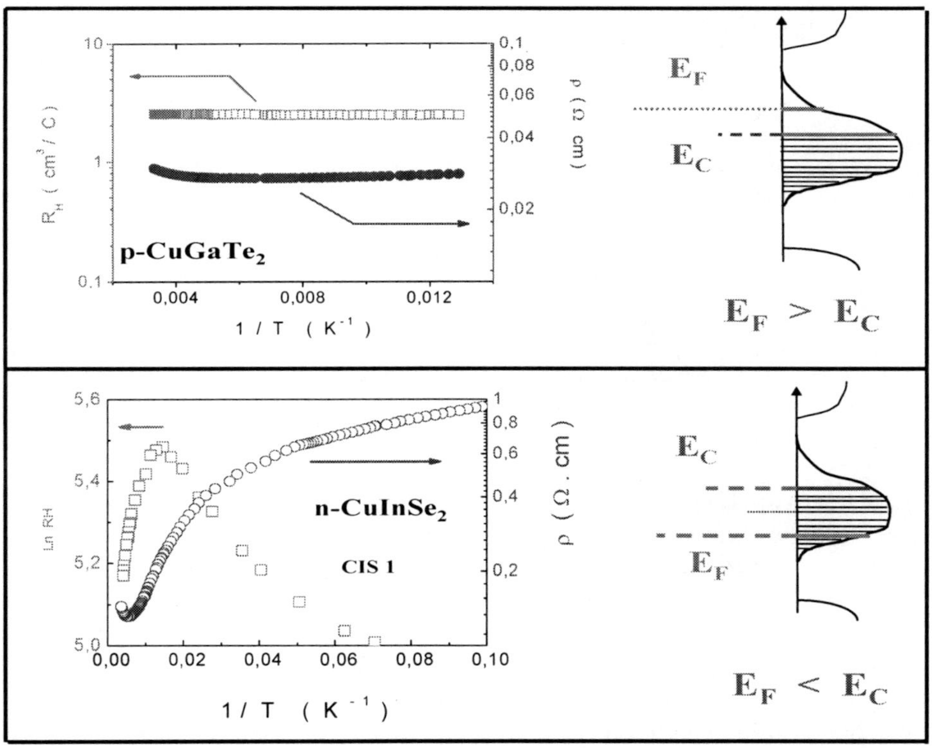

Figure 1. Variation of the electrical resistivity ρ and Hall coefficient R_H as a function of temperature in p-$CuGaTe_2$ and n-$CuInSe_2$. The relative position of the Fermi level E_F and the mobility edge E_C in the impurity band are indicated.

sample of $CuGaTe_2$ and another of $CuInSe_2$.The temperature dependent behavior of these parameters are indicative that the conductivity is metallic in nature in $CuGaTe_2$. In the case of $CuInSe_2$, the resistivity shows a weak minimum around 190 K and below this it increases with decreasing temperature. It is found to be very nearly exponential in two different temperature ranges. On the other hand, the temperature dependence of R_H shows a maximum around 70 K. This is a characteristic sign that at higher temperatures, above the maximum, the contribution to the electrical conduction is predominantly from electrons in the conduction band and at lower temperatures, below the maximum, from the impurity band formed by shallow intrinsic defects levels. The variation of electron mobility μ with temperature is given for $CuInSe_2$ in Fig. 2. At 250 K, it is about 1,250 cm^2/Vs. It increases with the decrease of temperature, reaches a maximum and then starts to decrease.

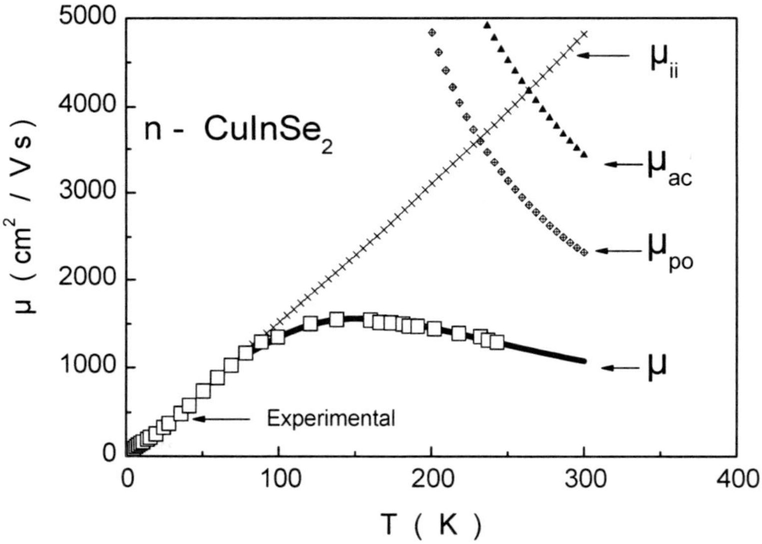

Figure 2. Experimental and theoretical electrical mobility in $CuInSe_2$ as a function of temperature. The calculated curves μ_{ii}, μ_{ac}, μ_{po} represent the contribution due to the scattering mechanism of electrons by ionized impurities, acoustic phonons and polar optical phonons, respectively. The total mobility is obtained from $\mu^{-1} = \mu_{ii}^{-1} + \mu_{ac}^{-1} + \mu_{po}^{-1}$.

The maximum mobility of 1,549 cm^2/Vs observed at 138 K is the highest reported[9] so far in n-type $CuInSe_2$ samples. In general, the temperature dependence of the electrical mobility originates from a combined effect of the contribution of scattering mechanism of electrons such as ionized and neutral impurities, acoustic phonons, polar and non-polar optical phonons and the space charge effect.[33]

The temperature dependence of the electrical resistivity for other materials like $CuIn_3Te_5$ and $CuGa_3Te_5$, together with $CuInSe_2$, is represented in Fig. 3. The data show that in the low temperature region, the predominant conduction mechanism is the variable hopping of Mott type. This we can see in Fig. 4 where the Mott's law $\rho = \rho_o \exp(T_o / T)^{1/4}$ is well verified in different temperature ranges.

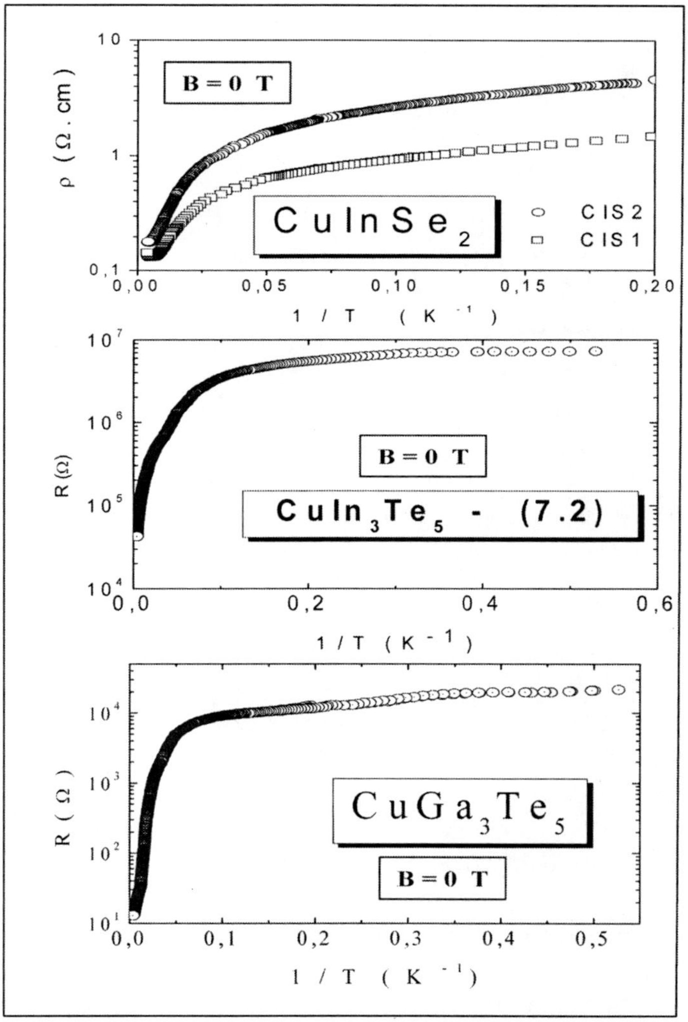

Figure 3. Variation of electrical resistivity as a function of temperature in two n-type samples CIS1 and CIS2 of $CuInSe_2$ and electrical resistance of $CuIn_3Te_5$ and $CuGa_3Te_5$ in the absence of magnetic field is shown.

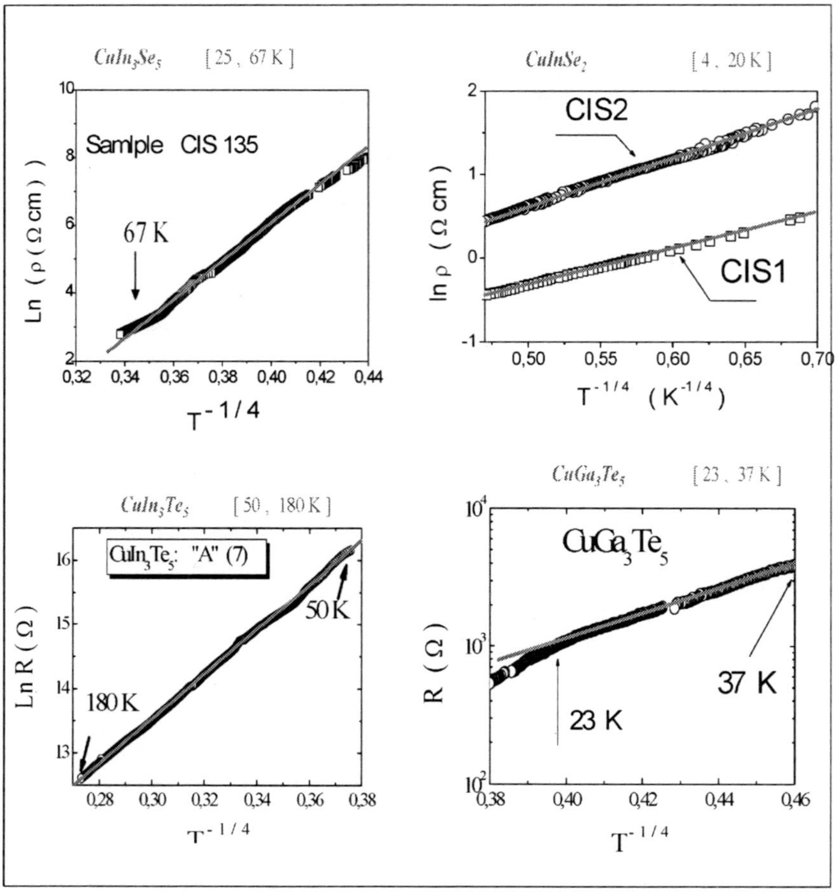

Figure 4. Variation of electrical resistivity as a function of $T^{-1/4}$ in two n-type samples CIS1 and CIS2 of $CuInSe_2$ and $CuIn_3Se_5$ and electrical resistance of $CuIn_3Te_5$ and $CuGa_3Te_5$ in the absence of magnetic field. Some of the representative temperature ranges where Mott's law is observed are indicated.

3.2 THE MAGNETORESISTANCE DATA

To further confirm the model proposed in the development of the theory of metallic conduction in $CuGaTe_2$ and VRH in other ternary compounds and to study the effect of magnetic field, the magntetoresistance was measured at different temperatures in a magnetic field up to 20 T. This is shown in Fig. 5 for $CuGaTe_2$ and in Fig. 6 for $CuInSe_2$.

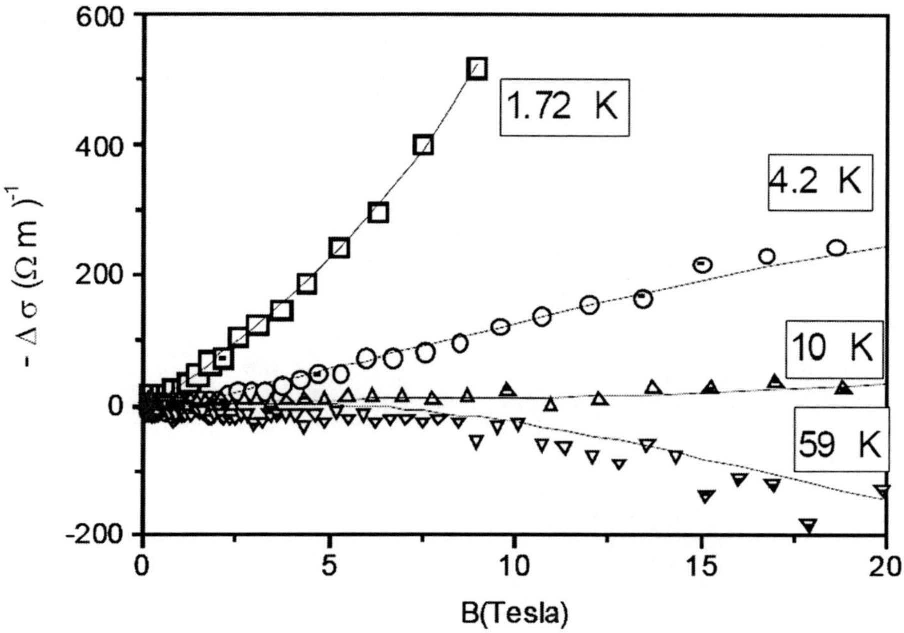

Figure 5. Variation of the magnetoconductivity as a function of magnetic field at representative temperatures in $CuGaTe_2$.

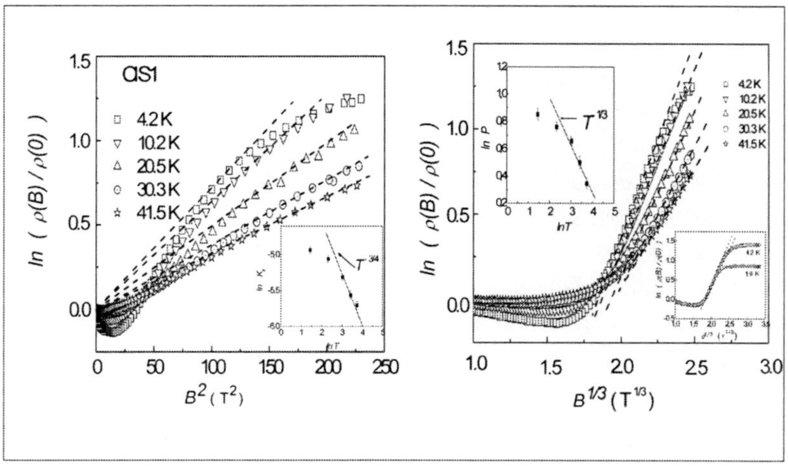

Figure 6. Variation of the magnetoresistance as a function of magnetic field at representative temperatures in $CuInSe_2$. In the insert, the variation of the slopes Ks and P as a function of $T^{-3/4}$ and $T^{-1/3}$, respectively are shown.

To analyze the experimental data for the metallic conductivity in $CuGaTe_2$, we use the model of weak localization and electron–electron interactions in its general form that takes into account the effect of the spin–orbit coupling. Good agreement is found over a wide range of magnetic field up to 20 T. The expression for the magnetoconductivity, used to fit the experimental data, originates from three different contributions. These are[31,32]:

$$\Delta\sigma = \sigma(B) - \sigma(0) \quad = \quad \Delta\sigma_1 + \Delta\sigma_2 + \Delta\sigma_3 \tag{9}$$

where $\Delta\sigma_1$ corresponds to weak localization, $\Delta\sigma_2$ arises from electron–electron interaction and $\Delta\sigma_3$ represents the contribution from weakly disordered electron gas arising from spin–orbit splitting of conduction electron energies.

The expression of the magnetoconductivity due to weak localization has been calculated by Kawabata[11] and Alt'shuler.[34] In the presence of strong spin–orbit coupling, the equation takes the following form:

$$\Delta\sigma_1 = \frac{e^{5/2}}{2\,\pi^2\,\hbar^{3/2}}\sqrt{B}\left\{\frac{1}{2}f_3\left(\frac{B}{H_i}\right) - \frac{3}{2}f_3\left(\frac{B}{H_{so}}\right)\right\} \tag{10}$$

In this relation

$$H_i = \frac{\hbar}{4\,e\,D}\,\tau_i^{-1}\,; \tag{11a}$$

and

$$H_{so} = \frac{\hbar}{4\,e\,D}\,(\tau_i^{-1} + 2\,\tau_{so}^{-1}) \tag{11b}$$

where D is the diffusion constant, τ_i the electron relaxation time for inelastic scattering which decreases with increasing temperature. τ_{so} is the electron relaxation time for spin–orbit scattering. The function f_3 is giving in reference.[11]

The contribution to magnetoconductivity arising from electron–electron interactions has been obtained by Alt'shuler et al.[34] taking into account the orbital part of the interaction. It is expressed as:

$$\Delta\sigma_2 = -\,\frac{e^{5/2}}{2\,\pi^2\,\hbar^{3/2}}\,\frac{1}{4}\,\sqrt{B}\;\;g(T,B)\;\varphi_3\left(\frac{2\,D\,e\,B}{\pi\,k_B T}\right) \tag{12}$$

The temperature and magnetic field dependence of the coupling term g(T,B) has been given by McLean and Tsuzuki.[35] For the function φ_3, we use the form presented in reference.[36]

For the case of three dimensional system, the magnetoconductivity of the weakly disordered electron gas arising from spin–splitting of conduction electron energies has been calculated by Lee and Ramakrishnan.[37] This is:

$$\Delta\sigma_3 = -\frac{e^2}{4\pi^2\hbar}\sqrt{\frac{k_B}{2\hbar}}\sqrt{\frac{T}{D}}\;F\;g_3\left(\frac{g\,\mu_B\,B}{k_B\,T}\right) \tag{13}$$

where F is the screening parameter for the Coulomb interaction. g is the Lande factor and μ_B the Bohr magneton. The analytical expression of the function g_3 is presented in reference.[36]

Using H_i and H_{so} as adjustable parameters, the total magnetoconductivity calculated through Equation (9), which represents the sum of Equations (10), (12) and (13) was fitted to the experimental data of Fig. 5. For this, following the method described in reference[38] for the case of $CuInSe_2$, the values of the diffusion constant D and the screening parameter F for the Coulomb interaction were deduced from hole concentration and mobility data. These were $D = 3.37\ 10^{-5}\ m^2/s$ and $F = 0.78$. As is shown in Fig. 5 by continuous curves, very good agreement is found at all temperatures between the calculated values and the experimental data over a wide range of magnetic field up to 20 T. The logarithmic variation of the corresponding values H_i and H_{so} with temperature is plotted in Fig. 7a. It is observed that H_i and thereby τ_i^{-1} has a tendency to vary as $T^{0.5}$ below 7 K and as T^2 above this temperature.

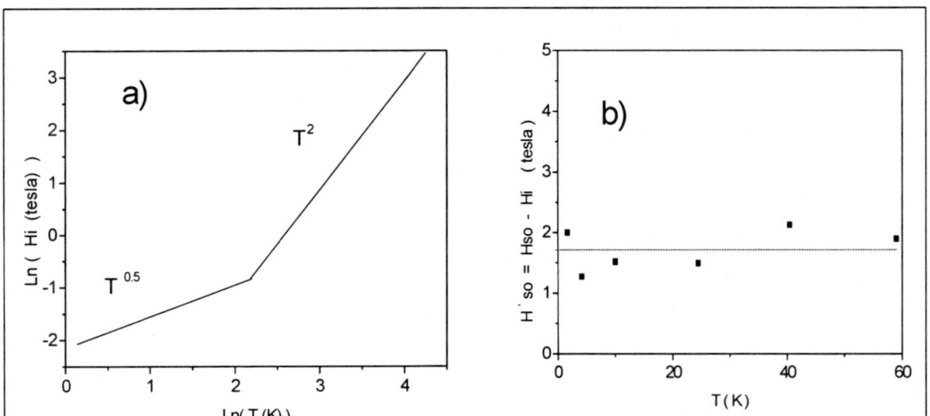

Figure 7. (a) The logarithmic variation of the parameter H_i with temperature. The straight lines showing $T^{0.5}$ and T^2 dependence serve as a guide to the eyes. (b) The variation of the parameter $[H'_{so} = H_{so} - H_i]$ with temperature. From the temperature independent straight line, τ_{so} is estimated to be $6.42\ 10^{-12}$ s.

This behavior ($\tau_i^{-1} \sim T^{0.5}$) has also been observed by Bieri et al.[39] and can be considered as a good approximation for the inelastic relaxation time at low temperatures. At higher temperatures, $\tau_i^{-1} \sim T^2$ is generally observed in the case of the diffusion by phonons.[39] It is interesting to observe, as shown in Fig. 7b, that the parameter $H_{so}^{'} = H_{so} - H_i$, which is proportional to τ_{so}^{-1}, is practically temperature independent. From this τ_{so} is estimated to be $6.42\ 10^{-12}$ s.

In Fig. 6, at low temperatures negative magnetoresistance (NMR) at lower fields and positive one (PMR) at higher fields can be observed. The PMR is well explained by the model proposed by Efros and Shklovskii.[14] The main mathematical expressions, already reported by us earlier,[40] are as follows:

$$\ln\left[\frac{\rho(B)}{\rho(0)}\right] = K_S\ B^2 \qquad \text{for low fields; } B < B_c \tag{14}$$

$$\ln\left[\frac{\rho(B)}{\rho(0)}\right] = \left[\frac{To(B)}{T}\right]^{1/3} = P\ B^{1/3} \text{ for high fields; } \quad B > B_c \tag{15}$$

In these expressions K_s is proportional to $T^{-3/4}$ and depends on the effective Bohr radius a and the zero field characteristic temperature T_o. The field dependent parameter T_o(B), in the high field region, is proportional to B and depends on the constant density of states $N(E_F)$ at the Fermi level and a. The critical field B_c that separates the low from high field regime[14] is expected to vary with the temperature as $T^{1/4}$.

From Equations (14) and (15) one should expect that at fixed temperatures $ln[\rho(B)/\rho(0)]$ should vary as B^2 and $B^{1/3}$ below and above B_c in the case of Mott type VRH conduction and at fixed magnetic fields it should be proportional to $T^{-3/4}$ and $T^{-1/3}$, respectively. Also, the variation of $ln\ K_s$ and $ln\ P$ with $ln\ T$ should give a respective slope of –3/4 and –1/3 (see the inserts in Fig. 6).

The low magnetic field negative magnetoresistance (NMR) in the VRH regime observed in Fig. 6 is explained by the theories of quantum interferences between many paths connecting the first and final state during the hopping process.[41] In our results on n-$CuInSe_2$[41] we have shown for the first time that $\Delta\rho/\rho$ is proportional to $T^{-\alpha}\ B^2$ in the low field region below about 0.35 T. Below 4 K, where $R_{hop}/\xi > 1.5$, we observed that $\alpha = 3/4$ and $B_M \propto T^{3/8}$. Above 4 K, we found $\alpha = 1.22$ and $B_M \propto T^{1.05}$ which is very similar to that reported in samples of n-GaAs[42] and n-CdSe.[43] In this case it is found that $R_{hop}/\xi < 1.5$. This leads us to suggest that for the correct application of the theories it is necessary that phase coherence should be maintained over a time scale larger than the hopping time and that the hopping length should be greater than the localization length.

4. Conclusion

In conclusion, it is established that the copper ternaries of the I-III-VI_2 and I-III_3-ϒ-VI_5 families, in addition to being SMART materials for their possible applications in opto-electronic devices such as solar cells are also of great academic interest. This is because they permit us to investigate physical phenomenon that can not be studied easily with elemental and binary compound semiconductors.

It is shown that Mott-type variable range hopping is the predominant electrical conduction mechanism at low temperatures on insulator side of the metal insulator transition (MIT). From the analysis of the experimental data, the magnetoresistance in n-$CuInSe_2$ is found to follow B^2 behavior at lower fields and $B^{1/3}$ at higher fields, in agreement with the theory of Efros- and Shklovskii. The hopping parameters K_s and P vary with temperature as $T^{-3/4}$ and $T^{-1/3}$, respectively. The presence of negative magnetoresistance (NMR) at lower magnetic fields in the VRH regime is explained by the theories of quantum interferences. In the case of $CuGaTe_2$ the presence of NMR on metallic side of MIT is observed. However, in this case, the theories of weak localization and electron-electron interactions must be considered to analyze the experimental data.

ACKNOWLEDGMENTS

A good part of the experimental work was carried out by L. Essaleh for his doctoral program at the Laboratoire de Physique des Solides et Service National des Champs Magnétiques Pulsés of Toulouse. The growth of the samples, their characterizations and the detailed analysis of the magnetoresistance data were supported by grants from CONICIT (Contract No. G-97000670), CDCHT-ULA (Contracts No. C917-98-05A; C918-98-05-E), EEC (Contract No. CI1*-CT-92-0099VE) and the Franco Venezuelan Co-operation Program through CONICIT and CEFI.

References

1. D.M. Chapin, C.S. Fuller, G.L. Pearson, J. Appl. Phys. 25, 676 (1954).
2. J.L. Stone, Phys. Today, 22–29, September (1993).
3. A. Rockett, R.W. Birkmire, J. Appl. Phys. 70(7), R81 (1991).
4. C.H. Champness, J. Mater. Sci.: Mater. Electronics 10, 605 (1999).
5. M.A. Conteras, B. Egaas, K. Ramanathan, J. Hiltner, A. Swartzlander, F. Hasoon, R. Noufi, Prog. Photovolt: Res. Appl. 7, 311 (1999).

6. M.A. Conteras, K. Ramanathan, J. Abushama, F. Hasoon, D. Young, B. Egaas, R. Noufi, Prog. Photovolt 13, 209 (2005).
7. D. Rudman, F.J. Haug, M. Kaelin, H. Zogg, A.N. Tiwari, G. Belger, Mat. Res. Soc. Symp. Proc. 668, H3.8.1 (2001).
8. S.M. Wasim, C. Rincon, G. Marin, Phys. Stat. Sol. (a) 194, 244 (2002).
9. S.M. Wasim, Sol. Cell. 16, 289 (1986).
10. B.I. Alt'shuler, A.G. Aronov, D.E. Khmel'nitskii, JETP Lett. 36, 195 (1982).
11. A. Kawabata, Solid State Comm. 34, 431 (1980).
12. E. Arushanov, L. Essaleh, J. Galibert, J. Léotin M.A. Arsene, J.P. Peyrade and S. Askenazy, Appl. Phys. Lett. 61, 958 (1992).
13. E. Arushanov, L. Essaleh, J. Galibert, J. Léotin and S. Askenazy, Physica B 184, 229 (1993).
14. B.I. Shklovskii and A.L. Efros, Springer, Berlin (1984).
15. V.L. Nguyen, B.Z. Spivak and B.I. Shklovskii, Sov. Phys.-JETP 62, 1021 (1985).
16. U. Sivan, Entin-Wohllman and Y. Imry, Phys. Rev. Lett. 60, 1566 (1988).
17. S.M. Wasim, C. Rincon, J.M. Delgado and G. Marin, J. Phys. Chem. Solids 66, 1990–1993 (2005).
18. S.B. Zhang, S.H. Wei, A. Zunger, H.K. Yoshida, Phys. Rev. B 57, 9642–9656 (1998).
19. L. Essaleh, S.M. Wasim, J. Galibert, Mat. Lett. 66, 1947 (2006).
20. S.M. Wasim, L. Essaleh and G. Marin. American Institute of Physics, Low Temperature Physics: 24th International Conference on Low Temperature Physics. Edited by Y. Takano, S.P. Hirchfield, S.O. Hill, P.J. Hirchfield and A.M. Goldmann, pp. 1468–1469 (2006).
21. B. Gudden and W. Schottky, Z. Tech. Phys. 16, 323 (1935).
22. A. Miller and S. Abrahams, Phys. Rev. 120, 745 (1960).
23. N.F. Mott, J. Non-Cryst. Solids 1, 1 (1968).
24. A.L. Efros and B.I. Shklovskii, J. Phys. C8, L49 (1975).
25. O. Entin-Wohlman, Y. Imry and U. Sivan, Phys. Rev. B 40, 8342 (1989).
26. W. Schirmacher, Phys. Rev. B 41, 2461 (1990).
27. M.E. Raikh, Solid State Comm. 75, 935 (1990).
28. B. Movaghar, Phys. Stat. Sol. (b) 97, 533 (1980).
29. G. Bergman, Phys. Rev. B 28, 2917 (1983).
30. Y. Isawa, J. Phys. Soc. Jap. 53, 2865 (1984); J. Phys. Soc. Jap. 53, 1415 (1984).
31. L. Essaleh, S.M. Wasim, G. Marin, E. Choukri and J. Galibert, Phys. Stat. Sol. (b), 225, 203 (2001).
32. L. Essaleh, J. Galibert , S.M. Wasim and G. Marin, Phys. Stat. Sol. (b), 240, 625 (2003).
33. L. Essaleh, S.M. Wasim and J. Galibert, J. Appl. Phys. 90, 3993 (2001).
34. B.L. Alt'shuler, A.G. Aronov, A.I. Larkin and D.E. Klmel'nitskii, Sov. Phys. JETP 54, 411 (1981).
35. W.L. Mclean and T. Tsuzuki, Phys. Rev. B. 29, 503 (1984).
36. J.C. Ousset, S. Askenazy, H. Rakoto and J.M. Broto, J. Physique 46, 2145 (1985).
37. P.A. Lee and T.V. Ramakrishnan, Phys. Rev. B 26, 4009 (1982).
38. L. Essaleh, J. Galibert, S.M. Wasim, and J. Léotin, Phys. Stat. Sol. (b), 177, 449 (1993).
39. J.B. Bieri, A. Fert, G. Creuzet, and J.C. Ousset, Sol. Stat. Commun. 49, 849 (1984).
40. L. Essaleh and S.M. Wasim, Mat. Lett. 61, 2491 (2007).
41. L. Essaleh, J. Galibert, S.M. Wasim, E. Hernandez and J. Léotin, Phys. Rev. B 50, 18040 (1994).
42. F. Tremblay, M. Pepper, D. Richie, D.C. Peacock, J.E.F. Frost and G.A.C. Jones, Phys. Rev B 39, 8059 (1989).
43. Y. Zhang, P. Dai, M. Levy and M.P. Sarachik, Phys. Rev. Lett. 64, 2687 (1990).

STRUCTURAL AND ELECTRICAL PROPERTIES OF NANO-CRYSTALLINE $LiCoO_2$ CATHODE MATERIAL SYNTHESIZED BY A SIMPLIFIED COMBUSTION METHOD

MUSTAPHA YAHIA[1,2], ISMAEL SAADOUNE[1*], ABDELMAJID ALMAGGOUSSI[2], ABDELHADI ABOUNADI[2] AND ABDELKADER OUTZOURHIT[3]
[1]*Materials and Environmental Chemistry Group, LP2M2E, FSTG, B.P. 549 University Cadi Ayyad, Marrakech, Morocco*
[2]*Optoelectronic Materials Group, LP2E2M, FSTG, B.P. 549 University Cadi Ayyad, Marrakech, Morocco*
[3]*Laboratoiry of Solide Physics and thin films, LPSCM FS Semlalia, B.P. 2390, Cadi Ayyad, Marrakech, Morocco*

Abstract. A new method based on combustion at low temperature is used to synthesize $LiCoO_2$ materials. The structural properties of obtained samples have been examined by X-ray diffraction associated to Rietveld refinement and by Scanning Electronic Microscopy. The formation of well-crystallized hexagonal particles with average size varying from 25 to 100 nm is confirmed. Transport charge properties are also investigated in the present study. Resistivity measurements were made over the temperature range of 10–300 K. The results suggest Mott-type hopping conduction at low temperature, while the data below 200 K are consistent with the Seto model. All these analysis show that the optimum synthesis condition is calcinations at 800°C for 1 h in air for 1.5 mol ratio of sucrose to nitrate. A relatively good cycling performance was obtained with a first discharge capacity of 157 mA h g^{-1} at 0.2°C .

Keywords: $LiCoO_2$, cathode materials, electronic materials, resistivity, hopping, thermo-emission.

---

* To whom correspondence should be sent: I. Saadoune, University Cadi Ayyad, FST Marrakech, ECME, Av. A. Khattabi, B.P. 549, Marrakech, MOROCCO: Email: saadoune1@yahoo.fr

1. Introduction

Much attention has been paid to the development of lithium secondary batteries for dispersed energy storage systems, such as laptop computers, cellular phones and electrical vehicles. A significant number of recent studies has focused on the synthesis, processing and electrochemical studies of cathode materials. Lithiated transition metal oxides, such as $LiCoO_2$, $LiNiO_2$ and $LiMn_2O_4$ have been studied widely as cathode materials because of their high operating voltage and energy density, large capacity and long cycle life.[1–3]

Because of the easy synthesis, stable discharge capacity and good rate capability, layered $LiCoO_2$ is still the most widely used as positive electrode material in commercial rechargeable lithium batteries, even though its raw materials are toxic and costly. Commercial $LiCoO_2$ is synthesized by solid state reaction at high temperatures (more than 900°C), mainly requiring pre-treatment steps at lower temperatures.[4–6]

In fact, high temperature firing for an elongated period to obtain the layered $LiCoO_2$ by solid state reaction method causes inhomogeneity, irregular morphology and a broader particle-size distribution of powder. To overcome these disadvantages, several solution routes have been developed such as sol-gel,[7] co-precipitation[8] Pechini process[9] and freeze drying methods.[10]

In this study, we have prepared ordered $LiCoO_2$ at relatively low temperature by the combustion method. The phase purity, crystal structure and particle sizes have been described. The electrochemical charging and discharging of the prepared sample have been carried out without using carbon as additive. The effect of this synthesis method on the electrical conductivity is reported trough the measurement of the temperature dependence of dc conductivity. We also discuss the mechanism of electric conduction of $LiCoO_2$ obtained from combustion method and solid state reaction in regard to either thermal activated mechanism or variable range hopping (VRH).

2. Methodology

2.1 SAMPLE PREPARATION

$LiCoO_2$ was synthesized by the combustion method, from the stoichiometric amounts of Li and Co nitrates which act as oxidants and sucrose as fuel according the following reaction:

$$LiNO_3 + Co(NO_3)_2{\cdot}6H_2O + C_{12}H_{22}O_{11} + 5/2O_2 \rightarrow LiCoO_2 + 6CO_2 + 3/2N_2 + 17H_2O$$

The reaction is extremely violent when using a stoechiometric amount of sucrose (1 mol; oxidant/fuel = 1). To avoid this, the oxidant/fuel ratio was chosen to be equal to 0.67 (1.5 mol of sucrose). Samples once obtained are treated for 1 h in air at 700°C (CS1), 800°C (CS2) and 900°C (CS3). Simultaneously, we have also prepared $LiCoO_2$ (S) by solid state reaction (900°C for 12 h in air).

2.2 STRUCTURAL CHARACTERIZATION

X-ray diffraction patterns of the products were obtained with a STOE STADI/P diffractometer (Mo-Kα_1 radiation). Structural refinements were carried out by using the Rietveld method.[11]

The morphology and microstructure of the powders were examined using a scanning electron microscope (SEM) (JEOL JSM-5500).

2.3 ELECTROCHEMISTRY AND CONDUCTIVITY

Electrochemical experiments were carried out in a multi-channel potensiostat (VMP2/Z; Ametek) by using a Swagelock type cell. Electrochemical cells were constructed using a two electrodes configuration where lithium was used as negative electrode. The samples, prepared by the combustion method, were mixed with only 3% of the binding agent, PVDF, and used as positive electrode. Indeed, small quantity of residual carbon is suspected to be present in the samples when prepared by the combustion method.

Measurements of conductivity were taken in the temperature range of 10–300 K, in DC current, according to the four points method using an assembly developed at the laboratory. The four points are aligned on the surface of the samples, and the contacts are carried out by silver glue.

3. Data and result

Figure 1 shows X-ray diffraction patterns of different kinds of $LiCoO_2$. For samples obtained by the soft combustion method, the XRD diagrams are characteristics of α-$NaFeO_2$ layered type structure (space group R-3m) even for the sample prepared at 700°C for 1 h (CS1). It is well known that the splitting of the (006)/(102) and (108)/(110) doublets clearly indicates the formation of well ordered layered structure. It should be noticed that by examining the XRD profiles, the peaks become more and more narrow indicating an increase of the degree of crystallinity by increasing the synthesis (heat or annealing) temperature.

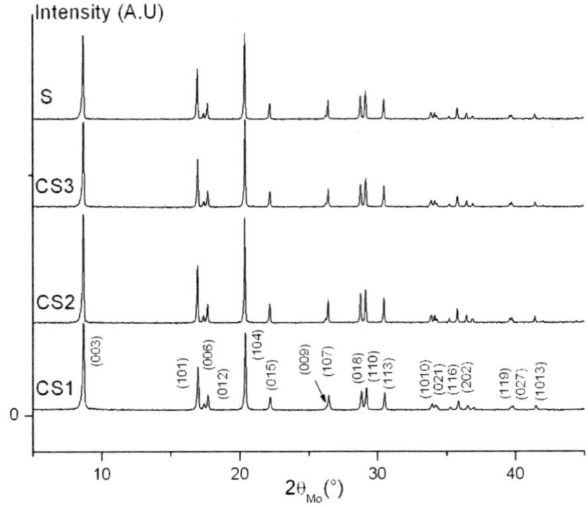

Figure 1. The XRD patterns of different $LiCoO_2$ compounds.

In order to determine accurately the structure of these materials, refinements by the Rietveld method of the XRD data were performed using the Fullprof program. In all case, we considered Li/Co = 1 even if we suspect that CS1 and CS2 samples are slightly over lithiated (Li/Co > 1). Li are located in the 3b (00½) site, Co in the 3a (000) and O in the 6c ($00z_{ox.}$). The occupancy ratios at the 3a, 3b and 6c were fixed while the isotropic atomic displacement parameters were refined. Figure 2 shows the comparison between the observed and calculated diffraction patterns for sample CS2, while Table 1 summarizes the Rietveld refinement results of the XRD patterns for all studied compounds. As shown, hexagonal unit cell parameters slightly increase from sample CS1 to CS2, while they remain constant from CS2 to CS3. The same evolution was observed for the c/a ratio which indicates that we reach a good 2D character for CS2 i.e. $LiCoO_2$ presenting high-layered character could be obtained at 800°C for 1 h by using the combustion method. Furthermore, the I_{003}/I_{104} diffraction lines ratio, which recognized as a criterion for the existence of cation mixing in the $LiMO_2$ oxide, decrease when sample was prepared at 900°C for 12 h (sample CS3). For sample CS2, the value of this ratio is 1.32 which is also greater than that of the sample prepared by the classical solid state reaction (sample S). According to these remarks, 800°C/1 h could be considered as the optimum condition for synthesizing $LiCoO_2$ by combustion method.

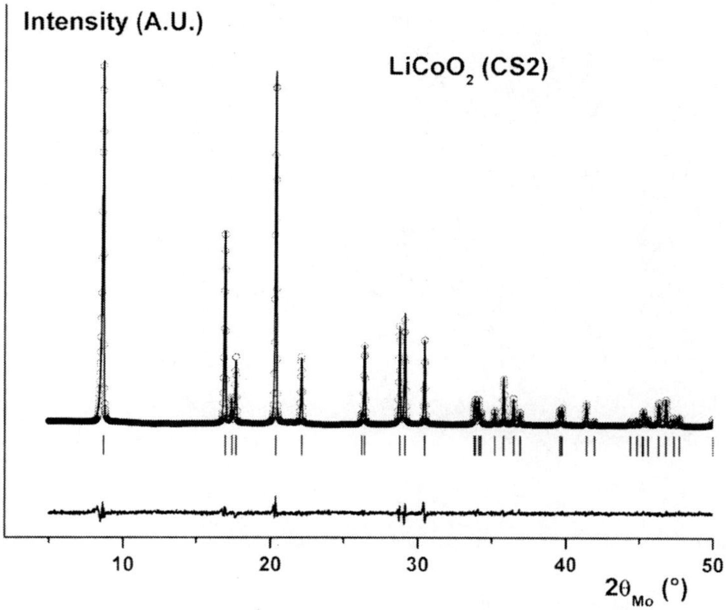

Figure 2. Rietveld refinement patterns of the XRD data for $LiCoO_2$ (CS2) sample.

TABLE 1. Results of the Rietveld refinement of the various kind of $LiCoO_2$.

Sample	a (Å)	c (Å)	c/a	I_{003}/I_{104}	$z_{ox.}$
$LiCoO_2$ (CS1)	2.8184(2)	14.059(1)	4.988	1.37	0.2606(6)
$LiCoO_2$ (CS2)	2.82246(7)	14.0855(5)	4.991	1.32	0.2599(5)
$LiCoO_2$ (CS3)	2.82214(7)	14.0869(5)	4.991	1.16	0.2606(5)
$LiCoO_2$ (S)	2.82138(5)	14.0775(5)	4.990	1.26	0.2603(5)

Figure 3 presents the morphology of $LiCoO_2$ annealed at various temperatures. From the SEM photographs, it is found that the particle size increases by rising temperature of heat treatment. Concerning the sample CS1, the SEM picture clearly shows that this compounds presents nano-size crystallites. Samples CS2 and CS3 consists of small and large particles with irregular shapes and their clusters. The particle size increases gradually from 1.5 μm to about 2.2 μm, while significant particle size growth is found for S sample (about 8 μm). A well defined lamellar shape can be seen for the CS2, this unique morphology is advantageous for electrode materials.

Figure 3. SEM images of the studied $LiCoO_2$ samples.

In Fig. 4, the Arrhenius plot (log(σ) versus 1,000/T) for all studied samples are presented. We observe an increase in conductivity with the temperature indicating a semiconductor behavior of these samples. This increase is spread out over several orders of magnitude showing the character thermically activated of conductivity. In all temperature range explored we remark that the sample Y02 give the best conductivity. Two distinguishable fields in temperature can be considered, giving place to two different mechanisms of conduction. When temperature is above 200 K, electrical conductivity as a function of temperature appears to fit with Arrhenius law. Conduction is assumed by thermionic emission resulting from carriers having enough energy to jump above the potential barrier at the grain boundary according to Seto's model.[15] The conduction is given by:

$$\sigma \propto T^{-1/2} e^{-\frac{E_b}{kT}} \quad \text{where } E_b = qV_B$$

At lower temperatures below 200 K, the thermal variations of conductivity are uniformly non-Arrhenius (Fig. 4). The thermal energy of carriers is not enough to surmount the potential barrier at the grain boundary. The charge carriers are semi-localized and the nature of conductivity dependence observed can be attributed to variable range hopping (VRH) of the electron over long distances at low temperatures. This appears in Fig. 5 where the logarithmic conductivities is plotted against $1/T^{1/4}$ on the basis of the well known Mott's law[16]:

$$\sigma \propto e^{-\left[\frac{T_0}{T}\right]^{1/4}}$$

where $T_0 = 1.5/\alpha^{-1}N(E_F)$, α^{-1} is the wave function decay length, which is generally taken around 10 Å, $N(E_F)$ is the density of states at the Fermi level.

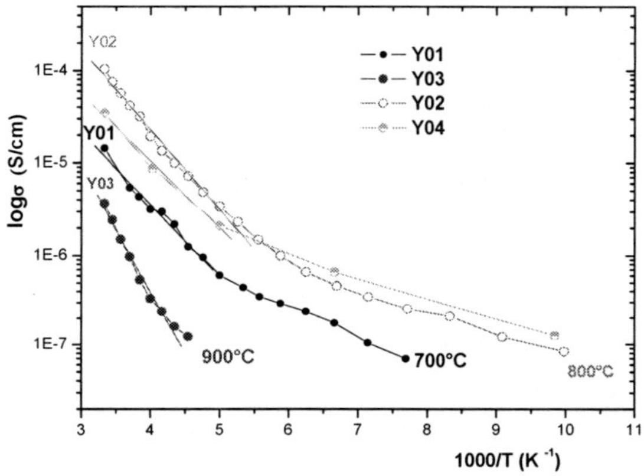

Figure 4. Arrhenius plots of the electrical conductivity of different samples $LiCoO_2$.

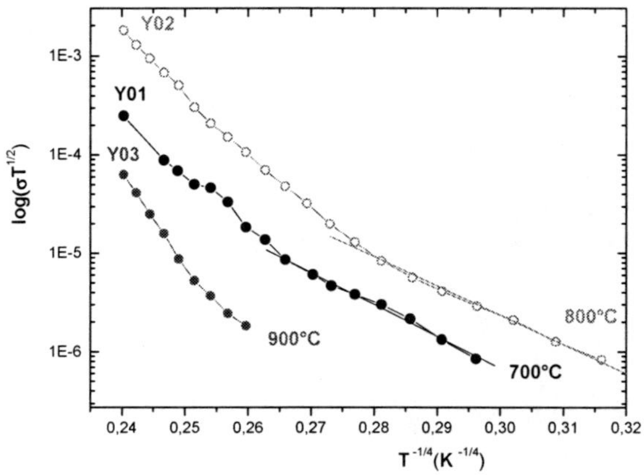

Figure 5. Conductivity plot according to the Mott's law.

As the sample CS2 presents similar structural features comparing to those prepared by the classical method (sample S), we have selected it ($LiCoO_2$ (CS2)) for the electrochemical properties study. Furthermore, as we suspect that combustion method leads to compounds containing small quantity of carbon, the electrochemical cycling has been carried out on the cathodic material composed only by $LiCoO_2$ and the binder (PVDF). Figure 6a shows the charge and discharge curves of a Li//$LiCoO_2$ electrochemical cell at a current density of 20 mA g^{-1} between 2.7 and 4.4 V. The similarity in the electrochemical behaviour between the charge and discharge curves shows a very good reversibility of the intercalation/deintercalation process in this material. Such reversibility is confirmed by the superposition of the second charge with the first charge. Furthermore, the cell polarization, given by the difference between charge and discharge curves, is small (about 80 mV). This behavior confirms that the cathodic material presents a good electronic conductivity as a result of the presence of small residual quantity of carbon.

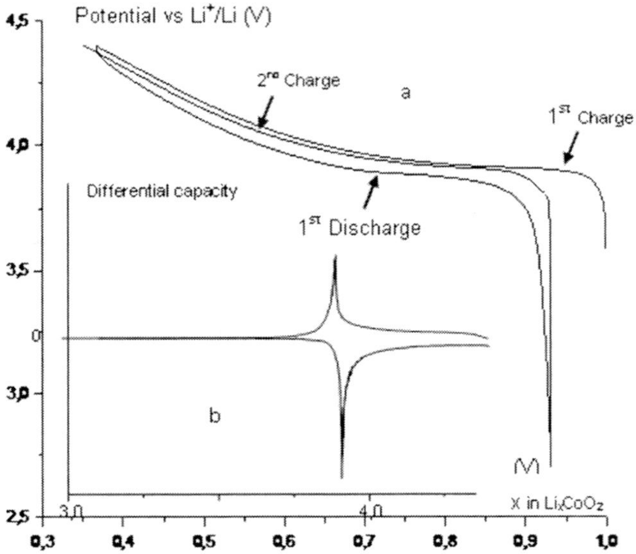

Figure 6. (a) Charge/discharge cycle properties of $LiCoO_2$ (CS2 sample), and (b) Derivative dx/dV as function of cell potential of a Li//$LiCoO_2$ (CS2) cell.

It should be noticed that the phase changes upon Li deintercalation from $LiCoO_2$ have been extensively studied.[12–13] By using electrical measurements and ^{7}Li NMR spectroscopy, we have evidenced that the two-phase domain at ca 3.93 V, is due to structural reasons. The metal-non metal transition is the driving force for this phase separation.[6] Furthermore, the electrochemical cycling curve of stoichiometric $LiCoO_2$ (Li/Co = 1) presents an hexagonal-monoclinc phase transition for x = 0.5 in Li_xCoO_2 (corresponding to ca 4.1 V). This phase transition disappears completely for the over-lithiated $Li_{1+z}CoO_2$.[14]

As shown in Fig. 6b and confirmed by the differential capacity given in Fig. 6, the studied sample presents no peak around 4.1 V indicating that no mono-clinic distortion occurs during lithium deintercalation. This indicates that the studied $LiCoO_2$ is over-lithiated as a result of the 5% lithium nitrate excess used for the sample preparation.

The CS2 sample has only one major reversible peak at 3.88 V, which is associated with a first-order transition from $LiCoO_2$ to $Li_{0.75}CoO_2$ as explained above. According to earlier works, we could also conclude that the studied sample shows only the hexagonal phase even up to 4.4 V.[14] It is well known that a phase transition from hexagonal to monoclinic phase during lithium deintercalation is the major cause of structural degradation in $LiCoO_2$.

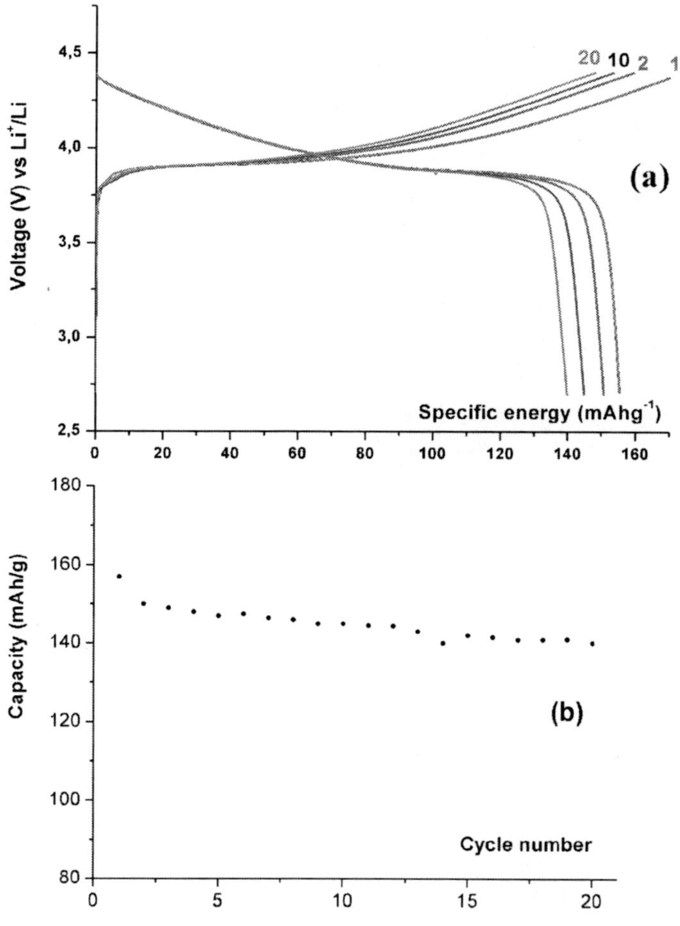

Figure 7a. (a) Comparison of the charge and discharge capacities for the first 20 cycles and (b) Discharge capacity as a function of cycle number for sample CS2.

Figure 7a shows the 20th charge-discharge voltage profiles for the Li//$LiCoO_2$ (without additional carbon) cell between 2.7 and 4.4 V at a constant current of 0.2 mA cm^{-2} which corresponds to C/5 rate. No changes in the potential versus capacity profile occur during the cycling indicating that structural change does not occur during the repetitive lithium extraction/insertion process. The studied positive electrode material presents a good structural stability even during high voltage cycling. The first charge-discharge curves show a discharging capacity of 157 mA h g^{-1}, larger than many of the literature samples prepared from low synthesis temperature (800°C for 1 h). As plotted in Fig. 7b, the specific capacity is reduced to 150 mA h g^{-1} from the second cycle indicating that the coulombic efficiency is 95%. After 10 cycles, the discharge capacity decreases slowly to 145 mA h g^{-1}, while after 20 cycles, the discharge capacity fading is equal to 11%. We are now investigating the rate capability of the studied samples and compared its electrochemical performances with those obtained by classical methods.

4. Conclusions

$LiCoO_2$ cathode materials were prepared by a simple sucrose-aided combustion procedure at different synthesis temperatures. Their structural features were compared to those of $LiCoO_2$ prepared by the classical solid state reaction. From the results of XRD, all prepared compounds were characterized by an α-$NaFeO_2$ type structure. The compound synthesized at 800°C for 1 h in air (Sample CS2) shows the largest I_{003}/I_{104} ratio (confirmed by the Rietveld refinement study) indicating better hexagonal ordering and less cation mixing. This sample has the best conductivity and shows a large first discharge capacity 157 mA h g^{-1} and good cycling performances. This value of capacity is larger than 136 mA h g^{-1}, reported in the literature for $LiCoO_2$ obtained by conventional solid state method. The electrical conductivity were found obeying to Mott type hopping at low temperature and controlled by the potential barrier at the grain boundary at relatively high temperature.

ACKNOWLEDGEMENTS

This work was supported by the French Foreign Affairs Ministry within the research program CORUS (REF 02 211 121).

References

1. A.K. Padhi, K.S. Nanjundaswamy, J.B. Goodenough, J. Electrochem. Soc. 144(4):1188 (1997).
2. S.-Y. Chung, J.T. Bloking, Y.-M. Chiang, Nat. Mater. 1:123 (2002).
3. Y.-S. Hong, K.S. Ryu, S.H. Chang, ETRI J. 25:412 (2003); W.A. Van Schalkwijk, B. Scrosati (Eds.), Advances in Lithium-Ion Batteries, Kluwar Academic/Plenum, New York, 2002.
4. Z. Chen, J.R. Dahn, Electrochim. Acta 49:1079 (2004).

5. I. Saadoune, C. Delmas, J. Mater. Chem. 6(2):193 (1996).
6. P. Barboux, J.M. Tarascon, F.K. Shokoohi, J. electrochem. Soc. 94:185 (1991).
7. X. Qui, X. Sun, W. Shen, N. Chen, Solid State Ionics 93:335 (1997).
8. W. Liu, K. Kowal, G.C. Farrington, J. Electrochem. Soc. 143:3590 (1996).
9. T. Lemercier, J. Gaubicher, E. Bermejo, Y. Chabre, M. Quarton, J. Mater. Chem. 9:567 (1999).
10. T. Roisnel, J. Rodriguez-Carjaval, Fullprof Manual, Institut Laue-Langevin, Grenoble, 2000.
11. J.N. Reimers, J.R. Dahn, J. Electrochem. Soc. 139:2091 (1992).
12. N. Imanishi, M. Fujii, A. Hirano, Y. Takeda, J. Power Sources 97–98:287 (2001).
13. S. Levasseur, M. Ménétrier, E. Suard, C. Delmas, Solid State Ionics 128:11 (2000).
14. J.Y.W. Seto, J. Appl. Phys. 46:5247 (1975).
15. G.C. Psarras I, Composites: Part A 375:1545 (2006).

IONIC TRANSPORT BEHAVIOR IN Na_2SO_4-Li_2O-MoO_3-P_2O_5 GLASSY SYSTEM

L. BIH[1*], D. MEZZANE[1], A. NADIRI[1], H. BIH[2],
M. MANSORI[2] AND M. AMALHAY[3]
[1]*UFR – PCMI – Céramiques et Verres, FST-Errachidia, B.P. 509, Boutalamine, Morocco*
[2]*FST-Guéliz-Marrakech, Morocco*
[3]*CERPHOS, Casablanca, Morocco*

Abstract. The need for electrolytes suitable for energy storage has spurred investigations into a number of lithium, silver and sodium ion conducting inorganic glass systems. It is apparent that two strategies have been used in the design of ion conducting electrolytes. The first is to use a combination of two anionic species which has been known to give increased ionic conductivity and is attributed to the so-called mixed-anion effect. The second strategy is to dissolve ionic cation salt in a conventional polymeric silicate, borate or phosphate glass containing a same cation. The increased conductivity is attributed to a volume increasing effect of the dissolved ionic salt. Several studies have been reported on lithium phosphate glasses to which lithium halides and lithium sulphate have been added. However, there has been no experimental results for the introduction of sodium sulphate into the lithium phosphate network. How specifically this feature – mixed alkali in the presence of sulphate ion – influences ion transport in the glasses is unclear. It is therefore necessary to investigate the ion transport behaviour of Na_2SO_4 containing lithium phosphate glasses in order to understand better the role of Na_2SO_4 in the ionic cation transport in these glasses. In this paper we report both dc and ac conductivity measurements performed on the $0.5[xNa_2SO_4\text{-}(1\text{-}x)Li_2O]\text{-}0.5[0.4\,(MoO_3)_2\text{-}0.6P_2O_5]$ glasses with x varying between 0 and 1. The complex impedance data is analyzed in conductivity and electric modulus formalism in order to throw light on transport mechanism. The glass composition relative to ($x = 0.5$) has exhibited a decreasing in the conductivity of about two orders of magnitude compared to the simple end glasses ($x = 0, 1$). This behaviour is associated with the mixed alkali effect (MAE). It could be attributed to the fact that the two types of alkali ions are randomly mixed and have distinct conduction pathways.

Keywords: Ionic conductivity, glass transition temperature, phosphate, molybdenum oxide, sulphate, glasses, mixed alkali effect.

* To whom correspondence should be sent: L. Bih; UFR – PCMI – Céramiques et Verres, FST-Errachidia, B.P. 509, Boutalamine, Morocco: Email: bihlahcen@yahoo.fr

I. A. Luk'yanchuk and D. Mezzane (eds.), *Smart Materials for Energy, Communications and Security*.

1. Introduction

Inorganic glasses doped with transition and rare-earth ions have attracted much attention because of their semiconducting properties and potential applications in optical devices such as solid-state lasers and optical fibers for communications.[1] Phosphate glasses possess technological importance because of their simple composition with strong glass forming character, low glass transition temperature and high thermal expansion coefficient. With the addition of appropriate doping agent, these glasses find a wide variety of applications in optoelectronic devices, laser host materials and as solid electrolytes in solid state ionic devices.[2] Fast alkali ion conducting glasses with high ionic conductivity are potential candidates for applications in solid-state electrochemical devices such as batteries, sensors, etc. due to their advantageous characteristics over the conventional crystalline materials.[3–5] Various attempts have been made to enhance the ionic conductivity and chemical durability of the phosphate glasses which include the addition of another glass former, modifier oxide, alkali halides and alkali sulphates.[3,6] It is well known that in phosphate glasses, the anionic skeleton is quite like an organic polymer with chains or cycles of connected PO_4 tetrahedra and therefore the non-bridging oxygen ions which are charged are attached to all the phosphorus atoms in the backbone.[7,8] Further addition of alkali oxide decreases the connectivity of the network structure and leads to the formation of shorter phosphate chains and ultimately to pyrophosphate or orthophosphate units.[7,8] Modified network glasses can be 'doped' with other lithium salts which contains the same cation (Li). Such modified and 'doped' glasses have larger ionic conductivities. However, to the best of our knowledge, there is no reports in the literature on Na_2SO_4 additives in lithium-conducting glasses. Studies on such a system could contribute to understand the mechanism of the conduction in glasses.

The aim of the present work is to investigate the effect of the substitution of Li_2O by Na_2SO_4 on the properties of Li_2O-Na_2SO_4-MoO_3-P_2O_5 glasses. The selected compositions for the present study behave more like ionic rather than electronic and/or mixed ion-polaron conductors. Therefore, our investigations will be focused on the system $0.5[xNa_2SO_4\text{-}(1\text{-}x)Li_2O]\text{-}0.5[0.4(MoO_3)_2\text{-}0.6P_2O_5]$, with $0 \leq x \leq 1$.

2. Methodology

The $0.5[xNa_2SO_4\text{-}(1\text{-}x)Li_2O]\text{-}0.5[0.4(MoO_3)_2\text{-}0.6P_2O_5]$, with $0 \leq x \leq 1$, compositions were prepared by mixing and finely grinding appropriate amounts of Li_2CO_3 (Merck, 99.5%), Na_2SO_4 (Merck, 99.5%), $NH_4H_2PO_4$ (Merck, 99.5%) and MoO_3 (Aldrich, 99.9%). The reactions in a glass melt are known to depend on the size of the melt, the sample geometry, the concentration of the total redox ions, the thermal history and the quenching rate. To keep these parameters constant, all glass samples were prepared under the same conditions as follows: about 10 g of mixture was initially heated at 573 K for about 12 h in order to remove volatile products. Next, the mixture was melted in a platinum crucible in air for 0.5 h at 1,100–1,200 K depending on the glass composition. The melts were then poured on a stainless steel plate preheated at 450 K, and the vitreous samples obtained were annealed at 650 K to relieve residual internal stress, and slowly cooled at room temperature.

The amorphous nature of samples was checked by XRD analysis, using a Seifert XRD 3000 diffractometer. Density measurements were carried out at room temperature, using the Archimede's method with diethyl orthophtalate as the immersion fluid. The relative error in these measurements was about ±0.03 g cm^{-3}. Glass transition temperatures (Tg) were determined by DTA analysis at a heating rate of 10 K min^{-1}, using a Seiko DTA thermal analyzer, these temperatures are reproducible to ±5 K. For measurements of electrical conductivity, silver electrodes were applied onto two opposite sides of the samples. The electrical conductivity was determined from the impedance/admittance spectroscopic method. The spectra were carried out on a Hewlett Packard Model 4284A precision LCR Meter in the frequency range 20–10^6 Hz with temperatures changing from 293 to 533 K. The EPR experiments were made at room temperature on glassy samples using a Brucker EMX spectrometer operating at X-band frequencies (9.5 GHz).

3. Data and result

The composition dependencies of physical and electrical properties of the glasses have been determined along the series $0.5(xNa_2SO_4\text{-}(1\text{-}x)Li_2O)\text{-}0.5[0.4(MoO_3)_2\text{-}0.6P_2O_5]$ (x varies from 0 to 1). All the compositions form glasses when cast onto a steel mould; these glasses were subjected to X-ray diffraction studies and no crystalline phases were detected. Figure 1 displays the density (d) and the molar volume (V) for these glasses.

When Li_2O is replaced by Na_2SO_4, it can be noted that the measured density decreases while the molar volume increases. The values of the density and the molar volume are consistent with the ionic size, atomic weight of elements and their amount in these glasses.

The reciprocal temperature dependence of the dc conductivity for the different glass compositions of the Li-Na containing tungsten phosphate glasses is shown in Fig. 2.

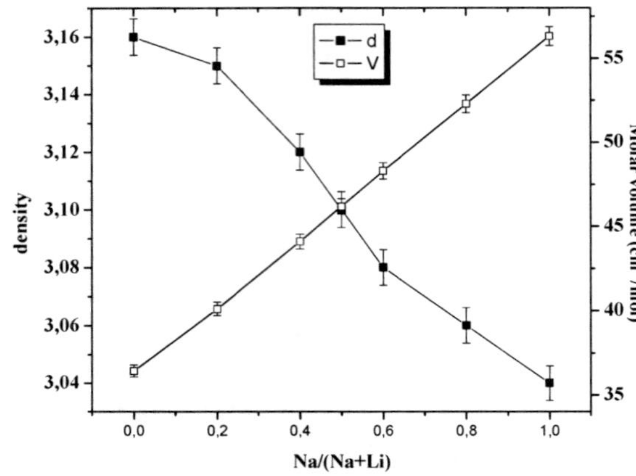

Figure 1. Plots of density and molar volume versus the molar ratio Na/(Na + Li) along the system $0.5[x(Na_2SO_4)\text{-}(1\text{-}x)Li_2O]\text{-}0.5[0.4(MoO_3)_2\text{-}0.6P_2O_5]$.

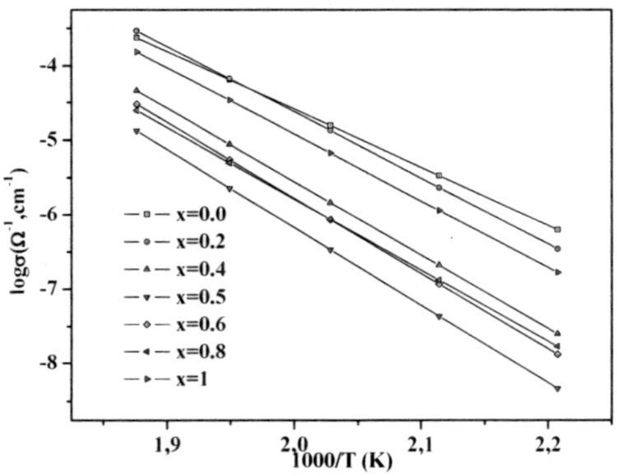

Figure 2. Thermal variation of the dc conductivity along the system $0.5[x(Na_2SO_4)\text{-}(1\text{-}x)Li_2O]\text{-}0.5[0.4(MoO_3)_2\text{-}0.6P_2O_5]$.

The figure shows that the plots are straight lines indicating that the conductivity obeys the Arrhenius law: $\sigma T = \sigma_0 \exp(-E_a/kT)$. This behaviour is close to that observed for Li_2O-NaF-MoO_3-P_2O_5 glasses.[9]

The compositional variation of the isothermal dc conductivity and its activation energy is shown in Fig. 3.

This figure shows a minimum and a maximum respectively at a particular glass composition ($x = 0.5$). It is worth to notice that this behaviour is usually observed in mixed alkali glasses.[10,11]

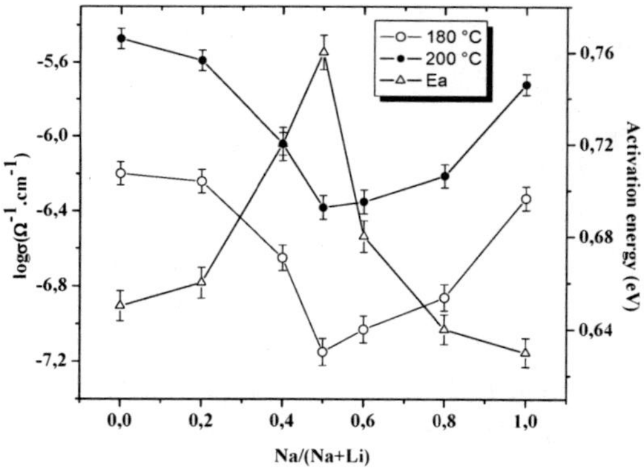

Figure 3. Plots of conductivity and activation energy versus composition along the glass system $0.5[x(Na_2SO_4)$-$(1$-$x)Li_2O]$-$0.5[0.4(MoO_3)_2$-$0.6P_2O_5]$. Line is drawn as guide to the eye.

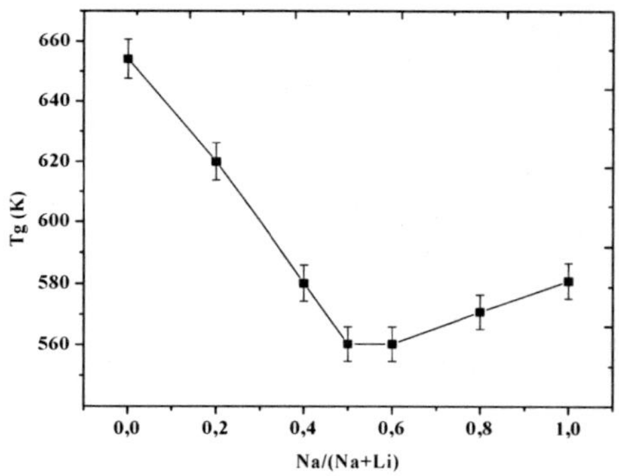

Figure 4. Composition change of the glass transition temperature for the glass system $0.5[x(Na_2SO_4)$-$(1$-$x)Li_2O]$-$0.5[0.4(MoO_3)_2$-$0.6P_2O_5$. Line is drawn as guide to the eye.

Figure 4 represents the variation of the glass transition temperature versus the composition. As can be seen, T_g decreases with increasing the ratio Na/(Na + Li) and reaches a minimum for the value Na/(Na + Li) = 0.5. Therefore, a similar trend associated with the conductivity MAE is observed for the glass transition temperature.

The frequency dependence of the real conductivity, as isotherms in a logarithmic scale, at different temperatures was investigated for all the compositions; as an example Fig. 5 shows the plot for the composition x = 0.4. The dynamic conductivity related to the real part of the complex conductivity showed a typical behaviour[12]: a frequency-independent plateau for low frequency range and a power-law increase at high frequencies. We observed a decrease in the power law exponent from 0.66 seen in the single-alkali end members to about 0.57 for the mixed alkali compositions. This decrease cannot be attributed to the average of the power law exponents found for two single Li- and Na-glasses, since each individually had exponent value larger than 0.57. In view of the Bunde and Maas model,[13] the existence of site speciation (Li-sites versus Na-sites) admits the possibility for formation of clusters of Li-sites and Na-sites within the glass matrix. Therefore, the reduced exponent seen for mixed-glasses could be associated with a reduction of the dimensionality of the conduction pathways over that present in single-alkali oxide glasses.

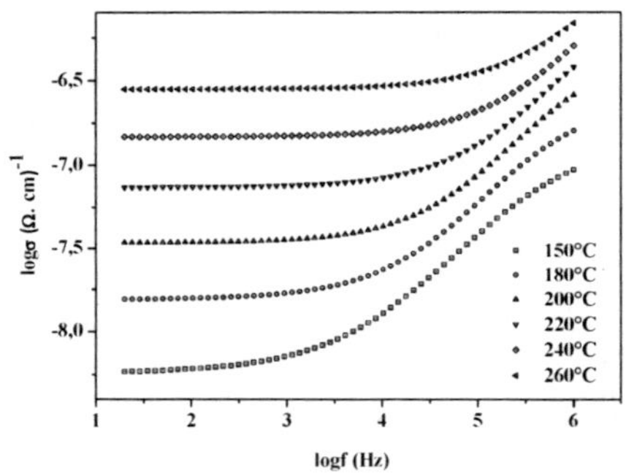

Figure 5. Typical example of the ac conductivity of the glass composition $0.5[x(Na_2SO_4)$-$(1\text{-}x)Li_2O]$-$0.5[0.4(MoO_3)_2$-$0.6P_2O_5]$ (x = 0.4).

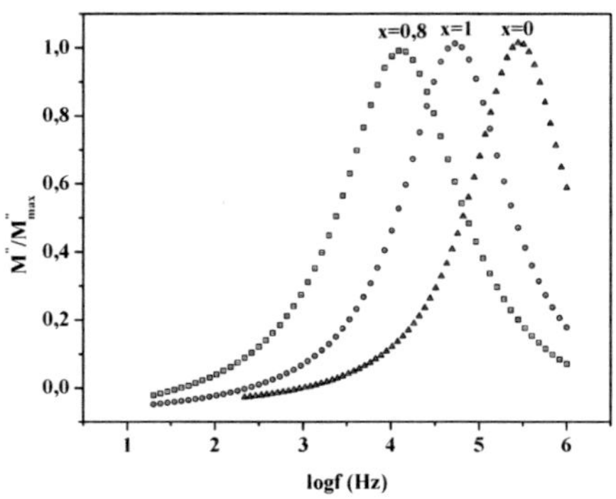

Figure 6. Frequency dependence of the imaginary part of the normalized modulus M''/M''_{max} for the two single Li and Na and mixed Li-Na (x = 0.8) glasses.

In order to neglect the electrode polarization effect,[14] the conductivity data are analyzed also by using the complex modulus formalism ($M^* = 1/\varepsilon^* = M' + iM''iwC_0Z^*$). Figure 6 shows a plot of the frequency dependence of the normalized imaginary part of M* (M''/M''_{max}) for single Li, Na and mixed Li-Na (x = 0.8) glasses at 473 K.

In each single alkali glass (Li or Na) the modulus peak maximum shifts to higher frequencies as temperature increases. The obtained M''/M''_{max} curves are not symmetric, proving a non-exponential behaviour of the conductivity relaxation which can be described by a Kohtrausch function $\Phi(t) = \Phi_0 \exp\text{-}(t/\tau_\sigma)^\beta$.[15–17] A more interesting observation obtained in this study is that the relaxation peak appears at a significantly higher frequency in single Li or Na glasses than in mixed Li-Na tungsten-phosphate glasses (Fig. 6).

4. Discussion

It is reported by Basan et al.[18] the presence of some conductivity anomalies in the system Li_2O-WO_3-P_2O_5. The authors suggested that negatively charged polarons (effectively the d-electron located on a W^{5+} centre) interact with mobile cations (Li^+) to form uncharged diffusing entities minimizing the conductivity. Such reduction of the conductivity is associated with the so-called "ion–polaron effect" (IPE).[18] In 1978, Kraevskii et al.[19] reported on a similar conductivity anomaly in the Na_2O-WO_3-P_2O_5 glassy system. This mixed ion-polaron effect (IPE) was also reported for other vitreous systems, e.g. xLi_2O-$yMoO_3$-zP_2O_5,[20] xLi_2O-$(1\text{-}x)V_2O_5$-$3TeO_2$,[21] xNa_2O-$(50\text{-}x)FeO$-$50P_2O_5$[22] and AgI-Ag_2O-V_2O_5-P_2O_5.[23] There is no consensus yet about the origin of such behaviour. Different explanations for IPE have been considered on the basis of the interaction[24] and trekking of the percolation paths.[25] In order to neglect such a IPE effect in the glasses under study, the authors have selected materials showing no Mo^{5+} ESR spectra. The total alkali oxide content is chosen in such a way that the electrical conductivity of the glasses under study is entirely ionic. Therefore, the stabilisation of the molybdenum in a high valence state Mo^{6+} is a fundamental key which has allowed us firstly to elaborate ionic glass conductors, and secondly to study the ionic transport, in the present mixed alkali glasses $0.5[xNa_2SO_4\text{-}(1\text{-}x)Li_2O]\text{-}0.5[0.4(MoO_3)_2\text{-}0.6P_2O_5]$, as a function of the relative composition while exchanging one type of ionic species with another one keeping both the total alkali content constant.

The experimental results shown in Fig. 3 are associated with the well known mixed alkali effect (MAE). This latter is a longstanding subject of investigation in oxide glasses. MAE corresponds to a nonlinear evolution of certain properties with the relative cation concentration in oxide glasses including two different alkali cations. Several models for MAE have been proposed in the literature and are compiled in several reviews.[26–28] Recently, V. Belostotsky[29] has presented a new defect model for the mixed mobile ion effect. The essential physical concept involved is that simultaneous migration of two unlike mobile ions in mixed ionic glass is accompanied by an expansion or contraction of the guest-occupied sites with distortion of the surrounding glass matrix.

The conductivity related to the mixed-alkali effect ($x = 0.5$) is lower than that of the original Li and Na analogue glasses. Indeed, deep minimum in isotherm of the conductivity increases with decreasing temperature (Fig. 2). Moreover, the explanation of MAE regularly reported in the literature can be considered either as based upon structural features (e.g., conduction pathways)[13,30,31] or as based upon differing cation interactions resulting from differences in the mass and/or size of the cation.[30,32] The promising model which takes into account the two features of the MAE is the Dynamic Structure Model (DSM) reported by Bunde et al. and Maass et al.[13,31] According to this model, the observed minimum of the conductivity in the title phosphate glasses could be attributed to the distinctly different local environment of the two alkali ions, which are preserved in the mixed glasses. The argument is that the atomic characteristics of Li and Na are very different and each cation may reside in a site formed by a local environment in the single glasses as well as in the mixed-alkali compositions. It is worth to mention that in oxide glasses, lithium and sodium cations are normally connected with non-bridging oxygen anions to satisfy the charge neutrality conditions. Since the activation energy associated with Na-glass is larger than that of single Li-glass, one can predict that the magnitude of the interactions and the polarisation effects related to the alkali-environment are different. In a single glass, where an alkali ion moves into a site previously occupied by the alkali, a sort of structural memory effect[13] favours its migration. However, in the mixed compositions the hopping dynamics of Li and Na cations are intimately coupled with the structural relaxations of the glass network. For instance, to accommodate the jump of Li cation into a site previously occupied by a sodium cation, the latter must undergo a local relaxation after which the cation can continue to migrate through the matrix. As consequence of such relaxation, a decrease in dc conductivity (MAE) is observed (Fig. 3). Statistically, it can be noted that a minimum may be pronounced for a composition corresponding to a maximum disorder of alkali elements. Accordingly, the conductivity decreases when we begin substituting lithium by sodium and vice versa, i.e., in the regions $0 \leq x \leq 0.5$ and $0.5 \leq x \leq 1$, and a flat plateau or an absolute minimum (Fig. 3) was reached for the composition with the ratio $Na/(Na + Li) = 0.5$. In addition to ionic conductivity, the glass transition temperature, Tg, which is not directly dependent upon ionic transport, shows a pronounced departure from linearity at intermediate mixed alkali ion compositions. We have also observed that the glass transition temperature is lower for mixed glasses than for the original compositions ($x = 0, 1$). Such behaviour could also be associated to the 'structural disorder' imposed by the presence of two kinds of cations. This kind of Tg variation was also reported for other vitreous systems, e.g. Li_2O-Na_2O-MoO_3-P_2O_5 [10], Li_2O-Na_2O-Al_2O_3-P_2O_5 [11], Li_2O-(Na, K, Rb)$_2$O-B_2O_3[33] and Li_2O-NaF-MoO_3-P_2O_5.[9]

It is also interesting to note that the cation localised motions, which give rise to the ac dispersion, show a small exponent for mixed glasses compared to that of single glasses (Fig. 5). In view of the Bunde et al. and Maas et al. model,[13,31] this decrease is in agreement with the notion of cluster formation occurring in the mixed alkali glasses; which is due to the fact that the alkali Li and Na cations occupy specialized sites in the glass matrix. Indeed, we can observe that the relaxation peak appears at higher frequency in single Li or Na phosphate glasses (Fig. 6). This behaviour is classical for mixed alkali glasses and confirms the slowing down of the ionic motions both on local and long ranges.

5. Conclusions

We have observed the mixed alkali effect in sulphate molybdenum-phosphate glasses along a series $0.5(xNa_2SO_4\text{-}(1\text{-}x)Li_2O)\text{-}0.5[0.4(MoO_3)_2\text{-}0.6\,P_2O_5]$ ($0 \le x \le 1$). The stabilisation of the molybdenum in a high valence state Mo^{6+} is a fundamental key which has allowed us firstly to elaborate ionic glass conductors, and secondly to study the ionic transport without interference with the ion-polaron effect (IPE). Thermal and electrical properties of the glass compositions showed that the dc conductivity and the glass transition temperature present minima when mixing Li_2O and Na_2SO_4.

References

1. E. Antić-Fidančev, B. Viana, P. Aschehoug, L. Bih, A. Nadiri, M. Taibi, J. Aride, Mater. Sci. Forum 518 (2006) 447.
2. R.K. Brow Jr., J. Non-Cryst. Solids 263–264 (2000) 1.
3. S.W. Martin Jr., Am. Ceram. Soc. 74 (1991) 1767.
4. C.A. Angell, Sol. St. Ionics 9–10 (1983) 3.
5. H.L. Tuller Jr., J. Non-Cryst. Solids 73 (1985) 331.
6. C.C. Hunter, M.D. Ingram, Solid State Ionics 14 (1984) 31.
7. J.R. Van Wazer, Phosphorus and Its Compounds, vol. I, Wiley-Interscience, New York, 1958.
8. S.W. Martin, Eur. J. Solid State Inorg. Chem. 28 (1991) 163.
9. L. Bih, S. Mohdachi, A. Nadiri, M. Mansori, M. Amalhay, O. Mykajlo, D. Kaynts, submitted to Opt. Mat. R. Commun.

10. L. Abbas, L. Bih, A. Nadiri, Y. El Amraoui, D. Mezzane, B. Elouadi, J. Mol. Structure, 876 (2008) 194.
11. A. Faivre, D. Viviani, J. Phalippou, Solid State Ionics, 1769 (2005) 325.
12. A.K. Jonscher, Nature 256 (1975) 566.
13. P. Maass, A. Bunde, M.D. Ingram, Phys. Rev. Lett. 68 (1992) 3064.
14. F.S. Howell, R.A. Bose, P.B. Macedo, C.T. Moynihan, J. Phys. Chem. 78 (1974) 639.
15. K.L. Ngai, J.N. Mundy, H. Jain, G.B. Jollen beck, O. Kamert, Phys. Rev. B 39 (1989) 6169.
16. K.L. Ngai, Phys. Rev. B 48 (1993) 13481.
17. K.L. Ngai, J. Non-Cryst. Solids, 203 (1996) 232.
18. J.C. Bazan, J.A. Duffy, M.D. Ingram, M.R. Mallace, Solid State Ionics 86–88 (1996) 497.
19. S.L. Kraevskii, T.F. Evdokimov, U.F. Solonov, E. Shimshmentseva, Fiz. Khim. Stekla 4 (1978) 839; S.L. Kraevskii, U.F. Solonov, Glass Phys. Chem. 32 (2006) 629.
20. L. Bih, M. El Omari, J.M. Réau, A. Nadiri, A. Yacoubi, M. Haddad, Mat. Lett. 50 (2001) 308.
21. G.D.L.K. Jayasinghe, M.A.K.L. Disanayake, P.W.S.K. Bandarayake, J.L. Souquet, D. Foscallo, Solid State Ionics 121 (1999) 19.
22. R.J. Barczynski, L. Murawski, J. Non-Cryst. Solids 307–310 (2002) 1055.
23. J.E. Garbarczyk, M. Wasiucionek, B. Wnetrzewski, W. Jakubowski, Phys. Status. Solidi (a), 156 (1996) 441.
24. P. Subbalakshmi, N. Veeraiah, J. Phys. Chem. Solids 284 (2003) 1027.
25. M.C. Ungureanu. M. Lévy, J.L. Souquet, Solid State Ionics 4 (1998) 200.
26. J.D. Isard, J. Non-Cryst. Solids 1 (1969) 235.
27. D.E. Day, J. Non-Cryst. Solids 21 (1976) 343.
28. M.D. Ingram, Phys. Chem. Glasses 28 (1987) 215.
29. V. Belostotsky, J. Non-Cryst. Solids 353 (2007) 1078.
30. G.N. Greaves, Philos. Mag. B 60 (1989) 793.
31. A. Bunde, M.D. Ingram, P. Maass, K.L. Ngai, J. Phys. A 24 (1991) 2881.
32. A.H. Dietzel, Phys. Chem. Glasses 24 (1983) 172.
33. Y. Gao, Chem. Phys. Lett. 417 (2006) 430.

Part III

PIEZO- AND ELECTRO-ACTIVE MATERIALS

ELECTRICAL CONDUCTION AND DIELECTRIC PROPERTIES IN PIEZOELECTRIC FIBRE COMPOSITES

HELMI HAMMAMI[1,2], MOURAD AROUS[1*],
MANUEL LAGACHE[2] AND ALI KALLEL[1]
[1]*Laboratoire des Matériaux Composites, Céramiques et Polymères, Faculté des Sciences de Sfax, 3018 Sfax, Tunisie*
[2]*Laboratoire d'Optimisation de la Conception et d'Ingénierie de l'Environnement, (LOCIE-ESIGEC)-Université de Savoie, 73376 Le Bourget du Lac, France*

Abstract. Smart structural composites are multifunctional structural materials which can perform functions such as sensing strain, vibration reduction and are essential because of their relevance to mitigation and structural vibration control. Piezoelectric fiber composites were developed to overcome drawbacks of typical monolithic piezoceramic (PZT) actuators. Piezoelectric fiber composites can improve the performance of various structures, and can be subject to wide temperature range where the thermoelastic behavior is important. A series of 1–3 connectivity PZT fibers/epoxy resin composites with different volume fraction is studied by means of dielectric spectroscopy in the wide frequency range 0.1 Hz–100 kHz and temperature varying from the ambient to 210°C. The conduction phenomenon is analyzed using the "universal power law" and its scaling is studied by the Jonscher's universal power law. At low frequencies ac conductivity tends to be constant, while in the high frequency region verifies the exponential law of conductivity. In the intermediate frequencies, the examined systems exhibit strong dispersion with frequency and the produced fitting curves deviate from the experimental data by not being able to describe the recorded relaxation and pointing out that in the vicinity of the relaxation peaks the power law is not applicable. Finally, dipolar relaxation mechanisms and interfacial or Maxwell-Wagner-Sillars relaxation were revealed in the frequency range and temperature interval of the measurements.

Keywords: Smart materials, composite materials, piezoelectric materials, dielectric properties, electrical properties.

---

*To whom correspondence should be sent: Email: mourad.arous@fss.rnu.tn

I. A. Luk'yanchuk and D. Mezzane (eds.), *Smart Materials for Energy, Communications and Security.* 169

1. Introduction

Structures using smart materials such as piezoelectric materials have potential applications in many areas, especially in controlling vibration, energy recovery on a vibrating structure and acoustic transducers.[1–3]

Monolithic piezoelectric materials have been widely used in electromechanical actuators and sensors, given their high piezoelectric performances. However, monolithic piezoelectric ceramic demonstrates a brittle behaviour and is very sensitive to cracking. Today, monolithic ceramics are increasingly substituted with 1–3 connectivity piezocomposite, which consist of aligned PZT fibres that are longitudinally polarized. Advantages of the 1–3 piezoelectric composites, over monolithic ceramics, include high sensitivity, low acoustic impedance (matching with water and human cells), a large range of mechanical properties, high strength and low density. The piezoelectric 1–3 composites have attracted a great deal of attention and have been used widely in engineering especially in the underwater hydrophone application and ultrasonic actuators and sensors for medical diagnostic devices.[4–8]

Several studies using piezoelectric composite materials were carried out during these last years concerning their manufacturing and characterizations from a mechanical point of view. Nevertheless, according to the literature, few works were interested with their dielectric behaviours or the mechanism of charge carriers transport under the influence of temperature and frequency of the applied electric field. These electric phenomena are essential to understand the mechanical behaviour. For example, the accumulation and detrapping of electric charges resulting from a mechanical or thermal load can be responsible to the failure of these materials. In precedent studies[9,10] a quantitative analysis was carried out in the aim to show that dielectric study is a good tool for the investigation of the dipolar and Maxwell-Wagner-Sillars (MWS)[11–13] relaxation processes under the influence of an electric ac field.

In the present work we report the electrical properties in unidirectional PZT fibres/epoxy resin composites, which offer valuable information about the behaviour of localized electric charge carriers. This can also lead to better understanding of the mechanism of electrical conduction and dielectric polarization in such systems.

2. Experimental

2.1 MATERIALS

A commercially available epoxy resin XB 3297 and a low reactivity formulated amine hardener XB 3298 (Vantico polymer specialities) with initial mixture viscosity 320–380 (MPa s) at 25°C, were used in this study. Ferroelectric ceramic fibres PZT-5A, commercialized by Advanced Cerametrics incorporated[14] with a density 7.7 g/cm^3 at 25°C, obtained by Viscous Suspension Spinning Process (VSSP),[15] were used with an average diameter of 250 μm.

The piezocomposite samples are elaborated similarly to the technique of the materials science and technology research institution in Dübendorf.[16] In this method, the fibres are placed manually on a grooved aluminium mould. These fibres are then attached with sticking paper on their sides and the film produced can then be easily transferred and coated by the epoxy resin in another mould. More details concerning the samples manufacturing can be found in our previous work.[9]

2.2 CHARACTERIZATION

In dielectric analysis, the sample is placed between two gold parallel plate electrodes (diameter of 24 mm), and the dielectric measurements were carried out perpendicularly to the fibres direction. The PZT fibres are totally embedded in the matrix, so there is no contact between fibres and electrodes. A sinusoidal voltage is applied, creating an alternating electric field. This produces polarization in the sample, which oscillates at the same frequency as the electric field, but with a phase angle shift δ. This phase angle shift is measured by comparing the applied voltage to the measured current, which is separated into capacitive and conductive components. Measurements of capacitance and conductance are used to calculate the real ε' and the imaginary ε'' parts of the complex permittivity ε^*.

Two kinds of dielectric experiments were conducted: isothermal runs for seven fixed temperatures from 80°C to 170°C with 10°C increment and varying frequency from 0.1 Hz to 100 kHz, and isochronal runs for seven constant frequencies (0.1, 1, 10, 100 Hz, 1, 10 and 100 kHz) with varying the temperature from ambient to 210°C with a heating rate of 2°C/min in a nitrogen atmosphere. The pure epoxy resin and three composites with square form, 25 mm length and 0.3–0.5 mm thickness were prepared with 8%, 20% and 26% volume ratio of PZT fibres and their dielectric relaxations were performed on a dielectric thermal analyzer DEA 2970 from TA Instruments.

3. Results and discussion

3.1 AC CONDUCTIVITY STUDY

The ac conductivity of all samples has been calculated from the dielectric losses according to the relation:

$$\sigma^* = j\varepsilon_0\omega\,\varepsilon^*(\omega) = j\varepsilon_0\omega\,(\varepsilon'\text{-}j\varepsilon'') = \varepsilon_0\omega\,\varepsilon'' + j\varepsilon_0\omega\,\varepsilon' \quad (1)$$

The real part of $\sigma^*(\omega)$ is given by:

$$\sigma_{ac}(\omega) = \varepsilon_0\omega\,\varepsilon''(\omega) \quad (2)$$

where $\varepsilon_0 = 8.85 \times 10^{-12}$ Fm^{-1} is the permittivity of the free space and $\omega = 2\pi f$ is the frequency (in this paper the term frequency always means angular frequency). Figures 1a–d show the dependence of ac conductivity from the frequency of the applied field at various temperatures and several PZT fibres content. It is evident that ac conductivity is both frequency and temperature dependent and increased, up to five orders of magnitude, with increasing frequency and temperature. It is remarkable that the frequency-temperature dependence of the ac conductivity is similar for variety of quite different disordered materials: ionic materials and ionic conducting polymers,[17–19] carbon black and carbon nanotube composites.[20–22] The plots exhibit the typical behaviour of ionic materials, i.e. $\sigma_{ac}(\omega, T)$ is found to be frequency independent in the low frequency region then bend off at some critical frequency ω_p and results for $\omega > \omega_p$ in a power law dependence.[23,24]

At low frequencies where the applied electric field, forces the charge carriers to drift over large distances, as temperature is increased, a tendency to retain almost constant values is recorded. When frequency is raised, the mean displacement of the charge carriers is reduced and the real part of conductivity, after reaching the critical frequency ω_p; follows the law $\sigma_{ac}(\omega,T) \sim \omega^s$ with $0 \le s \le 1$ characterizing hopping conduction.[24,25] The term hopping refers to sudden displacement of a charge carrier from one position to another neighbouring and in general includes both jumps over a potential barrier and quantum mechanical tunnelling.[26,27]

Jonscher[28] proposed the following equation for the frequency-temperature dependence of the ac conductivity:

$$\sigma_{ac}(\omega, T) = \sigma_{dc}(T) + A(T)\ \omega^S \quad (3)$$

where $\sigma_{dc}(T)$ is the $\omega \to 0$ limit value of $\sigma_{ac}(\omega, T)$ and corresponds to the dc conductivity values. A and s are parameters depending on temperature and filler content.[29–31] The parameter s describes the slope of power-law behaviour, usually varies between 0.6 and 1.0 and increases toward unity as temperature decreases.[32–34] The Equation (3) is often called the "ac universal power law" since it has been found to satisfactorily describe the ac response of numerous different types of materials.[24,35,36]

The frequency temperature superposition shows the presence of three distinct regions in the plots of Figs. 1a–d: the exponential part above ω_p, the tendency to reach constant values when frequency is low and the intermediate dispersion

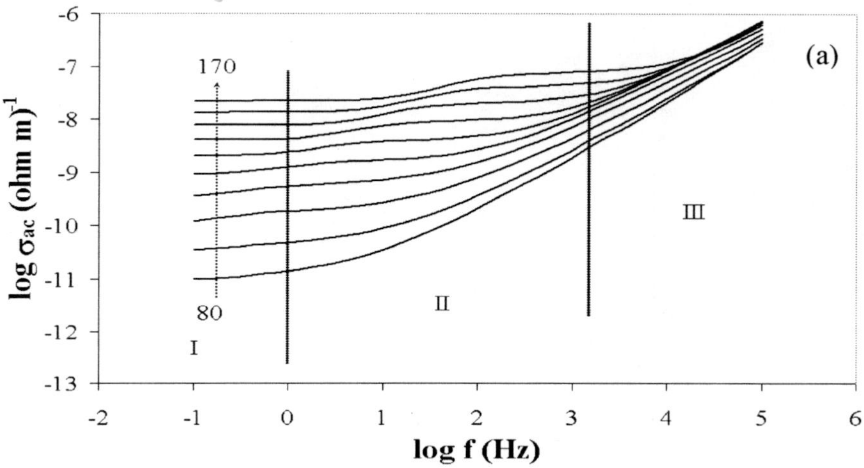

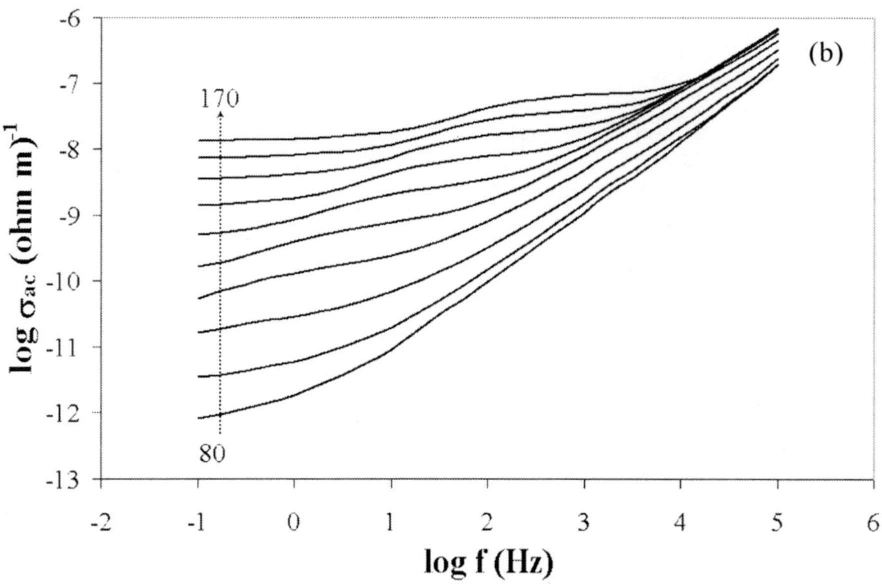

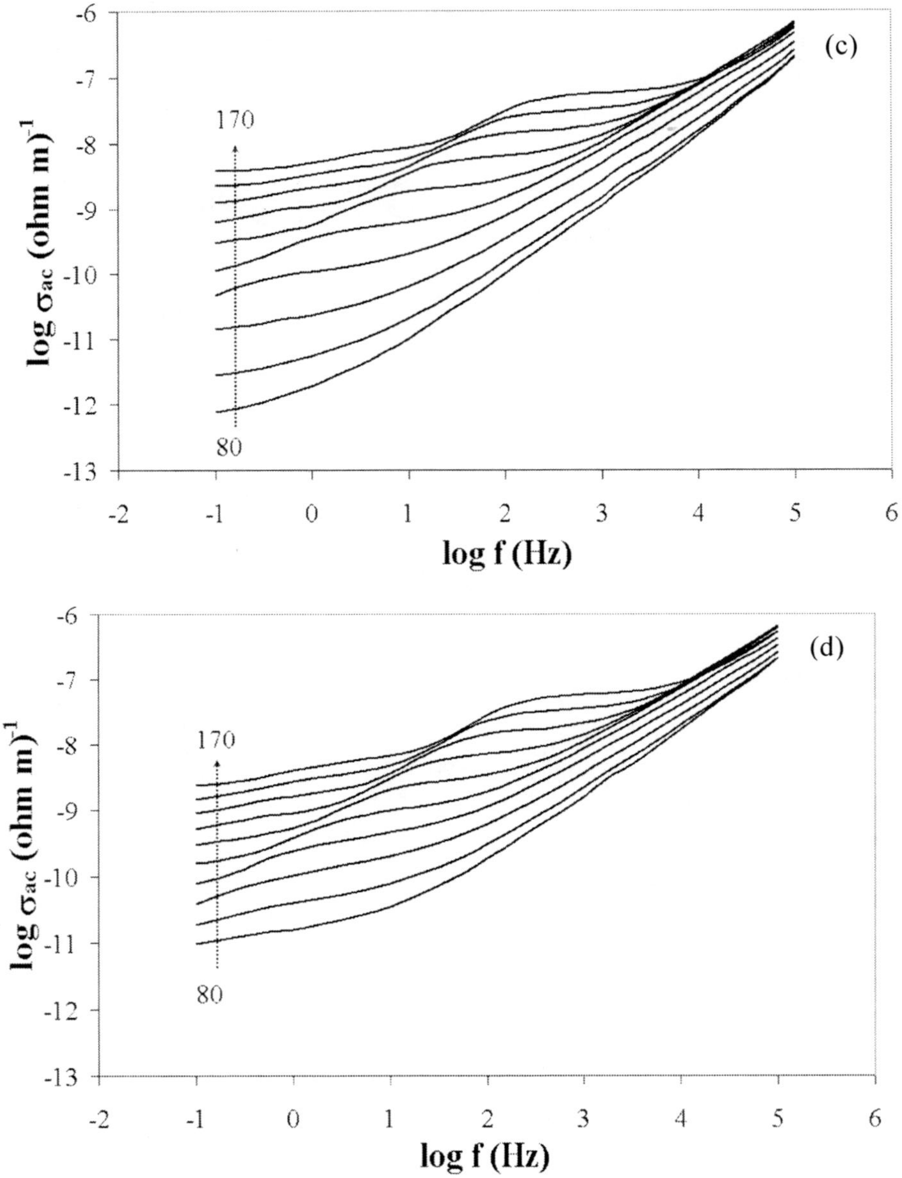

Figure 1. The ac conductivity (σ_{ac}) versus log f on the: (a) epoxy resin matrix and the piezocomposites with (b) 8%, (c) 20% and (d) 26% volume ratio of PZT fibres at various temperatures (°C).

which is more evident for high temperatures. Equation (3) was found to describe experimental data at the low and high frequency regions and to deviate for the intermediate frequencies where the relaxation processes are appeared. In fact, the ac conductivity includes both conduction and dielectric relaxation processes.

3.1.1 Region I

In the low frequency edge (region I, Figs. 1a–d) conductivity tends to remain constant and corresponds to the dc value. The evaluated values of dc conductivity for all examined samples were giving in Table 1 for temperatures going from 130°C to 170°C. These values were extracted from the frequency independent plateau by extrapolating data to the static frequency. We notice that the dc conduction effect shift to the lower frequencies with increasing the PZT fibres content. Thus the dc conductivity plateau is not reached for the composite with 26% fibres volume ratio (Fig. 1d). The dc conductivity, originated by the charge carriers, decreased when decreasing temperature and increasing PZT fibres content and obeys to the Arrhenius behaviour:

$$\sigma_{dc}(T) = \sigma_0 \exp[\frac{-E_{dc}}{KT}] \tag{4}$$

Here σ_0 is the conductivity at infinite temperature, E_{dc} is the dc activation energy and T is the temperature in Kelvin. Figure 2 shows the Arrhenius plots for the dc conduction obtained from all examined samples. The data can be described by straight lines, from which the activation energies E_{dc} were extracted (values listed in Table 1). The activation energy values for the piezocomposites were found to be close of the activation energy in the pure epoxy resin matrix, however in the piezocomposites E_{dc} decreased when increasing PZT fibres volume ratio.

3.1.2 Region II

Region II in Figs. 1a–d corresponds to the dielectric dispersion attributed to the dipolar relaxation mechanisms. In Figs. 3a–d, we represent the variation of the real M' and the imaginary M" parts of the electric modulus at several frequencies and varying temperature from the ambient to 210°C for the epoxy resin matrix and the piezocomposite samples.

$$M' = \frac{\varepsilon'}{\varepsilon'^2 + \varepsilon''^2} \text{ et } M'' = \frac{\varepsilon''}{\varepsilon'^2 + \varepsilon''^2} \tag{5}$$

Because the high values of ε' and ε'' obtained in this study, the electric modulus formalism was choosing in order to interpret the dielectric relaxation processes. The electric modulus has recently been adapted for the investigation of the dielectric response of composite polymeric system[37–44] and also been proposed for the description of systems with ionic conductivity.[45–47]

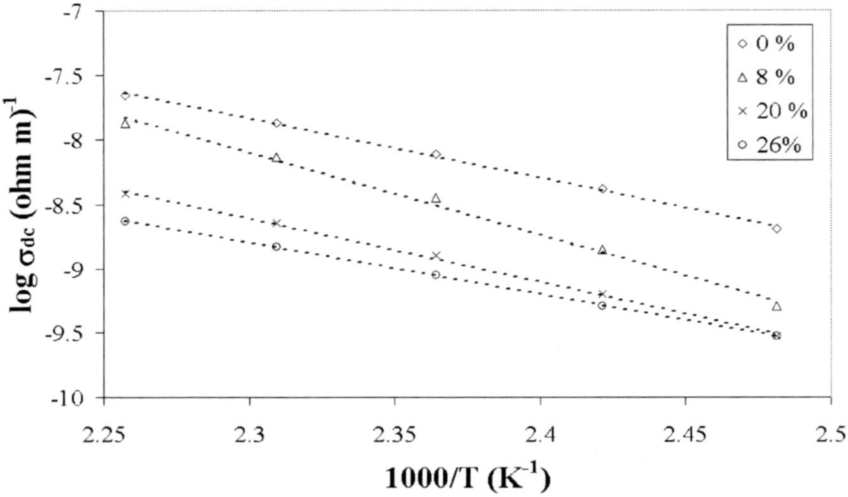

Figure 2. Arrhenius plots of the dc conductivity σ_{dc} in the all examined samples.

TABLE 1. The σ_{dc} values in the studied samples at different temperature and the evaluated activation energy.

PZT fibres content (%)	T (°C)	$\sigma_{dc} \times 10^{-10}$ (ohm m)$^{-1}$	E_{dc} (eV)
0	170	219.90	0.907
	160	134.77	
	150	77.24	
	140	41.79	
	130	20.58	
8	170	134.59	1.260
	160	73.93	
	150	35.49	
	140	14.07	
	130	5.13	
20	170	38.73	0.984
	160	22.91	
	150	12.67	
	140	6.33	
	130	2.99	
26	170	23.54	0.797
	160	14.82	
	150	8.92	
	140	5.15	
	130	2.97	

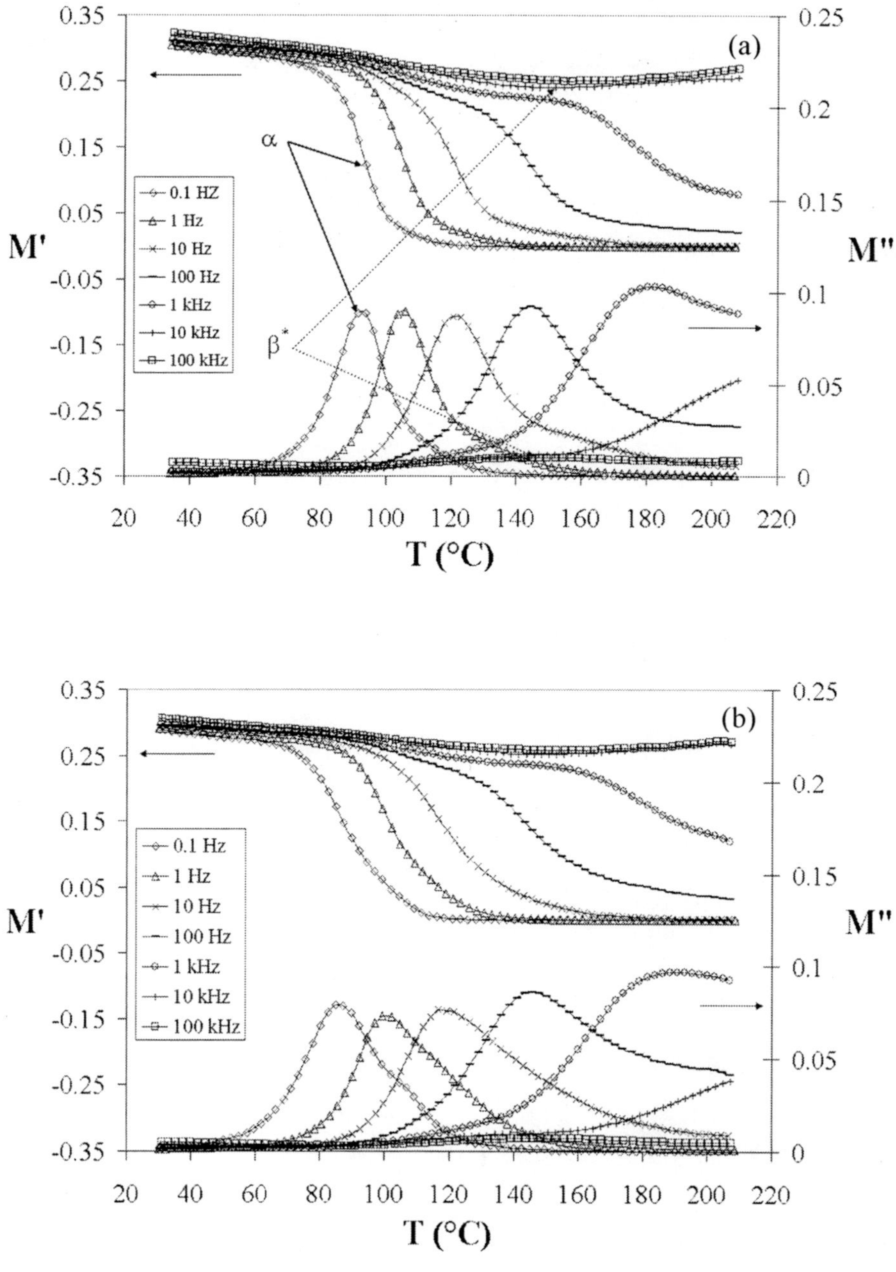
(a)
(b)
α
β*
0.1 HZ
0.1 Hz
1 Hz
10 Hz
100 Hz
1 kHz
10 kHz
100 kHz
M'
M"
T (°C)

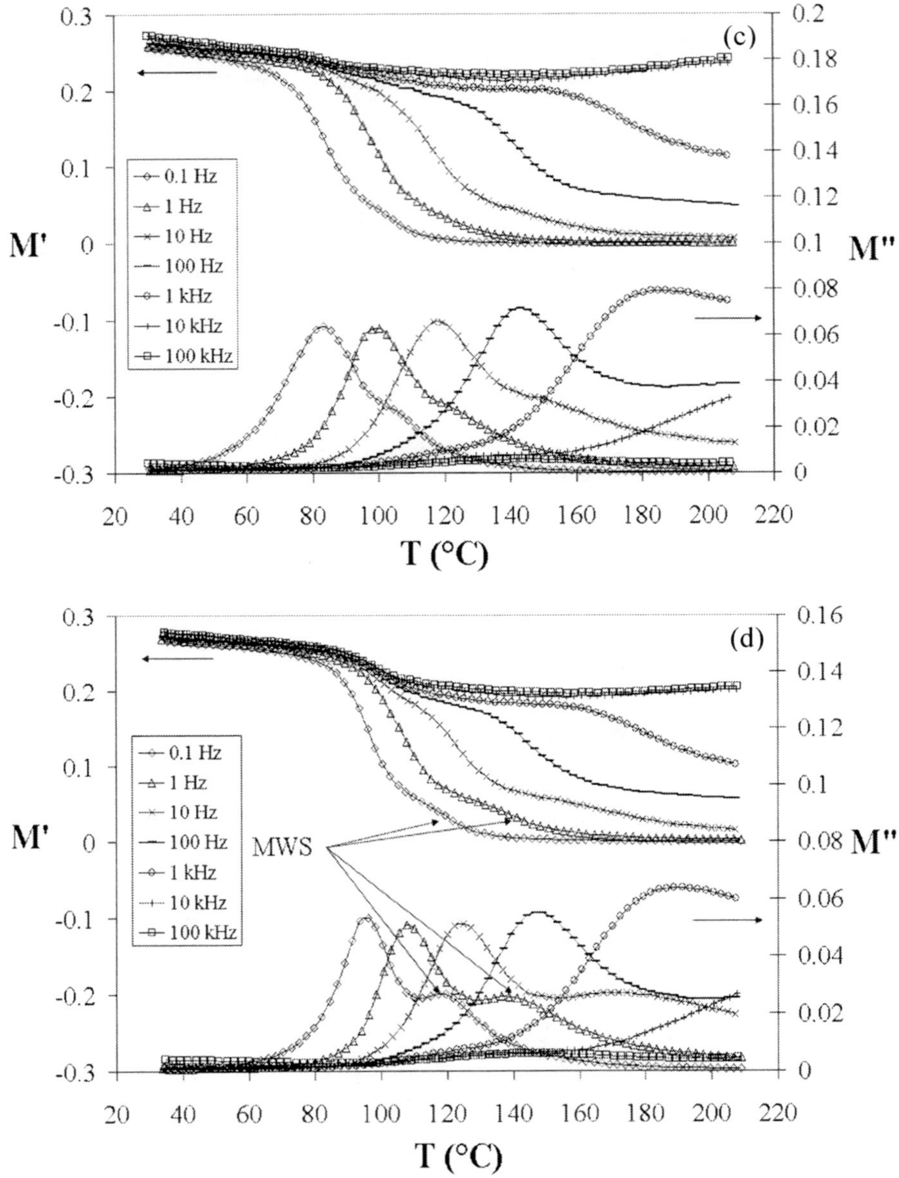

Figure 3. Variation of the real M' and the imaginary M'' parts of the electric modulus versus temperature in the: (a) epoxy resin matrix and the piezocomposites with (b) 8%, (c) 20% and (d) 26% volume ratio of PZT fibres, at various frequencies.

As it can be seen (Figs. 3a–d), M' take a rather constant value in the low temperature domain (30–80°C), then decreased when temperature increases at a constant frequency, but values of M' are ineffective at higher frequencies (from 10 kHz). In the temperature range of measurements, peaks in the plots of M" versus temperature, are developed, indicating the relaxation processes. Increasing the frequency involves a displacement towards the high temperatures of all the relaxations and values of M' and M" decreased when increasing the PZT fibres volume ratio. The analysis of these experimental results shows the presence of three distinct mechanisms. A large relaxation process begins approximately at 95°C and is noticeable only for frequencies higher than 100 Hz. We can notice that a peak maximum is located at 147°C for a frequency of 100 kHz in the epoxy resin matrix for example (Fig. 3a). This mechanism is related to the classical β^* relaxation of the epoxy resin, which is obscured for the lower frequency. Numerous paper[48–51] concerning epoxy matrix composites reveal such as an intermediate relaxation and other authors[52–54] consider this relaxation as a proof of the structural or molecular rearrangement within the network results from the moisture sorption. A second relaxation, which appeared for the whole examined frequencies, is allotted to the glass transition of the epoxy resin matrix and called α process. From a dipolar point of view it results from large-range motions "dipole-elastic process" corresponding to micro-Brownian segmental motion of chain.[55] This mechanism is the dominant one in the lower frequency and obscure the β^* process. By comparing the results obtained for the pure epoxy resin (Fig. 3a) and those of the piezocomposites (Figs. 3b–d), one notices that the β^* and α mechanisms always exist in all samples. However, a new mechanism, which is missed in the pure resin matrix, appears in the piezocomposites and increases with PZT fibres volume ratio. This mechanism is connected to the interfaces fibres/matrix and is called interfacial polarization or MWS effect. This phenomenon appears in heterogeneous systems due to the trapping of charges at the interfaces because of the discrepancy of permittivity and conductivity between constituents. Interfacial relaxation occurs in the low frequency region due to the inertia of the formed dipoles. More details concerning dielectric relaxation processes occurring in PZT fibres/epoxy resin piezocomposites can be found in previous studies.[9,10] The same behaviour in the intermediate region (region II, Figs. 1a–d) was experimentally observed by Psarras et al.[25] in iron particles loaded polymer composites. For these authors, region II corresponds to the dielectric dispersion attributed only to the interfacial relaxation process.

3.1.3 Region III

Region III in Figs. 1a–d corresponds to the exponential part of ac conductivity (second term of Equation (3)). This observed conductivity relaxation at high frequencies could be due to the probability of the correlated forward-backward hopping together with the relaxation of the ions.[33]

3.2 SCALING THE AC CONDUCTIVITY

The ability to scale different data sets as to collapse all to one single curve indicates that the process can be explained with a common physical mechanism only modified by thermodynamic scale.[56] In other words, systems that exhibit scale invariant behaviour lend themselves to simplified and often universal explanation.[57]

Equation (3) has been used to fit the ac conductivity data for all the examined samples in the temperature interval from 130°C to 170°C. In the fitting procedure, A and s values have been varied simultaneously to get the best fit. The obtained fitted curves were shown by dotted lines in Figs. 4a–d. As it can be seen, the fitting procedure qualitatively follows the dispersion of ac conductivity. In the low frequency edge reveals the tendency of conductivity to achieve a constant value, while in the high frequency region verifies the exponential law of conductivity. In the intermediate frequencies (region II of Figs. 4a–d), the produced curves deviate from the experimental data by not being able to describe the recorded relaxation and pointing out that in the vicinity of the relaxation peaks the power law is not applicable.[25,46,58–61] Table 2 summarizes the fitted values of A and s for the examined temperatures. The values of A were practically constant for all the samples, while the frequency exponent s is found to increase with decreasing temperature from 130°C until 110°C. Accordingly to Dyre et al.[24] and other authors,[34,62] in a fixed frequency range s increases as temperature decreases and $s \rightarrow 1.0$ for $T \rightarrow 0$. When there is no measurable dc conductivity s is found to be close to 1.0. We notice that s increased with increasing temperature from 140°C to 170°C and remarks said above are not respected. In fact, in the high temperature/high frequency domain s values were affected by the appearance of the β^* process.

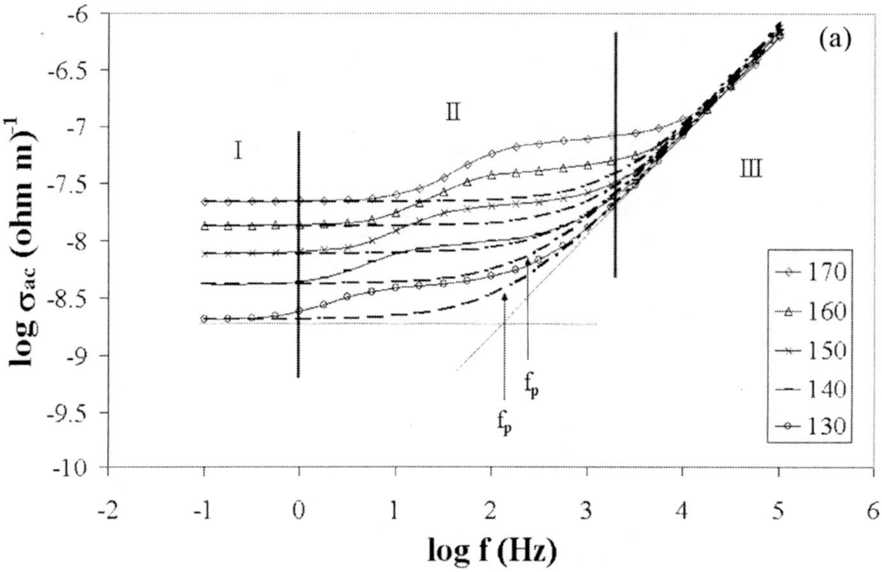
(a)
I
II
III
f_p
f_p
170
160
150
140
130
$\log \sigma_{ac}$ (ohm m)$^{-1}$
log f (Hz)

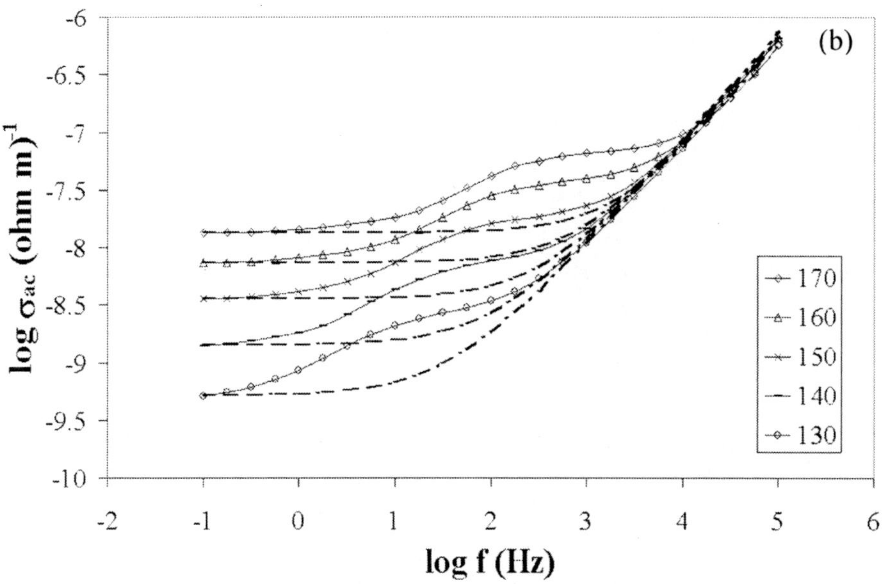
(b)
170
160
150
140
130
$\log \sigma_{ac}$ (ohm m)$^{-1}$
log f (Hz)

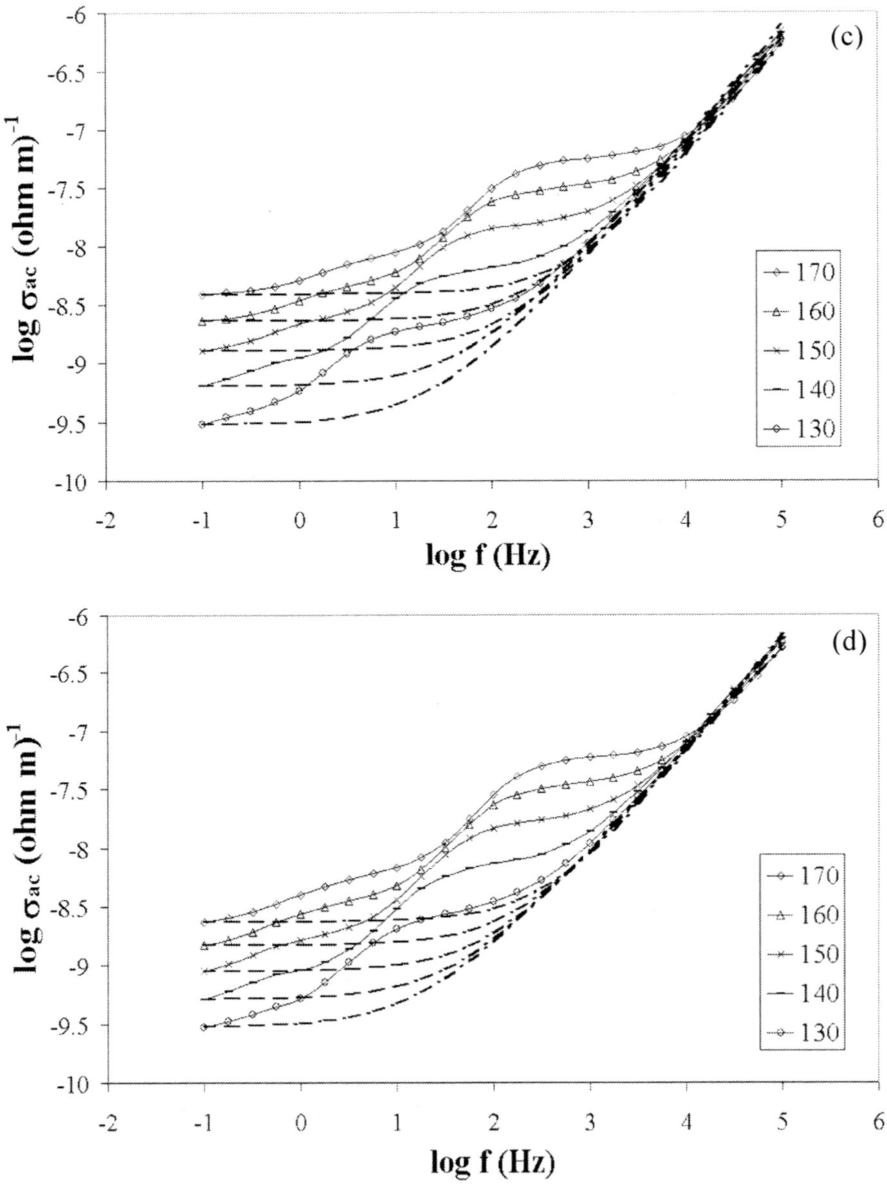

Figure 4. Analysis of the frequency dependence of ac conductivity on: (a) the epoxy resin matrix and the piezocomposites with (b) 8%, (c) 20% and (d) 26% of PZT fibres volume ratio, at various temperatures (°C). The dotted lines are produced by best fitting data to Equation (3). f_p denotes the crossover frequency.

TABLE 2. Fitting parameters of ac conductivity data to the Jonscher equation for the epoxy resin and the piezocomposites with 8%, 20% and 26% of PZT fibres content at various temperatures.

PZT fibres content (%)	T (°C)	s	A $(\text{ohm cm})^{-1}$ $\text{rad}^{-s} \times 10^{-12}$
0	170	0.976	1.737
	160	0.968	1.908
	150	0.949	2.447
	140	0.909	4.207
	130	0.898	4.197
	120	0.913	–
	110	0.960	–
8	170	1.000	1.007
	160	0.963	1.874
	150	0.941	2.592
	140	0.909	3.772
	130	0.898	4.069
	120	0.921	–
	110	0.981	–
20	170	1.000	0.899
	160	0.980	1.558
	150	0.950	1.919
	140	0.906	3.423
	130	0.892	3.431
	120	0.921	–
	110	0.981	–
26	170	1.000	1.060
	160	0.961	1.759
	150	0.932	2.430
	140	0.888	3.800
	130	0.867	4.840
	120	0.873	–
	110	0.914	–

The characteristic frequency corresponding to the onset of the power law conductivity variation is known as hopping frequency[63] and the values have been extracted from the plot of the ac conductivity at different temperatures (Figs. 3a–d) or from the universal law by:

$$\omega_p = \left(\frac{\sigma_{dc}(T)}{A}\right)^{1/s} \quad (6)$$

ω_p increased when increasing temperature and generally shifts to the higher frequencies with PZT fibres content. The evaluated hopping frequencies were found to be temperature dependent obeying the Arrhenius relation:

$$\omega_p = \omega_0 \exp(-\frac{E_p}{KT}) \tag{7}$$

where ω_0 is the hopping frequency at infinite temperature and E_p is the activation energy for hopping. From the plot of log f_p, ($f_p = \omega_p/2\pi$) versus 1000/T giving in Fig. 5, the hopping activation energies, calculated from the slopes of straight lines, were 1.09 eV for the epoxy resin sample, 1.58, 1.29, and 1.13 eV respectively for the composites with 8%, 20% and 26% PZT fibres content. These values are close to the dc conductivity activation energy. This result suggests that the charge carriers have to overcome the same energy barrier while conducting as well as relaxing.[64]

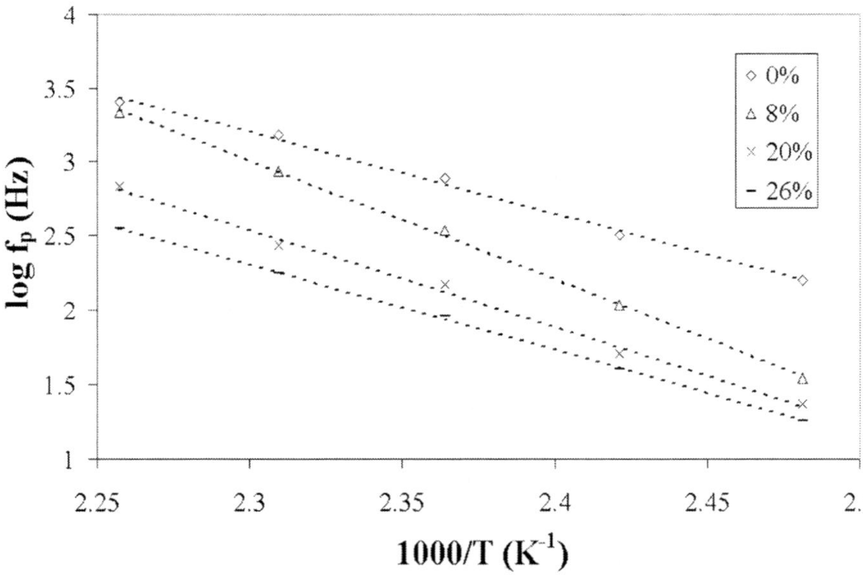

Figure 5. Arrhenius plots of the hopping frequency f_p in the all examined samples.

In recent publications,[24,56,65–71] the scaling behaviour in ac conductivity data has been studied by hopping frequency, ω_p, or by the directly measurable or accessible quantities such as the temperature, the dc conductivity, the concentration, the dielectric strength $\Delta\varepsilon = \varepsilon_s - \varepsilon_\infty$ and maximum magnitude of dielectric loss. In this work, the frequency axis is scaled with respect to hopping frequency which automatically taken into account the permittivity change and correlation effects between successive hops in the disordered lattice.[71] The formalism of Almond-West[63,72] is introduced here to study the scaling the conductivity spectra at different temperatures. In this approach, the hopping frequency ω_p, is used as the scaling frequency for the frequency axis and dc conductivity for the ac conductivity axis. In the Almond-West formalism, the scaling behaviour of ac conductivity is given by:

$$\sigma_{ac}(\omega) = \sigma_{dc}\ [1 + (\frac{\omega}{\omega_p})^s] \tag{8}$$

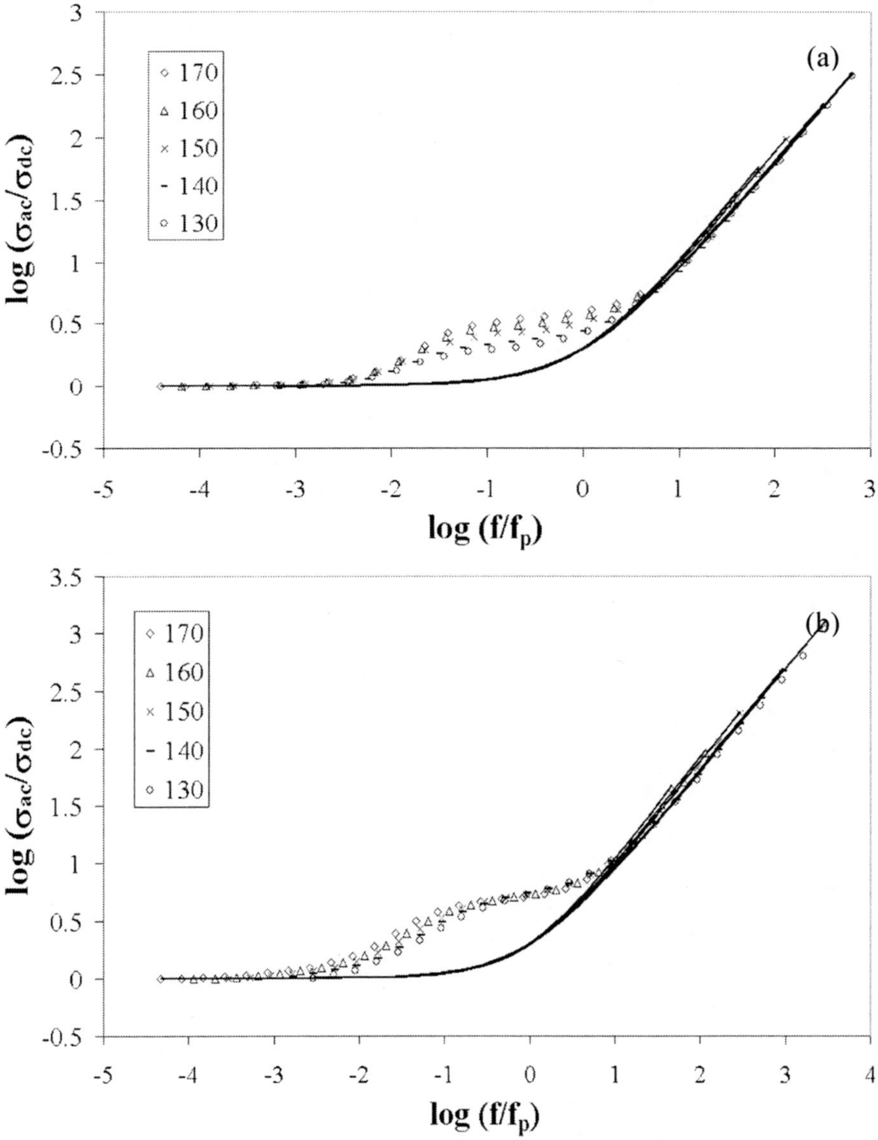

Figure 6. log (σ_{ac}/σ_0) as a function of log (f/f_p) for: (a) the epoxy resin matrix and (b) the piezocomposite with 8% of PZT fibres volume ration, at different temperatures (°C).

Figures 6a–b show the scaled conductivity spectra for the epoxy resin matrix and the piezocomposite with 8% of PZT fibres volume ratio at different temperatures under the Almond-West formalism. The same results were obtained also for the piezocomposites with 20% and 26% of PZT fibres content. From Figs. 6a–b, it is observed that the conductivity curves are superimposed into a single master curve defined by Equation (8). The magnitudes of the frequency exponent of the master curves are found to be in the range from 0.86 to 1. Finally, the critical frequency ω_p and the values of dc conductivity σ_{dc} were found to fulfil the Barton-Nakajima-Namikawa (BNN) relation[73]:

$$\sigma_0 = p\varepsilon_0\Delta\varepsilon\omega_p \tag{9}$$

where $\Delta\varepsilon = \varepsilon_0 - \varepsilon_\infty$, ε_0 is the dielectric permittivity of vacuum and p is a numerical constant of order 1. Experimental evidence for the BNN relation is reproduced in Fig. 7.

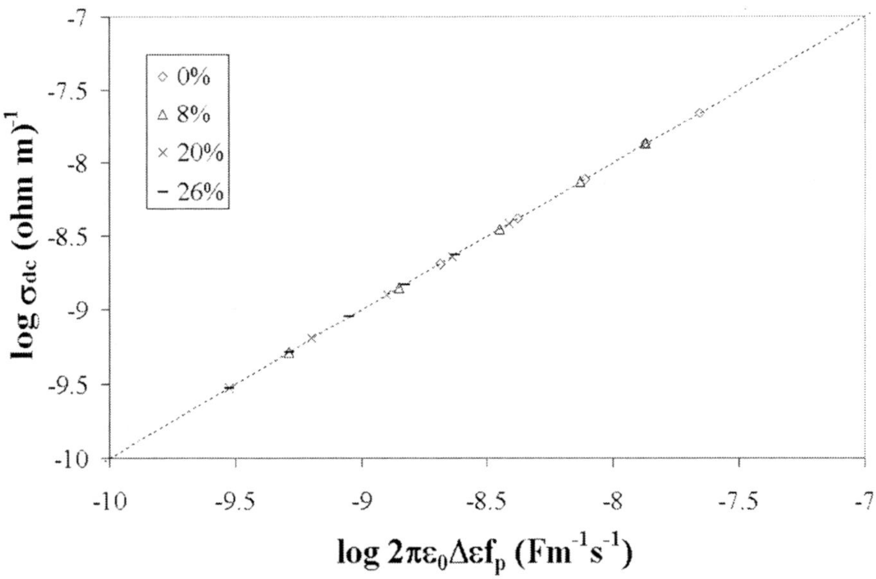

Figure 7. Test of the Barton-Nakajima-Namikawa relation for the studied samples. The dotted line gives the best fit to data.

4. Conclusions

The dielectric response of polymer composites consisting of an epoxy resin matrix with aligned PZT fibres has been studied in the frequency range 0.1 Hz–100 kHz and temperature interval from the ambient to 210°C. In these piezocomposite materials, ac conductivity exhibits a strong dispersion with the frequency of the applied electric field. Three domains have been highlighted. First, for the low frequencies, the ac conduction reaches a plateau corresponding to dc conduction. For the high frequencies, conductivity follows a classical power law. For these two domains, experimental data satisfied the "universal power law" proposed by Jonscher. For these two ranges of frequencies, the Almond-West formalism was used to scale the frequencies and conductivity in order to obtain a master curve which gives a synthetic description of the conductivity behaviour versus frequency. For the intermediate frequencies, ac conductivity cannot be modelled by the "universal power law". The discrepancy between the model and the experimental value reveals the existence of dipolar and MWS relaxation processes.

ACKNOWLEDGEMENTS

The authors wish to thank the Institut Français de Coopération et d'Actions Culturelles (French embassy-Tunisia) for its financial support.

References

1. R.E. Newnham and A. Amin, *Chem. Tech.* 29:38 (1999).
2. K. Uchino, *Piezoelectrics and Ultrasonic Applications, Kluwer, Deventer, MO* (1998).
3. S. Trolier-McKinsty and R.E. Newnham, *Mater. Res. Bull.* 18:27 (1993).
4. W.A. Smith, *The role of piezocomposites in ultrasonic transducers*, Proceedings of the 1989 IEEE Ultrasonic Symposium, pp. 755–766 (1989).
5. P. Challende, *IEEE Trans. Ultrason. Ferroelect. Freq. Control* 37:135 (1990).
6. S. Sripada, J. Unsworth and M. Krishnamurty, *Mat. Res. Bull.* 31(6):731 (1996).
7. C. Richard, L. Goujon, D. Guyomar, H.S. Lee and G. Grange, *Ultrasonics* 40:895 (2002).
8. A. Fernandes and J. Pouget, *Eur. J. Mech. A/Solids* 21:629 (2002).
9. H. Hammami, M. Arous, M. Lagache and A. Kallel, *Comp.: Part A* 37(1):01 (2006).
10. H. Hammami, M. Arous, M. Lagache and A. Kallel, *J. Alloys Compd.* 430:01 (2007).
11. J.C. Maxwell, *Electricity and magnetism, Clarendon, Oxford* (1892).
12. K.W. Wagner, *Arch. Elektrotech., Berlin* (1914).
13. R.W. Sillars, *J. Inst. Eng.* 80:378 (1937).
14. Advanced Cerametrics Incorporated, http://www.advancedcerametrics.com, P.O. Box 128, Lambertville, NJ 08530-0128.
15. R.B. Cass and I.A. Cornejo, *Adv. Cerametr., Intl. Ceramic Ind. Mag.* (2001).
16. EMPA, *Materials Science and Technology*, http://www.empa.ch/, Dübendorf (2003).
17. D.L. Sidebottom, *Phys. Rev. Lett.* 83(5):983 (1999).
18. B. Roling, A. Happe, K. Funke and M.D. Ingram, *Phys. Rev. Lett.* 78(11):2160 (1997).
19. S. Capaccioli, M. Lucchesi, P.A. Rolla and G. Ruggeri, *J. Phys.: Condens. Matter* 10:5595 (1998).
20. D.S. McLachlan and M.B. Heaney, *Phys. Rev. B* 60(18):12746 (1999).

21. M.T. Connor, S. Roy, T.A. Ezquerra and F.J. Balta Calleja, *Phys. Rev. B* 57(4):2286 (1998).
22. P. Pötschke, S.M. Dudkin and I. Alig, *Polymer* 44:5023 (2003).
23. F. Kremer and A. Schönhals, *Broadband Dielectric Spectroscopy, Springer, Heidelberg, Germany* (2002).
24. J.C. Dyre and T.B. Schroeder, *Rev. Mod. Phys.* 72(3):873 (2000).
25. G.C. Psarras, E. Manolakaki and G.M. Tsangaris, *Composites: Part A* 34:1187 (2003).
26. H. Böttger and U.V. Bryskin, *Hopin Conduction in Solids, vol. 41, Verlag Akademie, Berlin,* pp. 169–213 (1985).
27. G.C. Psarras, *Composites: Part A* 37:1545 (2006).
28. A.K. Jonscher, *Nature* 267:673 (1977).
29. G.M. Tsangaris, G.C. Psarras and E. Manolakaki, *Adv. Composites Lett.* 8(1):25 (1999).
30. R. Vijayalakshmi Rao and M.H. Shridhar, *Mater. Lett.* 55:34 (2002).
31. R. Vijayalakshmi Rao and M.H. Shridhar, *Mater. Sci. Engin. A* 35:73 (2002).
32. T.B. Schröder and J.C. Dyre, *Phys. Chem. Chem. Phys.* 4:3173 (2002).
33. P.S. Anantha and K. Hariharan, *Materials Sci. Engin. B* 121:12 (2005).
34. S.A. Saafan, A.S. Seoud and R.E. El Shater, *Physica B* 365:27 (2005).
35. A.K. Jonscher, *Universal Relaxation Law, Chelsea Dielectrics, London* (1992).
36. J.C. Dyre, *J. Appl. Phys.* 64(5):2456 (1988).
37. W. Woward, J.R. Starkweather and P. Avakian, *J. Polym. Sci.: Part B.* 30:637 (1992).
38. G.M. Tsangaris, G.C. Psarras and A.J. Kontopoulos, *J. Non-Cryst. Solids* 131/133(2):1164 (1991).
39. M. Arous, A. Kallel, Z. Fakhfakh and G. Perrier, *J. Phys. Soci. Japan* 66(11):3665 (1997).
40. G.M. Tsangaris, G.C. Psarras and N. Kouloumbi, *J. Mater. Sci.* 33:2027 (1998).
41. G.M. Tsangaris and G.C. Psarras, *J. Mater. Sci.* 34:2151 (1999).
42. M. Mudarra, R. Diaz-Calleja, J. Belana, J.C. Canadas, J.A. Diego, J. Sellarès and M.J. Sanchis, *Polymer* 42:1647 (2001).
43. M. Arous, F. Karray, H. Hammami and A. Kallel, *Phys. Chem. News* 10(1):5 (2003).
44. A.G. Charnetskaya, G. Polizos, V.I. Shtompel, E.G. Privalko, Yu Yu Kercha and P. Pissis, *Europ. Polym. J.* 39:2167 (2003).
45. L. Okrasaan, G. Boiteux, J. Ulanskia and G. Seytre, *Polymer* 42:3817 (2001).
46. C.T. Moynihan, *J. Non-Cryst. Solids* 172–174(2):1395 (1994).
47. K.L. Ngai and C. Leon, *Solid State Ionics* 195:81 (1999).
48. M. Schimbo, M. Ochi and M. Yoshizumi, *J. Polym. Sci. Polym. Phys*. 25:1817 (1987).
49. M. Ochi, M. Shimbo, M. Saga and N. Takashima, *J. Polym. Sci.* 24:2185 (1986).
50. H.J. Ploehn and J.Y. Wang, *J. Appl. Polym. Sci.* 59:345 (1996).
51. G.A. Pogany, *Br. Polym. J.* 1:177 (1969).
52. J.D. Keenan, J.C. Seferis and J.T. Quinlivan, *J. Appl. Polym. Sci.* 24:2375 (1979).
53. I.D. Maxwell and R.A. Pethrick, *J. Appl. Polym. Sci.* 28:2363 (1983).
54. K. Doukkali and Y. Segui, *J. Appl. Polym. Sci.* 41:1533 (1990).
55. A. Schönhals, *Novocontrol application notes, Dielectrics* 1 (2003).
56. D.L. Sidebottom, *Phys. Rev. Lett.* 82:3653 (1999).
57. D.L. Sidebottom, P.F. Green and R.K. Brow, *Phys. Rev. B* 51(5):2770 (1995).
58. G.M. Tsangaris, G.C. Psarras and E. Manolakaki, *Adv. Composites. Lett.* 8(1):25 (1999).
59. G.C. Psarras, E. Manolakaki and G.M. Tsangaris, *Composites: Part A* 33:375 (2002).
60. J. Malecki and B. Hilczer, *Ferroelectr. Polym. Ceram-Polym. Compos.* 92–99:181 (1994).
61. A. Bel Hadji Mohamed, J.L. Miane and H. Zangar, *Polym. Int.* 50:773 (2001).
62. M.D. Migahed, M. Ishra, T. Fahmy and A. Barakat, *J. Phys. and Chem. Solids* 65:1121 (2004).
63. D.P. Almond and A.R. West, *Solid States Ionics* 9–10(1):277 (1983).

64. J.M. Réau, Xu Y. Jun, J. Senegas, Ch. Le Deit and M. Poulain, *Solid States Ionics* 95:191 (1997).
65. K.L. Ngai and C. Leon, *Phys. Rev. B* 60:9396 (1999).
66. T.B. Schröder and J.C. Dyre, *Phys. Rev. Lett.* 84:310 (2000).
67. D.L. Sidebottom, P.F. Green and R.K. Brow, *Phys. Rev. Lett.* 74:5068 (1995).
68. B. Roling, *Solid State Ionics* 105:185 (1998).
69. D.L. Sidebottom and T. Zhang, *Phys. Rev. B* 62:5503 (2000).
70. B. Roling and C. Martiny, *Phys. Rev. Lett.* 85:1274 (2000).
71. A. Ghosh and A. Pan, *Phys. Rev. Lett.* 84:2188 (2000).
72. D.P. Almond and A.R. West, *Solid State Ionics* 23:27 (1987).
73. H. Namikawa, *J. Non-Cryst. Solids* 18:173 (1975).

NEW METHOD FOR PREPARATION OF POLYCRYSTALLINE LANGASITE FOR GAS SENSORS: STRUCTURAL STUDIES

KHALID OUZAOUIT[1], ABDELJALIL BENLHACHEMI[2*], SYLVIE VILLAIN[1], ABDELLATIF ESSOUMHI[1], HOURIA BENYAICH[2] AND JEAN-RAYMOND GAVARRI[1]

[1]*Laboratoire Matériaux and Microélectronique de Provence, UMR CNRS 6137, Université Sud Toulon-Var, B.P. 20132, 83957 La Garde – France*

[2]*Laboratoire Mat2riaux et Environnement, Faculté des Sciences, Université Ibn Zohr, B.P. 8106 Cité Dakhla, 80 000, Agadir – Maroc*

Abstract. Langasite $La_3Ga_5SiO_{14}$ (noted LGS) is a piezoelectric material presenting a strong electromechanical deformation at high temperature (up to 1,000–1,200°C). Classical solid state reaction at high temperature (1,450°C) is generally needed to synthesize the LGS phase. In this study, we present a new synthesis way for polycrystalline langasite (LGS), involving a series of specific thermal cycles at moderate temperatures (<1,200°C). The final langasite phase is then characterized by X-ray diffraction, scanning electron microscopy and infrared spectroscopy. FTIR observed vibration modes are commented in relation with literature results.

Keywords: Langasite, $La_3Ga_5SiO_{14}$, solid state reaction, sensors, X-ray diffraction, SEM, FTIR.

1. Introduction

Langasite $La_3Ga_5SiO_{14}$ (LGS) was discovered in 1982[1] in the general framework of material researches for laser radiation applications. Then, LGS appeared as an interesting piezoelectric material and was investigated for its electromechanical applications from 1983. The most pertinent property resides in its high temperature piezoelectricity: LGS presents strong mechanical responses

*To whom correspondence should be sent: Email: a.benlhachemi@gmail.com

I. A. Luk'yanchuk and D. Mezzane (eds.), *Smart Materials for Energy, Communications and Security.*

even above 1,000°C. Industrial surface acoustic wave (SAW) devices, based on LGS, and working at high temperature, were developed.[2] Today, langasite is involved in industrial resonators and is used in a wide range of applications such as microbalance at high temperature, bulk and surface acoustic wave detector devices, new communication systems, filters.[3,4] LGS might be operational up to its melting point at 1,470°C, without any phase transition leading to a loss of piezoelectric properties.[5] This is in contrast with the conventional piezoelectric quartz that exhibits a destructive alpha-beta modification at 573°C, prohibiting any use at high temperatures (above 300°C). Due to its strong electrochemical coupling, LGS and related compounds exhibit better piezoelectric properties than quartz, specifically at high temperatures: electro-mechanical coupling coefficient should be three times larger than that of quartz.[6–8]

Our own interest resides in the potential applications in gas detection field: we try to use LGS as a resonant (oscillating) substrate on which catalytic layers or inclusions will be deposited to interact with gases (CH4, CO, NH3…). Due to its exceptional properties, it was previously investigated as an alternative material for gas sensing: in resonant frequency devices, mass adsorption can modify resonant frequency and langasite might be highly sensitive to mass perturbation.[9,10]

Classically, LGS ceramics can be prepared from solid state reaction processes at 1,450°C (just below melting point). Our present study deals with a new synthesis way of polycrystalline LGS involving lower temperatures and lower costs. Intermediate and final phases were identified making use of X-ray diffraction, SEM microscopy and EDS element analysis. Infrared spectroscopy analyses are presented and commented. LGS crystallizes in a hexagonal (trigonal) lattice and can be considered as a layered compound (Fig. 1).

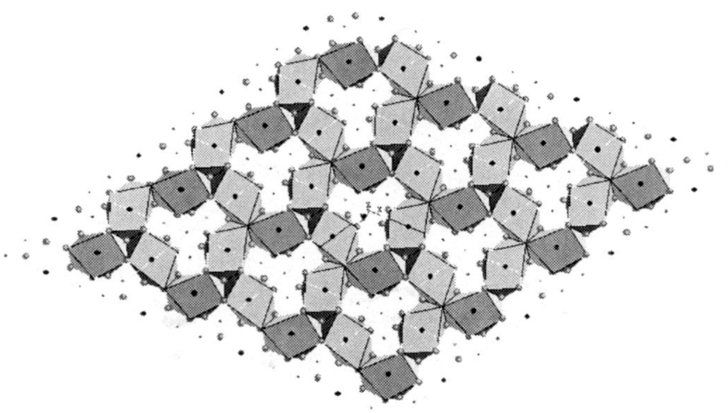

Projection of $La_3Ga_5SiO_{14}$ structure along the [001] direction.

La (3e) O (2d, 6g, 6g) G : octahedral sites (1a, 3f, 2d)

Si: half of the tetrahedral Sites (2d)

Figure 1. Structure of piezoelectric LGS phase.

2. Experimental techniques

2.1 SYNTHESIS OF POLYCRYSTALLINE SAMPLES

The classical route consists in mixing initial powders of La_2O_3, Ga_2O_3 and SiO_2 then heating at 1,450°C. The basic reaction is:

$$3\ La_2O_3 + 5\ Ga_2O_3 + 2\ SiO_2 \rightarrow 2\ La_3Ga_5SiO_{14} \quad (1)$$

This high temperature requires specific furnaces, high-energy costs. Following the chosen thermal process, new intermediate phases (solid or liquid) might appear as undesired intermediate phases. The available binary phase diagrams give hypothetic information on the apparition or not of intermediate phases above 1,400°C.[11] The final product resulting from high temperature reaction depends on initial homogeneity of this ternary system. Reacting with the two other phases, SiO_2 oxide can melt during the heating process: then, its migration in liquid form can fully modify the effective composition of the solid system. At lower temperatures (1,000°C instead of 1,450°C), as the initial phases in powder form may be distributed in a heterogeneous way, thus, crystallization of undesired phases should occur and block the formation of the desired LGS phase. Presently, we describe a new synthesis way allowing easy preparation of polycrystalline langasite. Two synthesis routes were experimented and compared.

From granulometry analyses (Malvern sizemeter equipment), particle size distributions of initial powders were determined. The three oxide powders were characterized by average particle (or agglomerate) sizes of 78 μm. Silicon dioxide was characterized by a broad distribution of sizes, while lanthanum and gallium oxides presented small size distribution widths.

First route: unique cycle. The three oxides La_2O_3, Ga_2O_3 and SiO_2 (ALDRICH powders 99.99%) were first mixed with stoichiometric composition $La_3Ga_5SiO_{14}$ of Equation (1). Then, the powders were milled in an agate mortar during 3 h. The resulting powder was compacted under a pressure of 4 kbar during 15 min. The obtained pellets were thermally treated according to the following cycle (Fig. 2a) obtained after optimization.

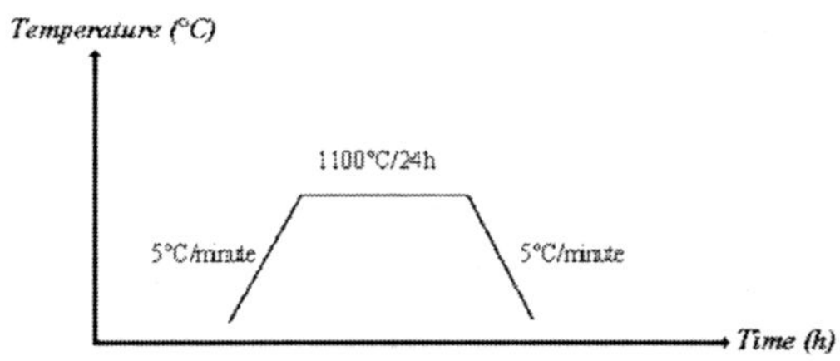

Figure 2a. Thermal cycle for first route.

Second route: new optimized three-steps process. This synthesis route was optimized after a series of experiments. Finally, two preparation steps allowed obtaining the desired LGS phase without any visible residual phase.

Preliminary step: step zero. Each initial oxide in powder form was first separately milled then heated at 600°C during 4 h: this preliminary step insured grain homogeneity, elimination of surface molecules (water, carbonates, adsorbed gases…) and more generally surface regeneration. Then, these oxides were mixed in stoichiometric proportions ($La_3Ga_5SiO_{14}$). The mixing process was operated in a large container during 15 min. The resulting powder was then milled in an agate mortar, during 2 h. Finally, pellets were compacted under a pressure of 4 kbars during 20 min. Each pellet was stocked in a dry box to avoid any hydration or gas adsorption.

Step 1. The obtained pellets (step 0) were thermally treated at 950°C during 5 h. The thermal cycle was conditioned by heating and cooling rates of 5°C/min. After cooling, the pellet was ground in agate mortar, with an additional mass of SiO_2 (5% of initial mass introduced in the initial system). A new pellet was then compacted in usual conditions.

Step 2. This new pellet was heated up to 1,200°C with heating rate of 5°C/min. The sample was heated during 6 h at 1,200°C, then cooled with the same rate and maintained under dry conditions. In fact, it was necessary to add small amounts of the three oxides to compensate mass losses. The best results were obtained by adding SiO_2 amount before the second final step.

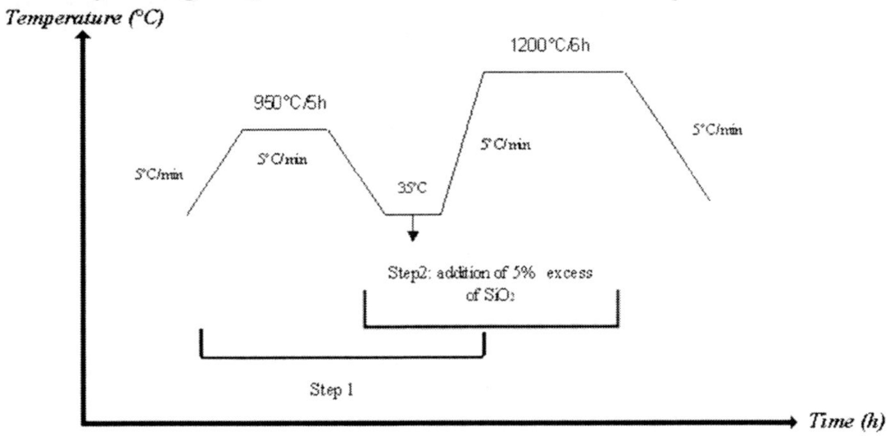

Figure 2b. Thermal cycle for final step of synthesis.

2.2 CHARACTERIZATION TECHNIQUES

X-Ray Diffraction (XRD). X-Ray diffraction patterns were recorded on Siemens-Brucker D5000 equipment working in a classical θ-2θ coupled mode, with copper X-Ray source (λ = 1.5406 Å), Soller slides, a secondary monochromator and a rotating sample holder. The patterns were recorded with a 2θ step of 0.02° and 26 s per step.

Scanning Electron Microscopy (SEM). SEM images were obtained from a Philips XL 30 SEM equipped with an energy dispersive spectrometry (EDS) microanalysis system.

Infrared spectroscopy. Transmittance spectra were recorded at different temperatures using Fourier Transform Infrared Mattson UNICAM spectrometer, in the wavelength range 2.5–25 μm (wavenumbers : ν = 400–4,000 cm^{-1}). The samples are KBr pellets having 1 wt% of La_2O_3, SiO_2, Ga_2O_3, $La_2Ga_3SiO_{14}$ (LGS) phases. Each phase is subjected to a preliminary heating treatment allowing elimination of surface or lattice water, hydroxides and carbonates. In the case of lanthanum phase, this heating process is carried out at 1,000°C during 5 h because of presence of $La_2O_{3-y}(OH)_y$ $(CO_3)_z$ phase in commercial product (Aldrich 99.999%).

3. Results

3.1 X-RAY DIFFRACTION

In Fig. 3 we have reported X-ray diffraction pattern of obtained powder from synthesis 1. We observe the presence of LGS phase which coexists with $LaGaO_3$ phase and residual started oxides.

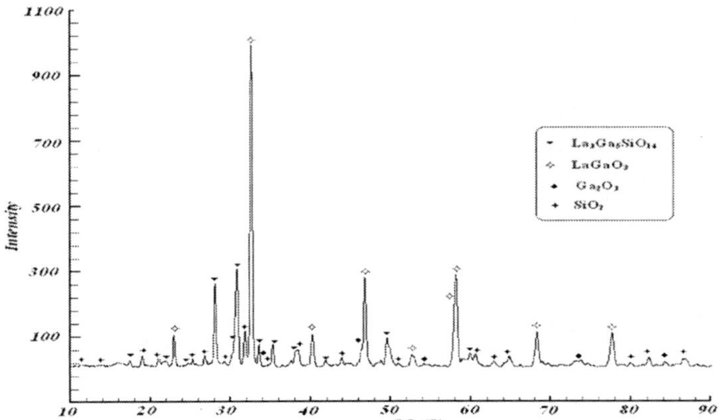

Figure 3. X-ray diffraction patterns from synthesis 1: presence of LGS desired phase, of $LaGaO_3$ phase, SiO_2 and Ga_2O_3 residues.

In Fig. 4a and b we have reported diffraction patterns corresponding to intermediate and final steps corresponding to synthesis 2. In intermediate steps (Fig. 4a) $LaGaO_3$, SiO_2, La_2O_3 and Ga_2O_3 phases coexist with LGS phase.

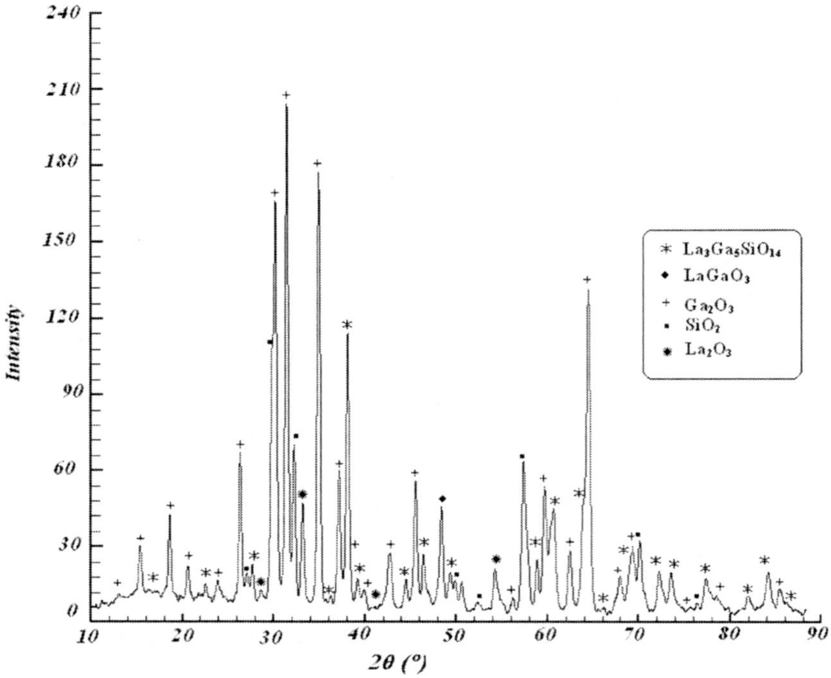

Figure 4a. X-ray diffraction patterns: step 1 of synthesis 2: presence of LGS phase, of $LaGaO_3$ phase, SiO_2 and Ga_2O_3 and La_2O_3 residues.

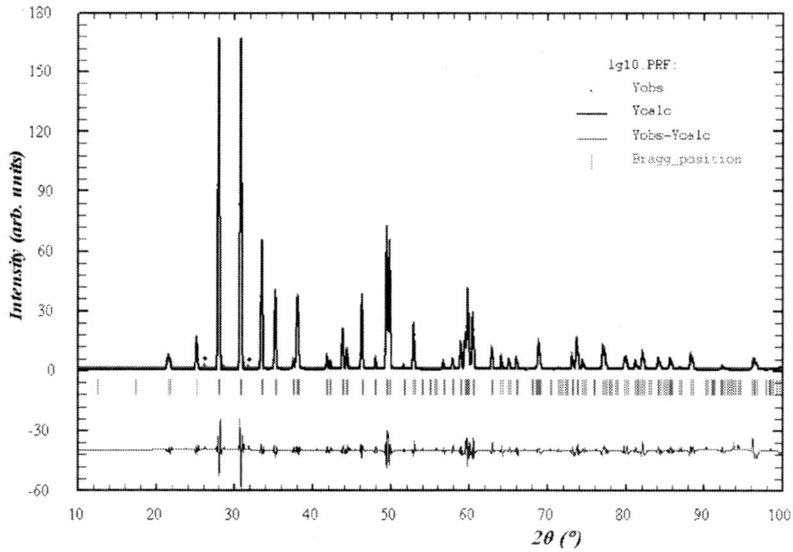

◆ SiO_2 (standard file JCPDS 89–7499)

Figure 4b. X-ray diffraction patterns: Calculated (FullProf) and observed profiles; identification of Bragg peaks.

The diffraction pattern obtained from the final sample (Fig. 4b) was used to determine the various structural parameters: lattice parameters (a and c), atom coordinates (x_i, y_i, z_i) and individual Debye-Waller factors $B_i = (8\pi^2/3) <\Delta X_i^2>$ where the term ΔX_i represents thermal vibrations or static distortions for one atom noted i. To refine these parameters, we used the specific FullProf software[12] that allows calculating X-ray diffraction intensity profiles and fitting these profiles to experimental data. The calculation was performed in the hexagonal setting of the P 321 space group. The refined parameters are listed in Table 1. In this table, the residuals for the weighted pattern R_{wp}, the pattern R_p and the goodness of fit χ^2 are mentioned.

TABLE 1. Calculated structural parameters of LGS.

Atom	Wyckoff Position	x	y	Z	B (Å^2)
La	3e	0.4159 (5)	0	0	1.02 (1)
Ga (1)	1a	0	0	0	1.16 (1)
Ga (2), Si[a]	2d	1/3	2/3	0.4623 (4)	0.87 (1)
Ga (3)	3f	0.8018 (2)	0	½	0.98 (1)
O (1)	2d	1/3	2/3	0.7871 (6)	1.44 (3)
O (2)	6g	0.4649 (4)	0.3483 (4)	0.6160 (4)	1.74 (3)
O (3)	6g	0.2849 (5)	0.0835 (5)	0.2460 (5)	1.95 (4)
a (Å) (10^4 σ)	8.1694 (4)				
c(Å) 10^4 σ)	5.0938 (4)				
Profile parameters: U = 0.08783 (2)			V = – 0.14599 (1)		
W = 0.24874 (8)					
Reliability factors	**R_p (%)**		**R_{wp} (%)**		**χ^2 (%)**
	2.09		2.42		1.52

The resulting refined cell parameters a = 8.1694 ± 0.0004 and c = 5.0938 ± 0.0004 Å are in good agreement with literature results.[13] From this Rietveld analysis, we clearly identified the LGS phase as being the standard LGS phase (JCPDS N° 72 22 49): the atom coordinates; occupation rates and thermal attenuation factors are in good agreement with the structural data of reference[14] obtained from a LGS crystal. The attenuation factors are somewhat greater ($\Delta B = 0.25$ Å^2) than the same factors for LGS crystal: this can be due to the fact that the profile refinement was performed on a polycrystalline sample, with a complex background. The R factors are of a high quality. This detailed analysis allowed us to evaluate the main residual phase: SiO_2 corresponding with file JCPDS 89-74.

Note: σ is standard deviation
Isotropic individual temperature factor $B_i = (8\,\pi^2/3) <\Delta X_i^2>$
$^a = 0.5$ Ga + 0.5 Si

$$R_p = 100 \times \frac{\Sigma|y_i - y_{ci}|}{\Sigma y_i}; \qquad R_{wp} = 100 \times \sqrt{\frac{\Sigma w_i |y_i - y_{ci}|}{\Sigma w_i y_i^2}} \qquad (2)$$

where y_i and y_{ci} are observed and calculated intensities at step i; w_i is the weighting factor.

3.2 SCANNING ELECTRON MICROSCOPY AND X-RAY EMISSION EDS ANALYSES

SEM analyses showed that powder samples were constituted of isolated small grains and of large agglomerates having diversified sizes ranging between 1 and 10 microns. In agglomerates, grain boundaries forming pseudo-hexagonal associations are visible in Fig. 5. Pores are systematically present with significant density and sizes. We also observed that, after sintering LGS ceramics at 1,200°C during 6 h, the observed porosity decreased.

Figure 5. SEM Images of LGS: (a) (× 2,500) 10 µm; (b) (× 10,000) 2 µm.

Local chemical analyses (EDS analyses) were carried out using a standard of langasite (single crystal). In Table 2, we report analyses from LGS single crystal and LGS powder. This gives a good agreement between compositions of standard and studied sample.

TABLE 2. Comparison of EDS Analyses of LGS pellet and commercial LGS single crystal.

	$La_3Ga_5SiO_{14}$	LGS wafer	Polycrystalline LGS
Element	Theo At %	At %	At %
Si (K)	11	16	16
La (L)	33	36	33
Ga (K)	56	48	51
Total	100	100	100

	LGS wafer	Polycrystalline LGS
Element	At %	At %
Si (K)	17	16
La (L)	35	33
Ga (K)	48	51
Total	100	100

3.3 INTERMEDIATE PHASES AND MECHANISM OF LANGASITE SYNTHESIS

To better describe the synthesis way delivering the final langasite phase, we used XRD analyses performed after successive preparation steps. The main steps are described below.

Synthesis 1. (unique cycle, incomplete reaction). This thermal process, operated at 1,100°C, was supported by XRD analyses and can be schematized as follows:

$$3\ La_2O_3 + 5\ Ga_2O_3 + 2\ SiO_2 \rightarrow (2-x)\ La_3Ga_5SiO_{14} + 3x\ \mathbf{LaGaO_3} + x\ Ga_2O_3 + 2\ SiO_2 \quad (3)$$

The term x is not a fixed value and can depend on all experimental conditions (temperature, powder granulometry, heterogeneity ...). This step may be the result of a fast reaction occurring between two types of individual La_2O_3 and Ga_2O_3 grains, directly in contact, while the reaction giving rise to the three-component LGS phase should be slower. An other origin will be the vaporization of Ga_2O_3 phase and migration of liquid-like phase SiO_2. This process was abandoned.

Synthesis 2. In synthesis 2, we can define two main steps.

Step 1. The temperature was fixed at 950°C and sample was heated during 5 h; in these conditions, we limited evaporation of some melt components and migrations of phases (Ga_2O_3 and SiO_2).

$$3\ La_2O_3 + 5\ Ga_2O_3 + 2\ SiO_2 \rightarrow La_3Ga_5SiO_{14} + 3LaGaO_3 + Ga_2O_3 + 2\ SiO_2 \quad (4)$$

In fact, we can express the intermediate reaction as resulting of initial fast reaction between lanthanum and gallium oxides:

$$\{\mathbf{3(La_2O_3 + Ga_2O_3)}\} + 2\,Ga_2O_3 + 2SiO_2 \rightarrow \mathbf{6\ LaGaO_3} + \{2\,Ga_2O_3 + 2SiO_2\} \quad (5)$$

So, in this step 1, we define a first reaction giving LGS and residues:

$$\{\mathbf{5\ LaGaO_3 + SiO_2}\} + LaGaO_3 + 2\,Ga_2O_3 + SiO_2 \rightarrow \{\mathbf{La_3\ Ga_5\ Si\ O_{14}} + La_2O_3\} + 2\,Ga_2O_3 + SiO_2 + LaGaO_3 \quad (6)$$

Step 2. In step 2, as temperature reaches higher values of 1,200°C (heating time: 6 h), a loss of SiO_2 is systematically observed, and we observed that adding some amount of SiO_2 (deigned through **s** coefficient in equation just below) could improve the final synthesis of a unique LGS phase (corresponding with file JCPDS N° 72 22 49):

$$\{\mathbf{La_3\ Ga_5\ Si\ O_{14}} + La_2O_3\} + 2Ga_2O_3 + LaGaO_3 + \mathbf{(1\text{-}s)\ SiO_2} + \{\mathbf{s\ additional\ moles\ of\ SiO_2}\} \rightarrow 2\ La_3Ga_5SiO_{14} \quad (7)$$

3.4 FOURIER TRANSFORM INFRARED SPECTROSCOPY ANALYSES

FTIR spectroscopy analyses as presented in figure 6 were carried out on LGS powder and thermally treated powders of SiO_2, La_2O_3 and Ga_2O_3 used as standards. In Table 3, the observed vibrational bands are listed and correlations are proposed by direct comparison between standards and LGS spectra. Let us recall that infrared absorption bands can be observed only for antisymmetric molecular vibration modes: in these modes, polar moments vary and absorption intensity increases with polarization amplitude variation.

We also compared the present results with Raman scattering experiments published by authors.[15] Let us recall that Raman active modes only depend on tensor of polarizability and that symmetric and antisymmetric vibration modes can be active in this technique. In their study, the authors presented a detailed analysis of vibrational modes and force constants. It is interesting to note that, as the higher Raman wavenumber is found at $\nu = 866\ cm^{-1}$, the higher IR wave-number is observed at a higher value of 1,099 cm^{-1}. This high frequency band (952–1,099 cm^{-1}) is a doublet having large intensity. Compared with the SiO_2 band of Table 3, this doublet can be ascribed to La-O (952 cm^{-1}) and Si-O (1,099 cm^{-1}) stretching vibration modes. As the 866 cm^{-1} Raman frequency was ascribed to a symmetric stretching mode of $[SiO_4]$ entity, we should admit that the 1,099 cm^{-1} IR mode could be an IR active antisymmetric stretching mode of the same entity. Other Si-O, La-O and Ga-O vibration bands can be clearly identified.

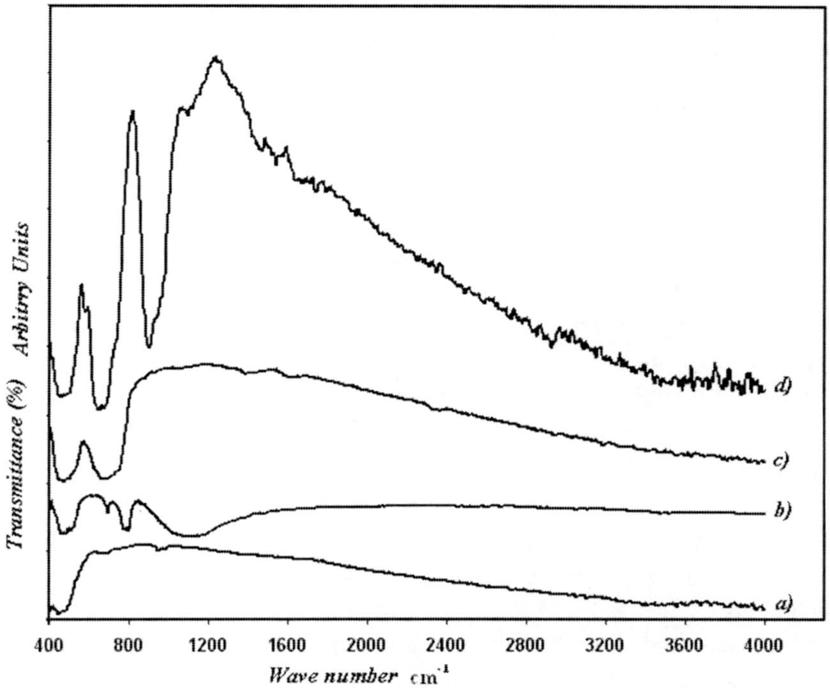

Figure 6. FTIR spectra of synthesized LGS, SiO_2, La_2O_3 and Ga_2O_3 phases: (a) = La_2O_3; (b) = SiO_2; (c) = Ga_2O_3; (d) = LGS.

TABLE 3. FTIR analyses: comparative study of SiO_2, La_2O_3, Ga_2O_3 and LGS phases. Samples in KBr pellets.

SiO_2	La_2O_3	Ga_2O_3	Raman wavenumbers for LGS crystal (literature results)[17]	LGS powder Present study	Vibration mode
500–550	450–500	454–497.5	450–522	480–525	O-La-O[16] O-Ga-O
695.8	650–680	671–757.8	640	650–700	La-O[16]
			712		Si-O
			755		Ga-O [17]
782.5–807				800	Si-O[16]
	954		866	952	La-O[16]
1060–1180				1099	Si-O Stretching

In their analysis of substitution effects in piezoelectric properties of LGS systems, authors[14] clearly showed that piezoelectric responses (piezoelectric modulus d_{11}) should be strongly related to ionic radius of atoms in the lanthanum (La^{3+}) crystallographic site. The authors claimed that these modifications in piezoelectric properties should be due to space modification of this site in a direction [100] of the hexagonal structure.

It is interesting to note that, in La_2O_3 IR spectrum, the vibration bands present very weak intensities. This might be ascribed to a high symmetry of this structure and a low polarization amplitude. In LGS spectrum, the associate bands present large intensities (650 and 952 cm^{-1}): this might be ascribed to a large variation of polar moment linked to La-O chemical bonds in LGS structure. This should be strongly related to the interesting piezoelectric properties of LGS.

4. Conclusion

Langasite was successfully prepared via a specific solid state route involving thermal cycles with maximal temperature of 1,200°C. The final thermally treated LGS phase was subjected to a Rietveld refinement giving structural parameters in full agreement with literature data. The ceramics phase was clearly identified from XRD analyses as being the phase giving rise to high performance piezoelectric responses. Successions of chemical steps are proposed to interpret the final result obtained at relatively low temperature. FTIR analyses allow attribution of IR vibrational modes. Large IR absorption bands are associated with polarized La-O bonds.

References

1. A.A. Kaminsky and S.E. Sarkissov. *Reports of the U.S.S.R, Academy of Science* 264(1):93–95(1982).
2. A.A. Kaminsky and I.M. Silvestrova. *Izv. of the U.S.S.R. Academy of Science Physics Series* 47(10):1903–1909(1983).
3. M. Schulz and H. Fritze. *Renewable Energy* 33:336–341(2008).
4. D. Damjanovic. *Solid State & Materials Science* 3:469–473(1998).
5. J.A. Thiele and M. Pereira da Cunha. *Sensors and Actuators B* 113:816–822(2006).
6. S. Huankiat, H.L. Tuller and H. Fritze. *Sensors and Actuators B* 93:169–174(2003).
7. C. Klemenz. *Journal of Crystal Growth* 237–239:714–719(2002).
8. J. Kskela, S. Lehtonen, V.P. Plessky and M.M. Salomaa. *Applied Physics Letters* 72:2665–2667(1998).
9. H. Fritze, D. Richter and H.L. Tuller. *Sensors and Actuators B* 111–112:200–206(2005).
10. H. Fritze, H.L. Tuller, H. She and G. Borchardt. *Sensors and Actuators B* 76:103–107(2001).
11. S.-Qi Wang and U. Satoshi. *Journal of Crystal Growth* 250:463–470(2003).

12. J.R. Carvajal and T. Roisnel. WinPLOTR, a graphic Tool For Powder Diffraction, Laboratoire Léon Brillouin (CEA/CNRS) 91191 Gif sur Yvette CEDEX, France. www-llb.cea.fr/fullweb/fp2k/fp2k.htm.
13. H. Zhang, N.B. Singh, A. Berghmans, J.D. Adam, S. Tidrow and C. Fazi. *Journal of Crystal Growth* 234:660–665(2002).
14. T. Iwataki, H. Ohsato, K. Tanaka, H. Morikoshi, J. Sato and K. Kawasaki. *Journal of the European Ceramic Society* 21:1409–1412(2001).
15. G.W. Lu, C.X. Li, W.C. Wang, Z.H. Wang, Z.M. Wang, D.R. Yuan and H.R. Xia. *Physical State Solid (b)* 241(2):439–446(2004).
16. F. Ruiz, J.R. Martínez and J. González-Hernández. *Journal of Molecular Structure* 641:243–250(2002).
17. M. Nieminen. *Inorganic Chemistry Publication Series*, Helsinki University of Technology, No.1(2001).

DIELECTRIC ANOMALIES AND RELAXATION BEHAVIOR IN HYDROTHERMALLY PROCESSED PLZT FERROELECTRIC CERAMICS

TAJEDINE LAMCHARFI[1,2], NOUR-SAID ECHATOUI[1,2], SALAHEDDINE SAYOURI[1], DAOUD MEZZANE[3], L. HAJJI[3] AND LAHCEN ELAMMARI[4]
[1]*LPTA, Département de Physique, Faculté des Sciences-DM, B.P. 1796, Fès-Atlas, Morocco*
[2]*Département Génie Electrique, FST, Route d'Immouzer, Fès, Morocco*
[3]*Faculté des Sciences et Techniques Guéliz, Marrakech*
[4]*Département de Chimie, Faculté des Sciences Agdal, Av. Ibn Battouta, B.P. 1014, Rabat, Morocco*

Abstract. $Pb_{1-y}La_y(Zr_{0.52}Ti_{0.48})O_3$ (PLZTy) powders where y = 0, 0.015, 0.03, 0.06, 0.08, 0.10, 0.12, 0.15 and 0.20 were prepared using the hydrothermal process and their structural and dielectric properties investigated. Increasing La content is shown to enhance crystallization of the raw samples and to transform the average symmetry to tetragonal one in the calcined ones. Two anomalies are observed on the real part of the permittivity on both undoped and doped samples, at relatively high temperatures (~180°C; ~260°C). The anomaly located at about 180°C was interpreted as a transition from ferroelectric-rhombohedral phase to ferroelectric-quadratic phase. A polynomial law was used to fit the thermal behavior of the permittivity.

Keywords: Smart materials, PLZT, dielectric anomalies, relaxation, MPB

1. Introduction

Lead titane ($PbTiO_3$; PT), a ferroelectric material, is known for its interesting properties; high Curie temperature, pyroelectric coefficient and spontaneous polarization, and low dielectric constant. These properties make it suitable for numerous applications: ultrason transducers,[1] thermistors, optical electronic devices and satellite detection,[2] etc. At room temperature, PT has a tetragonal perovskite structure,[3] and combined with other oxides it forms materials such as (Pb, La)TiO_3, (PLT), Pb (Zr, Ti)O_3, (PZT), and (Pb, La)(Zr,Ti)O_3, (PLZT) whose

I. A. Luk'yanchuk and D. Mezzane (eds.), *Smart Materials for Energy, Communications and Security.*

very wide range of applications is well known: nonvolatile random access memory (NVRAM) and dynamic random access memory (DRAM) devices,[4] infrared sensors,[5] etc. In fact, these materials belong to what is called Smart materials, materials (and structures) which may give information about their environment to an observer or monitoring device, especially those exhibit-ing two or more ferroic features such as ferromagnetism/magnetostriction, ferroelectricity/piezoelectricity or ferroelasticity/shape memory effects due to their unusual responses, which are benefiting of a growing interest due to the possibility of their applications and integrations into industrial systems including civilian, industrial, medical and military applications. PLT ceramics belong to the family of ferroelectric relaxors: frequency dispersion of the dielectric permittivity, ε_r, and a large maximum of ε_r at a temperature, T_m, which is shifted towards high values with increasing frequency; the corresponding phase transition is called diffuse phase transition.

Both experimental and theoretical works have been devoted to the study of the relaxor behavior in ferroelectric materials. From theoretical point of view, different approaches and models have been established to clarify the mechanisms involved in the normal-to-relaxor behavior.[6–11] Different experimental methods have also been used to elucidate the origin of relaxor ferroelectricity.[12–17] Structural investigation of the ferroelectric phase transition in La-doped lead titanate (PLT) revealed that for La contents in excess of 5 at%, local polarization persists until temperatures significantly above that of the average phase transformation.[18] Moreover, transmission electron microscopy (TEM) studies showed the presence of sub-domain modulation along the mechanically soft tetragonal c axis.[18] As the dopant content was further increased between 5 and 20 at%, the degree of modulation increased suggesting a relaxation of the lattice. At a La-content of 25 at%, the ferroelectric domain structure was found difficult to define, rather the bright-field image consisted of a cross-hatched or "tweed" microstructure; the latter has also been observed in the incipient relaxor state of rhombohedral[19] and tetragonal PLZT materials.[20] Besides, it has been suggested that La modification of PT gives rise to the development of a spatially varying polarization within the normal microsized ferroelectric domain structure.[18] These structural changes appear to be directly related to the nature of the phase transition, the latter transforms from a conventional first order to a diffuse transition as the La content is increased between 5 and 10 at%.[21] As the La content is increased to above 20 at%, an extreme broadening of the permittivity has been reported.

Ferroelectric lead zirconate titanate $Pb(Zr_{1-x}Ti_x)O_3$ (PZT) are known to exhibit high piezoelectric and pyroelectric properties.[22–24] These materials are also known for their unusual phase boundary, called the morphotropic phase boundary (MPB),[25] which occurs at around 50% Zr substitution for Ti in $PbTiO_3$, and which separates two structural phases rhombohedral (Zr-rich region) and tetragonal (Ti-rich region) structures; in particular, a very high piezoelectric response is recorded in the MPB. The occurrence of the MPB depends mainly on the procedure of preparation of the samples and the grain size.[26,27] Substitution of Lanthanum for Lead (PLZT) in these materials changes their macroscopic properties from normal ferroelectric to relaxor ferroelectric types; relaxation phenolmenon affects deeply PLZT properties (anomalies). In particular, high values of dielectric permittivity and electromechanical and electrooptic co-efficients are recorded, which make these materials suitable for various applications such as capacitors, optoelectronic modulators....[28–30] These anomalies manifest themselves within a broad temperature region around T_m; the phase transition is called 'diffuse phase transition' (DPT). Several models and approaches have been developed to interpret this DPT involving different mechanisms: chemical heterogeneities, superparaelectric behaving due to mesoscopic heterogeneities....[31–35] However, details concerning the physical process of the DPT remains not completely clear. As mentioned above, a maximum response of different characteristics, among them the electromechanical coupling coefficient, is obtained due to the coexistence of ferroelectric tetragonal and ferroelectric rhombohedral phases near the MPB. Based on this coexistence, different phase diagrams have been proposed.[36,37] Moreover, this enhancement of the electromechanical response is suggested to be linked with the instability in the vicinity of the rhombohedral to tetragonal transition.[37]

Low temperature dielectric measurements have revealed the presence of two anomalies in the real part of the dielectric constant of PZT materials near the MPB.[38] The occurrence of the anomaly corresponding to the lowest temperature was imputed to a tetragonal to monoclinic phase transition which was supported by XRD studies of Noheda et al.[39] The presence of such anomaly has also been reported in dielectric studies of PZT materials near the MPB.[40] With the aim of studying the effect of La addition on the structural and dielectric properties of the morphotropic phase boundary composition $Pb(Zr_{0.52}Ti_{0.18})O_3$ (PZT), we have prepared, using the hydrothermal process, $(Pb_{1-y}La_y)(Zr_{0.52}Ti_{0.48})O_3$ (PLZTy) powders, where y = 0, 0.015, 0.03, 0.06, 0.08, 0.10, 0.12, 0.15 and 0.20. Indeed, addition of La (La belongs to the so-called soft-dopants additives) is known to improve permittivity and piezoelectric coupling coefficient of PZT materials.[41] Moreover, hydrothermal processing, contrary to other methods of preparation, in particular the solid-state reaction method, permits a good control of composition and morphology of the powder, does not necessitate any milling and calcinations operations and has lower agglomeration effects (the latter having dramatic effects on the consolidation and densification behaviour of the powder).[42,43] The results obtained from our studies were analysed and compared to those of the literature.

Variations of the permittivity with temperature are usually approached by the following law[44]:

$$\frac{\varepsilon'_{max}}{\varepsilon} = 1 + \left[\frac{(T - T_m)^{\gamma}}{2\delta^2}\right] \tag{1}$$

where γ is a parameter such as $\gamma = 1$ in the case of a Curie Weiss behaviour, and $\gamma = 2$ in the case of a relaxor behaviour. δ represents the diffuse character of the transition. We have fitted the results obtained from our study with the help of Equation (1) where we have changed the factor δ^2 by δ^{γ} to preserve the dimensionless character of the term $\left[\frac{(T - T_m)^{\gamma}}{2\delta^2}\right]$ appearing in this equation, so Equation (2) becomes:

$$\frac{\varepsilon'_{max}}{\varepsilon} = 1 + \left[\frac{(T - T_m)^{\gamma}}{2\delta^{\gamma}}\right] \tag{2}$$

Moreover, we have used the following polynomial law to study thermal behavior of the permittivity:

$$\frac{\varepsilon_m}{\varepsilon} = 1 + \frac{(T - T_m)}{B_1} + \frac{(T - T_m)^2}{B_2} \tag{3}$$

and have in particular observed that the coefficient B_1 appearing in this expression may indicate a critical value of the percentage of the dopant element for which the relaxation phenomenon may take place.

2. Experiment

The procedure of preparation of the samples, or two-steps procedure, was conducted in conformity with the flow diagram of Fig. 1. In the first step, aqueous pH was raised to 11 by adding 2M KOH, was put into an autoclave (Filling ratio superior to 80%) and heated at 200°C during 12 h. The resultant PLZT suspension was filtered and the remaining product was washed with distilled water for several times and finally dried at 70°C in an oven. Annealing of the powders, at different temperatures, was done in air solutions of $ZrOCl_2$, $8H_2O$ 0.1M and $TiOCl_2$-xHCL (Ti: 15%) were mixed, with stirring, and 2M KOH was added until precipitation to obtain a precipitate (ZTO) which was then washed with distilled water for several times to obtain its neutrality. ZTO was put into an aqueous solution of 0.2M lead acetate and 0.5M lanthanum acetate. Thereafter, the solution obtained, which pH was raised to 11 by adding

2M KOH, was put into an autoclave (Filling ratio superior to 80%) and heated at 200°C during 12 hours. The resultant PLZT suspension was filtered and the remaining product was washed with distilled water for several times and finally dried at 70°C in an oven. Annealing of the powders, at different temperatures, was done in air. For dielectric measurements, PLZT powders were pressed into discs (diameter ~13 mm, ~1 mm thick) using poly (vinyl alcohol) as a binder.

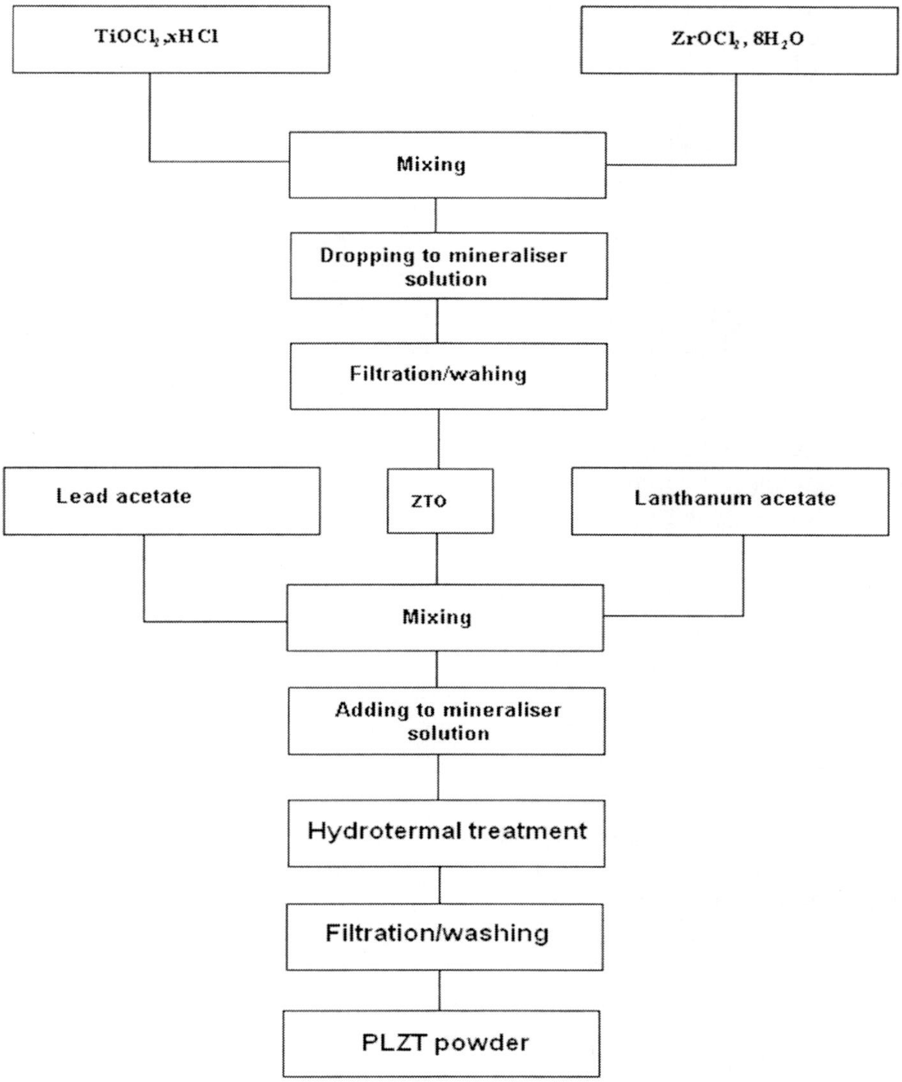

Figure 1. The flow chart of the procedure of fabrication of PLZT powders using the hydrothermal technique.

Dielectric constant (ε_r) and dissipation factor (tanδ) were measured using an impedance analyzer (LCR) (HP, Model 4284 A) in the frequency range 20 Hz–1 MHz and for temperatures ranging from 30°C to 450°C.

3. Results and discussion

Figures 2 and 3 display XRD spectra of PLZTy samples, with y = 0, 0.015, 0.03, and 0.10, in their raw state and calcined at 700°C, respectively. Figure 3a shows the coexistence of rhomohedral and tetragonal phases which is consistent with the fact that the compositions are near the MPB. Moreover, these two figures clearly show the influence of addition of La on the structure of the samples prepared under the same conditions. Indeed, increasing La content seems to favor crystallization of the raw powders (Fig. 2) and tends to transform the average symmetry to tetragonal one (Fig. 3) as indicated in the PLZT phase diagram.[25] Besides, no secondary phases (pyrochlore) have been detected. Dielectric measurements performed on samples annealed at 850°C, 1,100°C and 1,200°C during 2 h led to the curves given on Figs. 4, 5 and 6, respectively. In regard to these figures, the following observations can be made.

(i) The transition temperature from ferroelectric (FE) phase to paraelectric (PE) phase, corresponding to the maximum of the permittivity, ε_r (Fig. 4), of the undoped sample (PLZT0) is in agreement with the FE-PE phase boundary in the PZT phase diagram proposed by Jaffe et al.[25]

(ii) Increasing the annealing temperature, T_a, of the samples ameliorates ε_r and gives a slight decrease of the temperature, T_m, of the maximum of ε_r (Figs. 5 and 6) reaching low values for the sample with y = 0.12 (Fig. 6d) and for T_a = 1,200°C (T_m inferior to 50°C, Table 1). However, annealing of the sample with y = 0.12 at 1,200°C during 4 h led to a deterioration of ε_r (Table 1) which points out the role played by both annealing temperature and the duration, t_a, of the annealing process.

(iii) A relaxation phenomenon takes place for y = 0.10, accompanied with a broad temperature region around T_m, together with a decrease of T_m (Figs. 5 and 6). This phenomenon is more pronounced in the samples heated at 1,200°C (Fig. 6).

(iv) ε_r increases with increasing La reaches a maximum for y = 0.12 and then decreases (Table 1). Gupta et al.[45] obtained a maximum of ε_r at y = 0.02, however relaxation occurs at a concentration (y = 0.08) close to ours.

(v) Weak values of dielectric losses (tanδ) (Figs. 7a, b, c) are recorded and the latter show anomalies at the same temperatures as for the permittivity. The increase of tanδ above T_m traduces the conducting character of the sample.

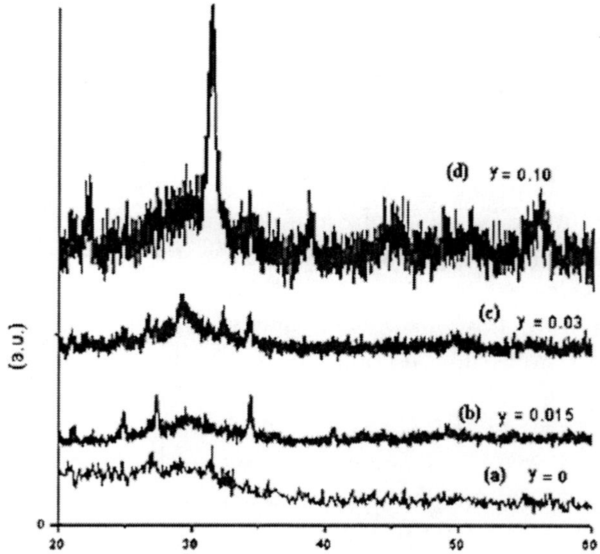

Figure 2. XRD spectra of PLZTy raw samples.

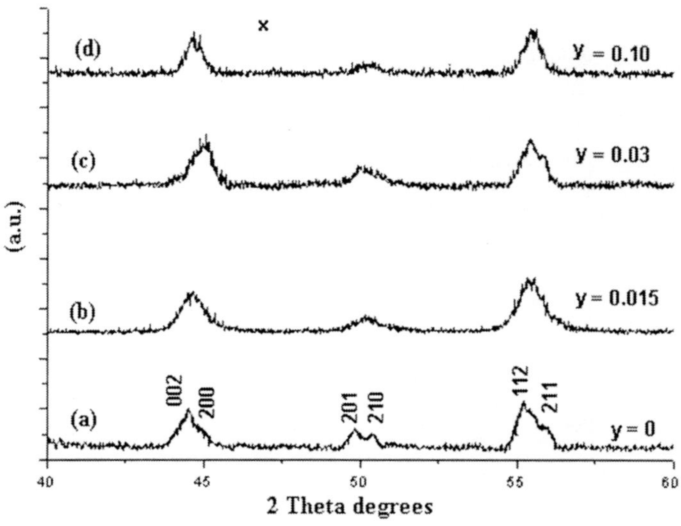

Figure 3. XRD patterns of PLZTy samples calcined at 700°C during 4 h showing the gradual transformation from the MPB phase to quadratic phase.

(vi) Two anomalies are observed at about 180°C and 260°C on PLZT0 (Fig. 4) and PLZT0.03 (Fig. 5a) heated at 850°C and 1,100°C, respectively. The anomaly at 260°C is not observed on Fig. 6a corresponding to the sample (PLZT0.03) heated at 1,200°C. The existence of such anomalies has been reported in the literature; however, other studies have not revealed them.[45] Ragini et al.[38] have observed two anomalies at low temperatures (197 and

253 K) on PZT samples ($PbZr_xTi_{1-x}O_3$, x = 0.52 and x = 0.515), which they imputed, based on the work of Noheda et al.,[46] to a transition from tetragonal to monoclinic phase for the first anomaly (located at about 197 K) and to a cell-doubling transition[46] for the second anomaly (located at about 253 K). Sheen and Kim[40] have also observed an anomaly on PZT thin films which they interpreted as a transition from .tetragonal to monoclinic symmetry with decreasing temperature ($PbZr_{1-x}Ti_xO_3$, x = 0.46, 0.48, 0.49, where the anomaly occurs at about 490, 325, and 235 K, respectively). Our results reveal the existence of anomalies at relatively high temperatures in comparison with those mentioned above. Bouzid studied the effect of Niobium and Potassium doping on the anelastic behavior and on the dielectric response of Pb ($Zr_{0.54}Ti_{0.46}$)O_3 (PZT54/46) samples prepared using the conventional mixed-oxide method.[47,48] From dielectric reasurements, Bouzid observed two anomalies, at relatively high temperatures, on undoped PZT54/46 samples and on those doped with Nb and K. These two anomalies were interpreted as being due to a rhombohedral to quadratic phase transition for that located at a temperature of about 180°C (undoped sample), and to a quadratic to cubic phase transition for that occurring at about 370°C (undoped sample). This interpretation was based on the correlation made between mechanical losses and dielectric losses.[47,48] Indeed, anelastic behavior (mechanical losses and elastic modulus) studies showed the existence, for the undoped and doped samples, of two minimums of the elastic modulus,[48] which indicates the occurrence of two phase transitions[48]: rhombohedral to quadratic phase and quadratic to cubic phase. Moreover, these observations are supported by the work of Gerthsen et al.[49] pointing out the correlation existing between mechanical and dielectric losses (Fig. 8).

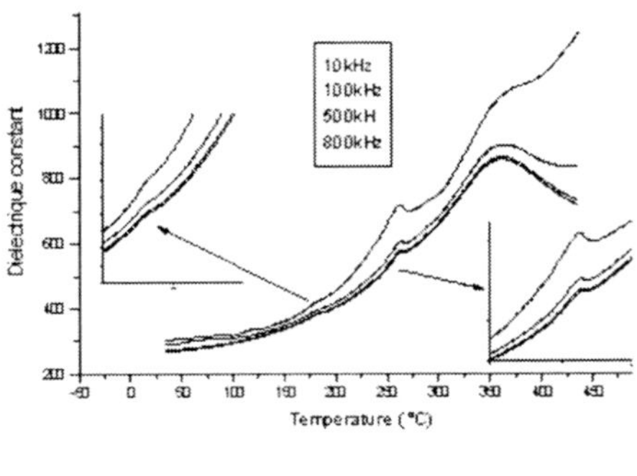

Figure 4. Dielectric constant of PLZT0 heat treated at 850°C during 2 h. Arrows indicate in more enlarged scale the two anomalies.

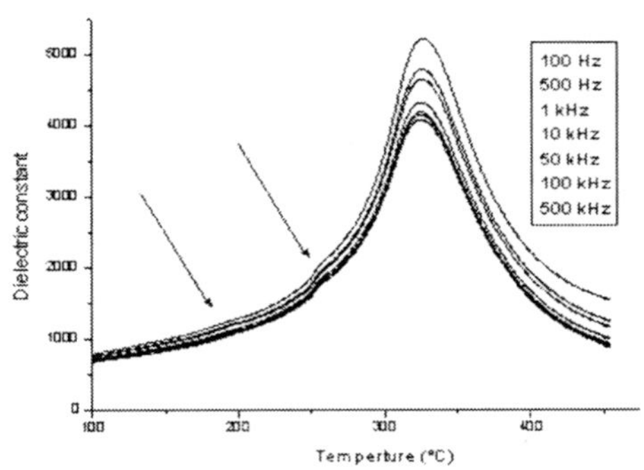

Figure 5a. Dielectric constant of PLZT0.03 heated at 1,100°C during 2 h. Arrows indicate the anomalies located at about 180°C and 260°C.

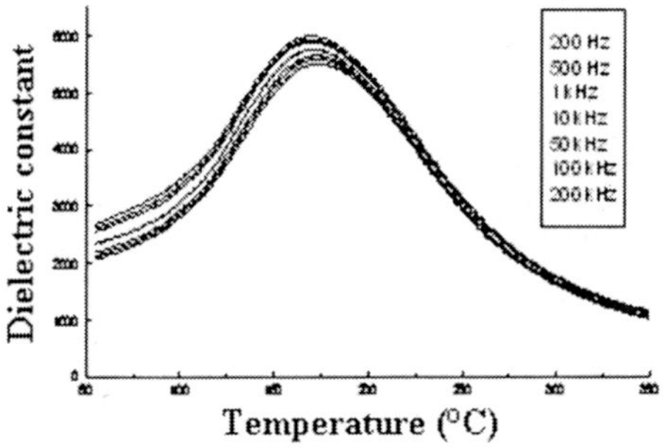

Figure 5b. Dielectric constant of PLZT0.10 heated at 1,100°C during 2 h: broad maximum and shift of the temperature of the maximum to higher values.

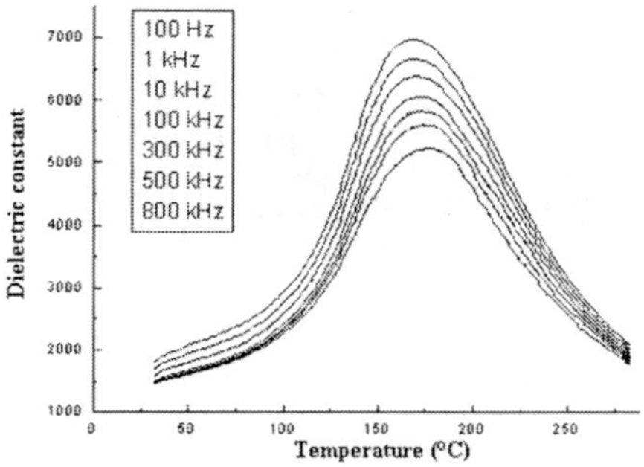

Figure 5c. Dielectric constant of PLZT0.12 heated at 1,100°C during 2 h.

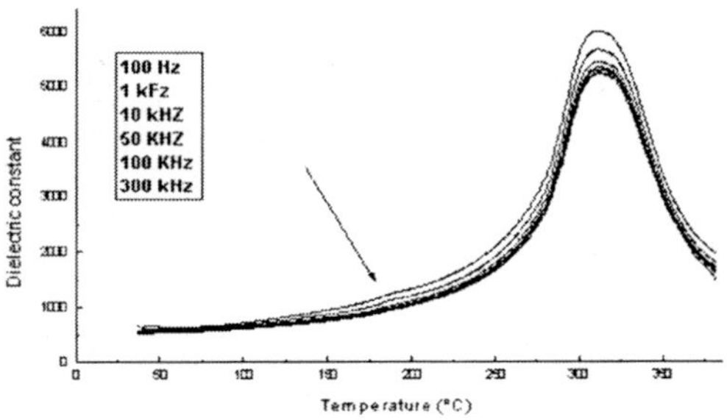

Figure 6a. Dielectric constant of PLZT0.03 heat treated at 1,200°C during 2 h. The arrow shows the anomaly located at about 180°C.

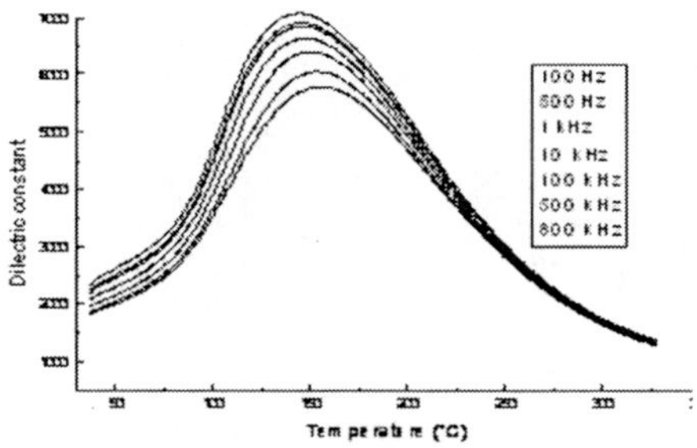

Figure 6b. Dielectric constant of the sample PLZT0.10 heated at 1,200°C during 2 h: more pronounced relaxation effects than the same composition heated at 1,100°C (Fig. 5b).

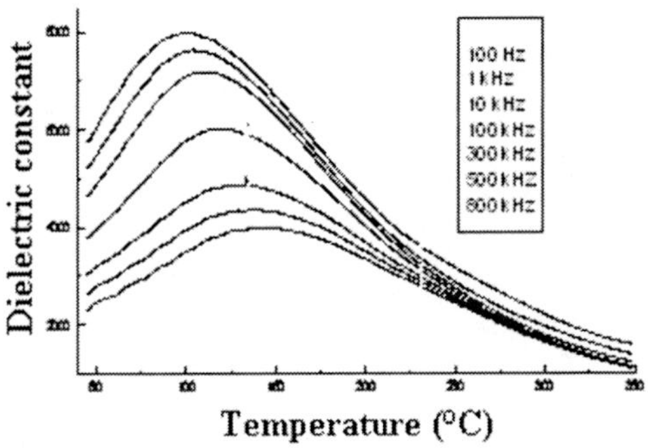

Figure 6c. Dielectric constant of the sample PLZT0.12 heated at 1,200°C during 2 h: broad maximum and strong dispersion with increasing frequency.

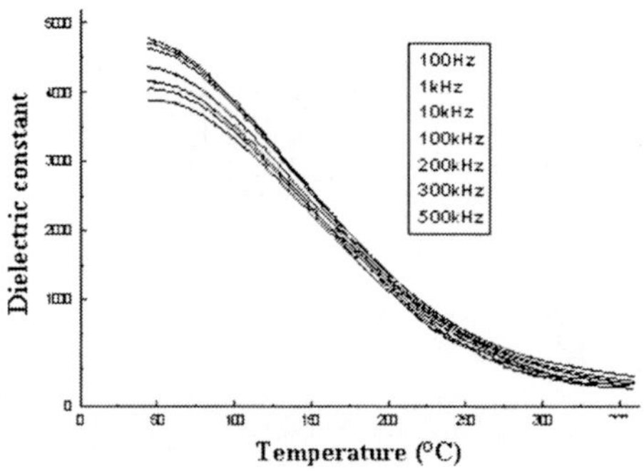

Figure 6d. Dielectric constant of PLZT0.20 heated at 1,200°C during 2 h.

TABLE 1. Maximum of the dielectric constant and the corresponding temperature, T_m, for PLZTy samples.

Y		0.03	0.10	0.12	0.20
1,100°C 2 h	$(\varepsilon)_{max}$	5208	6000	6976	2629
	T_m (°C)	327	169	168	93
1,200°C 2 h	$(\varepsilon)_{max}$	6057	7098	7989	4813
	T_m (°C)	311	144	100	< 50
1,200°C 4 h	$(\varepsilon)_{max}$	–	–	3739	–
	T_m (°C)	–	–	108	–

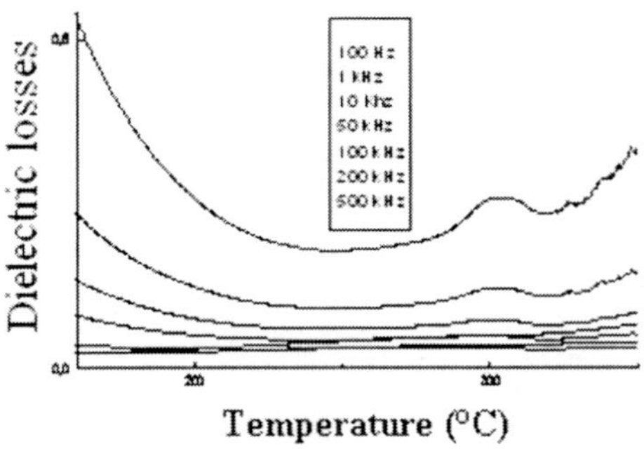

Figure 7a. Dielectric losses recorded for PLZT0.03 heated at 1,200°C during 2 h.

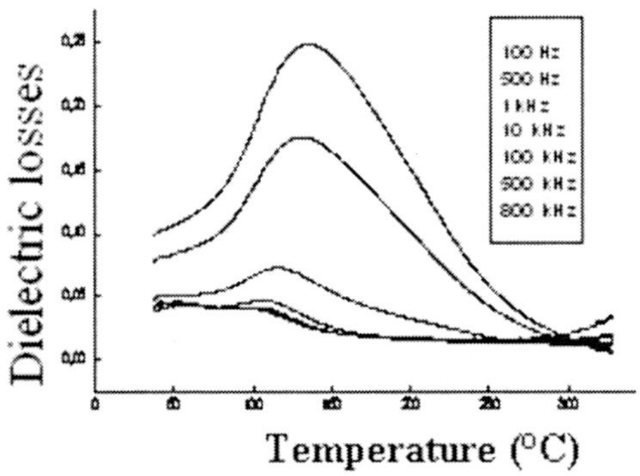

Figure 7b. Dielectric losses of PLZT0.10 heated at 1,200°C during 2 h.

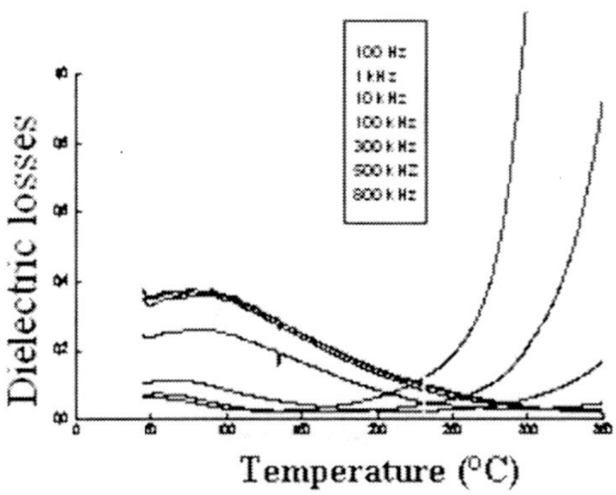

Figure 7c. Dielectric losses of the sample PLZT0.12 heated at 1,200°C during 2 h.

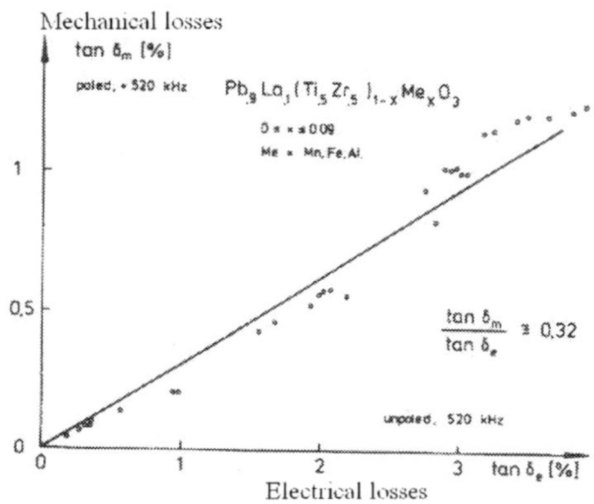

Figure 8. Correlation between mechanical losses and electrical losses in PLZT compounds.[27]

Concerning the two anomalies we have observed from dielectric studies of our samples, one may infer from comparison of our results to those mentioned above that the two anomalies located at about 180°C and 370°C are due to a transition from ferroelectric-rhombohedral phase to ferroelectric-quadratic phase and to a ferroelectric-quadratic phase to a paraelectric-cubic phase, respectively; it can be noticed that the former has not been reported in other works on PZT ceramics prepared by other methods[44]. The anomaly located at about 260°C, which disappears in the sample heated at 1,200°C, may be due to an "intrinsic" phenomenon and needs more investigation to clarify its origin. We can noticed that in a recent work performed on sol gel processed (Pb, La)($Zr_{0.52}Ti_{0.48}$)O_3 ceramics,[50] we have observed a dielectric anomaly located at about the same temperature of 180°C.

To get more information about the establishment of relaxation and diffuseness characters, we have studied the thermal variation of the permittivity by using two laws to fit this thermal behavior. Table 2 gathers values of the two parameters γ and δ, deduced with the help of Equation (2). It is clear from these values that relaxation behavior ($\gamma = 2$) takes place for y = 0.10, and that the diffuse character (indicated by values of δ) increases until a concentration in La equal to 0.08, shows a slight decrease at the occurrence of the relaxation effect and then increases. Figure 9 displays the variation of $\varepsilon_m/\varepsilon$ as a function of $(T - T_m)^\gamma$, and a clear distinction between normal behavior ($\gamma < 2$) and relaxor one ($\gamma = 2$) is indicated by the slopes of the corresponding lines. We have obtained similar results, related to the appearance of the relaxation phenomenon, on the same materials as in this study prepared using the sol gel process;[28] indeed, relaxation clearly appears for y = 0.10, and the permittivity behaves in the same manner as in the present case.

TABLE 2. Values of the two parameters γ and δ deduced from fitting to Equation (2).

Samples	δ(K)	γ
PZT	20.8	1.7
PLZT y = 0.03	37	1.8
PLZT y = 0.06	63	1.6
PLZT y = 0.08	63.8	1.8
PLZT y = 0.10	58.5	2
PLZT y = 0.12	53.6	2
PLZT y = 0.15	68	2
PLZT y = 0.20	85	2

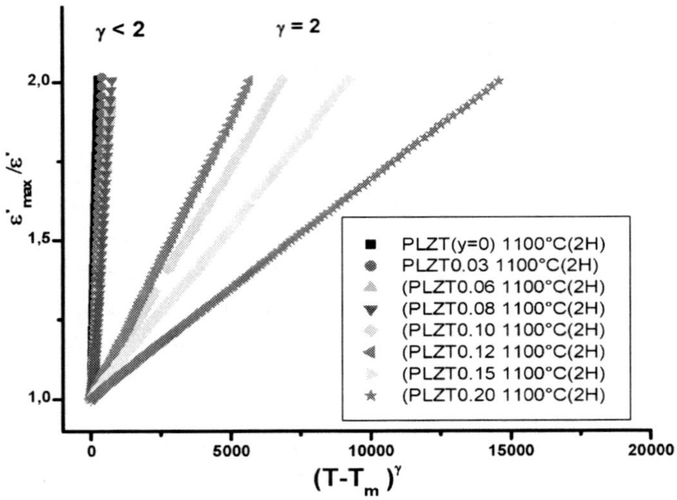

Figure 9. Variations of $\varepsilon_m/\varepsilon$, as a function of $(T-T_m)^\gamma$, for different concentrations in La.

We have used relation (3) to simulate the behavior of the permittivity. Excellent fits are obtained with Equation (3) as shown on Fig. 10, representing the variation of $\varepsilon_m/\varepsilon$ as a function of $(T - T_m)$. Figure 11 shows the behavior of the two coefficients B_1 and B_2 as a function of La concentration, the former reaching a maximum at $y = 0.10$ and then may be an indicator of the occurrence of the relaxation phenomenon. The parameter B_2 increases and may indicate the evolution of the diffuse character of the permittivity.

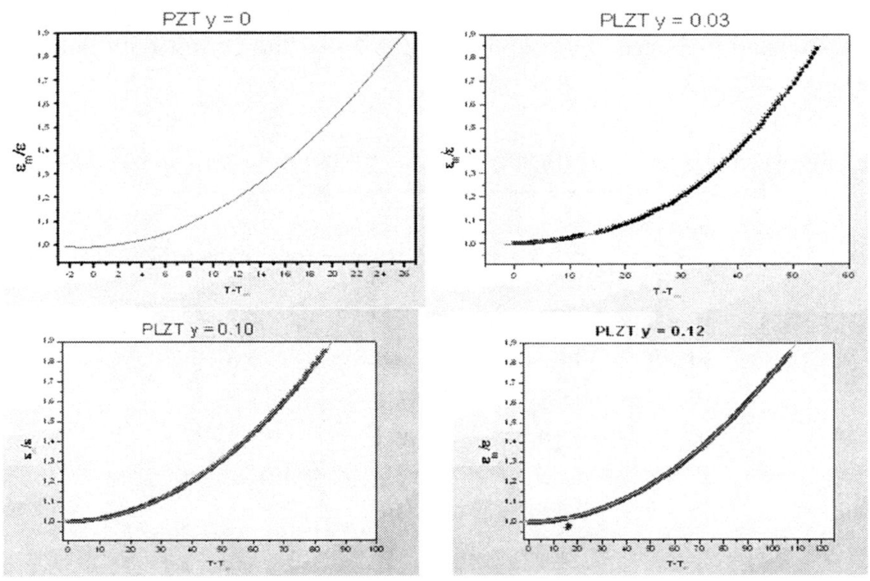

Figure 10. Fit of $\varepsilon_m/\varepsilon$, as a function of $(T-T_m)$, to the polynomial law given by Equation (3).

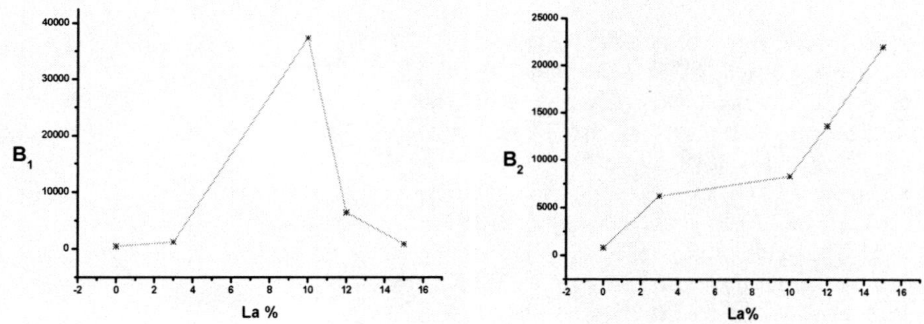

Figure 11. Evolution of the coefficients B_1 and B_2, appearing in Equation (3), as functions of La content.

4. Conclusion

PLZT ceramics with compositions near the morphotropic boundary phase have been prepared and their structural and dielectric properties investigated. Addition of La seems to enhance crystallization of the raw powders and moves the structural phase towards tetragonal one. Moreover, no secondary phase has been detected. Dielectric measurements revealed the existence of two anomalies at relatively high temperatures T ≈ 160°C and ≈ 260°C, the former, which is consistent with previous work, is interpreted as the temperature at which the transition from ferroelectric-rhombohedral phase to ferroelectric-quadratic phase occurs. The apparition of such anomalies has not been systematically observed in other works, studying the same materials, and does not be inherent to the method of preparation of the samples. Relaxation occurs in our samples at about y = 0.10, and the dielectric constant reaches its maximum for y ≈ 0.12, which value is different from other studies where this maximum occurs at low values of y. The temperature, T_m, of the maximum of the dielectric constant reaches low values (T_m < 50°C) for the sample with y = 0.2. Fitting the thermal behavior of the permittivity with the help of Equation (2) has given satisfactory results in terms of appearance and evolution of diffuseness and relaxation phenomena. Moreover, we have successfully applied a polynomial law which may be an indicator of the occurrence of the relaxation phenomenon. Indeed, the coefficient B_1 appearing in Equation (3) increases reaches a maximum before decreasing; this maximum (for the series prepared) coincides with the value at which the relaxation phenomenon takes place.

References

1. A. Mansingh, Ferroelectrics 102 (1990) 69–84.
2. A. M. Glass, Science 235 (1987) 1003–1009.
3. G. Burns, B. A. Scott, Phys. Rev. B 7 (1973) 3088.
4. S. J. Lee, K. Y. Kang, S. K. Han, M. S. Jang, B. G. Chae, Y. S. Yang, S. H. Kim, Appl. Phys. Lett. 72(3) (1998) 299–300.
5. K. Iijima, Y. Tomita, R. Takayama, I. Veda, J. App. Phys. 60 (1986) 361.
6. B. E. Vugmeister, H. Rabitz, Phys. Rev. B 57 (1998) 7581.
7. A. E. Glazounov, A. K. Takantsev, A. J. Bell, Phys. Rev. B53 (1996) 11281.
8. B. J. Helperin, C. M. Varma, Phys. Rev. B 14 (1976) 4030.
9. W. Kleemann, Int. J. Mod. Phys. B7 (1993) 2469.
10. R. Pirc, R. Blinc, Phys. Rev. B 60 (1999) 13470.
11. R. Fisch, Phys. Rev. B 67 (2003) 094110.
12. E. Prouzet, E. Husson, N. de Nathan, A. Morell, J. Phys.: Condens. Matter 5 (1993) 4889.
13. G. A. Rossetti Jr., W. Cao, C. A. Randall, Ferroelectrics 158 (1994) 343.
14. L. A. Bursill, P. Julin, Q. Hua, N. Setter, Physica B 205 (1995) 305.
15. J. Zhao, A. E. Glazounov, Q. M. Zhang, B. Toby, Appl. Phys. Lett. 72 (1998) 1048.
16. S. Vakhrushev, A. Nabereznov, S. K. Sinha, Y. P. Feng, T. Egami, J. Phys. Chem. Solids 57 (1996) 1517.
17. M. D. Glinchuk, V. V. Laguta, I. P. Bykov, S. Nokhrin, V. P. Bovtum, A. A. Leschenko, J. Rosa, L. Jastrabik, J. Appl. Phys. 81 (1997) 3561.
18. G. A. Rossetti Jr., Ph.D. thesis, The Pennsylvania State University (1993).
19. J. F. Li, X. H. Dai, A. Chow, D. Viehland, J. Mater. Sci. 10 (1995) 926.
20. X. H. Dai, Z. Xu, D. Viehland, Philos. Mag. B 70 (1994) 33.
21. K. G. Keiser, G. J. Lansink, A. J. Burgaaf, J. Phys. Chem. Solids 39 (1978) 59.
22. J. F. Scott, C. A. Aranjo, Science 246, 1400 (1989).
23. P. K. Larsen, R. Cuppens, G. A. C. M. Spiering, Ferroelectrics 128, 65 (1992)
24. J. T. Evans, R. Womack, IEEE J. Solid State Circuits 23, 1171 (1988).
25. B. Jaffe, W. R. Cook, H. Jaffe, Piezoelectric Ceramics (Academic, London, 1971).
26. K. Kakegawa, O. Matoumaga, T. Kato, Y. Sasaki, J. Am. Ceram. Soc. 78, 1071 (1995).
27. J. F. Meng, R. S. Katiyar, G. T. Zou, X. H. Wang, Phys. Stat. Sol. A 164, 851 (1997).
28. G. A. Smolenskii, V. A. Bokov, V. A. Isupov, N. N. Krainik, R. E. Pasynkov, A. I. Sokolov, Ferroelectrics and Related Materials, p. 763 (Gordan and Breach, New York, 1984).
29. Thin Film Ferroelectric Materials and Devices, ed. by R. Ramesh (Kluwer, Boston, MA, 1997).
30. G. H. Haertling, Ceramic Materials for Electronics, ed. by Relva C. Buchman, New York, p. 139 (1986).
31. V. V. Kirillov, V. A. Isupov, Ferroelectrics 5, 3 (1973).
32. G. burns, F. Dacol, Phys. Rev. B 28 , 527 (1983).
33. L. E. Cross, Ferroelectrics 76, 241 (1987).
34. S. Li, J. A. Eastman, R. E. Newnham, L. E. Cross, Phys. Rev. B 55, 12067 (1997).
35. D. Viehland, J. F. Li, S. J. Jang, L. E. Cross, M. Wuttig, Phys. Rev. B 43, 8316 (1991).
36. S. K. Mishra, A. P. Singh, D. Pandey, Appl. Phys. Lett. 69, 1707 (1996).
37. S. K. Mishra, D. Pandey, Philos. Mag. B 76, 227 (1997).
38. Ragini, S. K. Mishra, D. Pandey, H. Lemmens, G. Van Tendeloo, Phys. Rev. B 64, 4101 (2001).

39. B. Noheda, D. E. Cox, G. Shirane, J. A. Gonzalo, S. E. Park, L. E. Cross, Appl. Phys. Lett. 74, 2059 (1999).
40. D. Sheen, J.-J. Kim, Phys. Rev. B 67, 144102 (2003).
41. Y. Xu, Ferroelectric Materials and Their Applications, pp. 101–210 (North Holland, Amsterdam, 1991).
42. P. Cousin, R. A. Cross, Mater. Sci. and Eng. A130, 119 (1990).
43. W. J. Dawson, Am. Ceram. Soc. Bull. 67(10) 1673 (1988).
44. K. Uchino and S. Nomura, Critical Exponents of Dielectric Constant in Diffused-Phase Transition Crystals, Ferroelectric Lett., 44, 55–61 (1982).
45. S. M. Gupta, J.-F. Li, D. Viehland, J. Am. Ceram. Soc. 81(5) 557 (1998).
46. B. Noheda, J. A. Gonzalo, L. E. Cross, R. Guo, S.6E. Park, D. E. Cox, G. Shirane, Phys. Rev. B 61(13) 8687 (2000).
47. A. Bouzid, M. Gabbay, G. Fantozzi, Defects and Diffusion Forum, Vols. 206–207, pp. 147–150 (Trans Tech Publications, Switzerland, 2002).
48. A. Bouzid, Ph.D. thesis, INSA-Lyon (Fr.) (2002).
49. P. Gerthsen, K. H. Härdtl, N. A. Schmidt, J. Appl. Phys. 51(2) 1131 (1980).
50. K. Bouayad, S. Sayouri, T. Lamcharfi, M. Ezzejari, D. Mezane, L. Hajji, A. Elghazouali, M. Filali, P. Dieudonné, M. Rhouta, Physica A 358, 75 (2005).

180° FERROELECTRIC DOMAINS IN THIN FILMS AND SUPERLATTICES

ANAÏS SENE[1], IGOR A. LUK'YANCHUK[*1] AND LAURENT LAHOCHE[2]
[1]*University of Picardie Jules Verne, Laboratory of Condensed Matter Physics, Amiens, 80039, France*
[2]*Roberval Laboratory, University of Technology of Compiegne, France*

Abstract. In the micro- and nanoscale ferroelectric samples, formation of periodic polarization domains is the efficient mechanism of reducing depolarization field that is produced by the surface bound charges. This makes the physics of these devices different from the bulk samples. We present the results of modeling of ferroelectric domains and domain textures in ferroelectric thin films and periodic paraelectric/ ferroelectric superlattices, basing on the self-consistent solution of the coupled electrostatic and Ginzburg-Landau equations. We go beyond the traditionally used low-temperature Kittel approximation (in which the polarization is assumed to be temperature independent and constant across domains) and explore the temperature evolution of the domain-induced properties. We study in detail the evolution of polarization profile $P(x, z)$ in the periodic domains structure as function of the temperature and of the film width and propose the simple interpolation formula that can recover all the regimes of the domain structure, from high temperatures and thin films to low temperatures and thick films.

Keywords: Ferroelectric domains, modeling, thin films, superlattices

1. Introduction

In recent years, a lot of attention was given to studies of finite size, surface and interface effects in ferroelectric thin films and superlattices[1] because of their potential applications in Ferroelectric Random Access Memories (FRAM) in which the up- and down-polarized domains serve as the binary information units. Stability of ferroelectric domains and conditions of their switching is therefore

* To whom correspondence has to be sent: Email: lukyanc@ferroix.net

I. A. Luk'yanchuk and D. Mezzane (eds.), *Smart Materials for Energy, Communications and Security.*

the crucial question for the reliability of ferroelectric devices. Generally two mechanisms of domain creation are possible. For macroscopic bulk samples domains appear as the metastable formations and their physics is mostly governed by switching kinetics. In micro- and naonoscopic thin films domains form periodic thermodynamically stable structures, provided by interplay of ferroelectric-condensation and electrostatic energies. It is the objective of the current article to elaborate the methods of domain modelling in thin films and discuss the results, relevant for their application in ferroelectric devices.

As we just mentioned, ferroelectric polarization domains are created in slabs and films to reduce the energy of the depolarization field produced by the space and surface charges with charge density $\rho(r)$ = div **P** that are provided by the discontinuity and non-uniformity of the polarization close to the crystal surface. This energy becomes smaller since depolarization field concentrates closer to the film surface and does not penetrate into the volume, but a new contribution due to domain wall energy appears. A delicate balance between depolarization and domain wall energies provides the equilibrium domain configuration.

In simple Kittel approach,[2–4] the domain texture is considered as a set of up- and down-oriented domains, having a flat polarization profile $P(x,z) = \pm P_0$ (hard domains). The domain walls are assumed to be infinitively thin and the boundary effects on the ferroelectric-paraelectric (or ferroelectric-vacuum) interface are neglected. This structure is assumed to be temperature independent and is valid for any width $2a_f$ of the ferroelectric film.

Being justified at low temperatures, this approach is not valid close to Ferroelectric-Paraelectric transition temperature T_c where the polarization gradient is very soft because of the divergency of the coherence length[4] and this approximation does not allow the analyze neither the temperature evolution of domain texture nor the influence of the surface and interface phenomena. Meanwhile these effects play the decisive role in thermodynamic and electric properties of the emerging nanoscopic ferroelectric devices.

We perform the modeling of domain structure in a whole temperature interval and study the temperature evolution of the related properties of the film based on the self-consistent solution of the coupled electrostatic and Ginzburg-Landau equations. The objective is to study in detail the evolution of polarization profile $P(x, z)$ in the periodic domains structure as function of the temperature and of the film width and to propose the simple interpolation formula that can recover all the regimes of the domain structure, from high temperatures and thin films to low temperatures and thick films.

2. Model description

The studied configuration of ferroelectric film with 180° periodic domain texture that is sandwiched between two paraelectric layers and placed into the short-circuited capacitor is analogous to the problem of ferroelectric/paraelectric superlattice with periodic (along c) boundary conditions, as shown in Fig. 1.

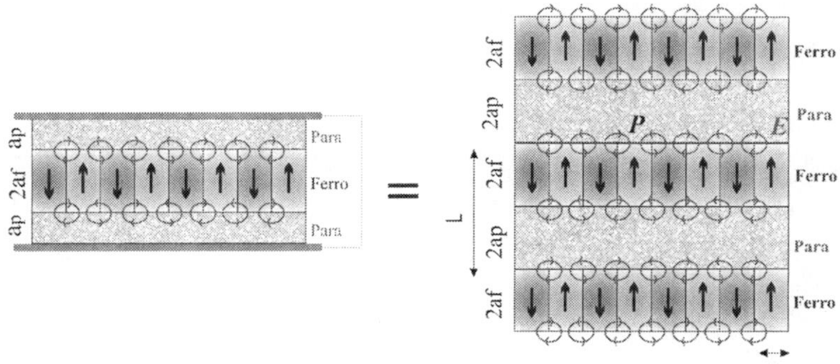

Figure 1. Capacitor-like and superlattice-like geometry of the model.

The physics of ferroelectric state is described by the system of electrostatic equations[6,7]:

$$div(E + 4\pi P) = 0, \qquad rot\, E = 0 \tag{1}$$

in which the polarization is related to the electric field by the "equation of state" **P = P (E)**.

The most appropriate tool to model the polarization distribution is the set of equations constructed from Ginzburg-Landau equations coupled with these electrostatic equations describing the depolarization field. These equations firstly derived by Chensky and Tarasenko[4,5] are obtained by variation of the general energy functional:

$$F = \int \Phi(P,E)dxdz, \qquad \Phi(P,E) = \Phi(P,0) - EP - \frac{1}{8\pi}E^2 \tag{2}$$

where $\mathbf{E} = (E_x, E_z)$, $\mathbf{P} = (P_x, P_z)$ and the field-independent part

$$\Phi(P,0) = \frac{4\pi}{\varepsilon_\perp}\frac{1}{2}P_x^2 + \frac{4\pi}{\varepsilon_{i\|}}\frac{1}{2}P_{zi}^2 + \frac{4\pi}{\chi_\|}f(P) \tag{3}$$

includes the transversal P_x, and non-polar longitudinal P_{zi} noncritical contributions ($\varepsilon_{\perp}$, $\varepsilon_{i||}$ »1). The nonlinear Ginzburg-Landau energy for the order parameter depends on the spontaneous z-oriented polarization P (assuming that $P_z = P_{zi} + P$) and is written as:

$$f(P) = \left[\frac{t}{2} P^2 + \frac{1}{n} P_0^{2-n} P^n + \frac{\xi_{0x}^2}{2} (\partial_x P)^2 + \frac{\xi_{0z}^2}{2} (\partial_z P)^2 \right] \tag{4}$$

where the reduced temperature t is expressed via the bulk critical temperature as: $t = T/T_{c0} - 1$, parameter $\chi_{||}$ is expressed via paramagnetic Curie constant C and via longitudinal permittivity $\varepsilon_{||}$ as: $\chi_{||} = C/T_{c0} \simeq (n-2)\varepsilon_{||}$ and the nonlinearity index n = 4 for second order transition in ferroelectrics and n = 6 for tricritical ferroelectrics[10] and coefficient P_0 is roughly equal to the saturated bulk polarization at $T \ll T_c$.

Variation of Equation (2) by P and electrostatic potential $\varphi(E = -\nabla \varphi)$ and exclusion of the non-critical variables P_x and P_{zi} gives the system of Chensky equations that describes the ferroelectric transition with the self-action of depolarizing field:

$$(t - \xi_{0x}^2 \partial_x^2 - \xi_{0z}^2 \partial_z^2) P + (P/P_0)^{n-2} P = -\frac{\chi_{||}}{4\pi} \partial_z \varphi \tag{5}$$

$$(\varepsilon_{i||} \partial_z^2 + \varepsilon_{\perp} \partial_x^2) \varphi = 4\pi \partial_z P$$

These equations should be completed by the Poisson equation for paraelectric media in which ferroelectric film is embedded:

$$(\partial_x^2 + \partial_z^2) \varphi^{(p)} = 0 \tag{6}$$

and by boundary conditions at the Para-Ferro interface

$$\varepsilon_{i||} \partial_z \varphi - \varepsilon_p \partial_x \varphi^{(p)} = 4\pi P, \qquad \varphi = \varphi^{(p)}, \qquad \partial_z P = \lambda P \tag{7}$$

that are also obtained as result of variation of Equation (2) where λ is the extrapolation length (neglected in this study).

Periodic conditions

$$P(x,z) = P(x + 2d, z), \qquad \varphi(x,z) = \varphi(x + 2d, z) \tag{8}$$

with variational parameter d are imposed to describe the periodicity of domain structure.

3. Model implementation

For numerical modeling we used the nonlinear iterative Newton-Raphson algorithm realized in the Comsol Multiphysics Finite-Element toolbox (Comsol Multiphysics, (c) COMSOL AB, 1994–2005).

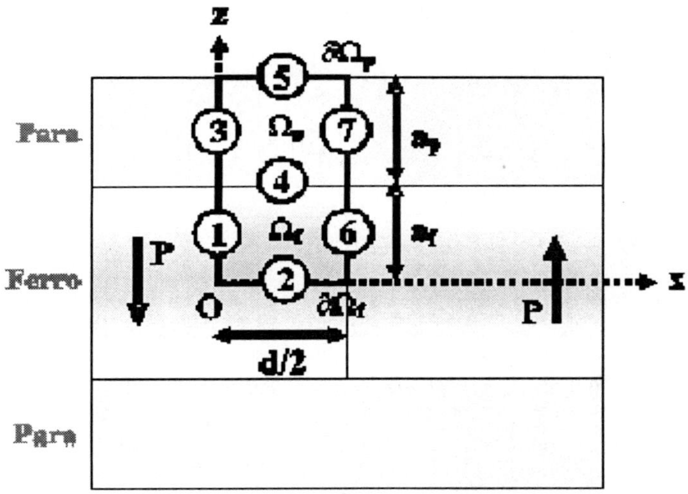

Figure 2. Elementary region used for the modeling.

The symmetry of the periodic domain structure let us to perform the calculations in the elementary region presented in Fig. 2. This region of the half domain width d/2 is composed from the paraelectric subregion Ω_p with boundary $\partial\Omega_p = (3) + (4) + (5) + (7)$ and height a_p and from the ferroelectric subregion Ω_f with boundary $\partial\Omega_f = (1) + (2) + (4) + (6)$ and height a_f.

It was convenient to map the elementary region Ω_f onto the unit square cell, using the coordinate rescaling $\tilde{x} = 2x/d$ and $\tilde{z} = z/a_f$ where $0 \le \tilde{x} \le 1$ and $0 \le \tilde{z} \le 1$.

We got the system of partial nonlinear differential equations (PDE) with three independent variables: $u = (P, \varphi, \varphi')$ that can be formalized as:

$$d_a \frac{\partial u}{\partial t} + \nabla.(-c\nabla u - \alpha u + \gamma) + \beta.\nabla u + au = f$$

where coefficients are given by

	d_a	C	α	β	γ	a	f
P in Ω_f	0	$<\frac{4}{d^2}\frac{1}{a_f^2}>$	0	0	0	$T + P^2$	$-\frac{\chi_{//}}{4\pi a_f}\frac{\partial\varphi}{\partial\tilde{z}}$
φ in Ω_f	0	$<\frac{4\varepsilon_\perp}{d^2}\frac{1}{a_f^2}>$	0	0	0	0	$-\frac{4\pi}{a_f}\frac{\partial P}{\partial\tilde{z}}$
φ' in Ω_p	0	$<\frac{4}{d^2}\frac{1}{a_f^2}>$	0	0	0	0	0

The corresponding boundary conditions are written either in Neumann:

$$n.(c\nabla u + \alpha u - \gamma) + qu = g - h^T\mu,$$

or in Dirichlet form: hu = r, where coefficients are given by:

Boundary	(1)		(2)		(3)	(4)			(5)	(6)	(7)	
Variable	P	φ	P	φ	φ'	P	φ	φ'	φ'	P	φ	φ'
q	0	0	0	0	0	0	0	0	0	0	0	0
g	0	0	0	0	0	0	$4\pi Pa_f + \varepsilon_p\frac{\partial\varphi'}{\partial\tilde{z}}$	0	0	0	0	0
h	1	1	–	1	1	–	0	1	1	–	–	–
r	0	0	–	0	0	–	0	φ	0	–	–	–

Considering now the domain width d as variational parameter we find numerically the equilibrium (most stable) structure, substituting solution *P(x,z)* into the generating Euler functional, that for its extremals is simplified to:

$$F = -\frac{1}{4}\int P^4 dxdz \tag{9}$$

and minimizing it over d.

For numerical calculations we selected the typical for the displacive ferroelectrics material parameters: $\varepsilon_{||} \simeq 500$, $\varepsilon_\perp \simeq 100$, $\varepsilon_p \simeq 10$. The coherence length was estimated as the low-temperature domain wall half-width $\xi_0 = 1$ nm. The ferroelectric film thickness $2a_f$ varied between 25 and 1,000 nm, the paraelectric layer thickness a_p was always selected larger then characteristic domain width to keep the sample in the multi-domain regime.

Basing on the equations we calculated numerically both the dependence of the domain structure period on temperature and the temperature evolution of the polarization profile.

Figure 3 presents the temperature dependence of domain width d for the films having different thickness $2a_f$. It appears that this parameter is almost temperature independent. Basing on scaling properties of this set of equations we can present the dependence $d(T)$ in the generalized Kittel form as:

$$d(T) = \sqrt{\gamma(T)\left([\varepsilon_\perp/\varepsilon_{||}]\right)^{\frac{1}{2}} 2a_f} \tag{10}$$

where the weakly temperature dependent numerical parameter $\gamma(T)$ can be calculated analytically in the limit cases of low temperatures (at $T = 0$ when $\varepsilon_p^2 \ll \varepsilon_\perp \varepsilon_{||}$), using the Kittel approximation[2,3,6] and in vicinity of T_c.[4,5]

$$\gamma(0) = \frac{2\sqrt{2}\pi^3}{21\zeta(3)} \simeq 3.53 \quad \gamma(T_c) = \pi$$

Since the period of domain texture is almost temperature independent, the contribution of domain wall motion to the temperature hysteresis of ferroelectric properties should be weak.

Discuss now the temperature evolution of polarization profile inside domains. Phase diagram of domain states in thin films is shown in Fig. 3. The transition from paraelectric to ferroelectric phase occurs directly into the shown in Fig. 4 periodically modulated soft domain structure in which the gradual polarization distribution is approximately written[4]

$$P(x, z) = P_{\max}(T)\sin\left(\pi x/d\right)\cos\left(\pi z/2a_f\right) \tag{11}$$

Since creation of domain structure costs the gradient and depolarization field energy, the transition temperature T_c is reduced with respect to transition temperature of the bulk short circuited sample T_{c0} as.[4,5]

$$T_c = T_{c0}\left(1 - 2\pi\sqrt{\frac{\chi_{||}}{\varepsilon_\perp}\frac{\xi_{0x}}{2a_f}}\right) \tag{12}$$

When temperature decreases below T_c the amplitude of polarization modulation P_{max} increases and, at the same time, the polarization profile in the domain centers becomes more flat. At lower temperatures the polarization profile inside domains becomes hard, as shown in Fig. 4.

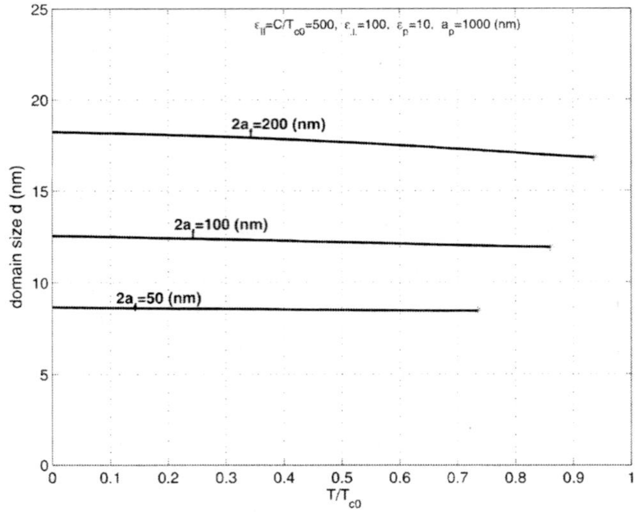

Figure 3. Temperature variation of equilibrium domain width d for films with different thickness.

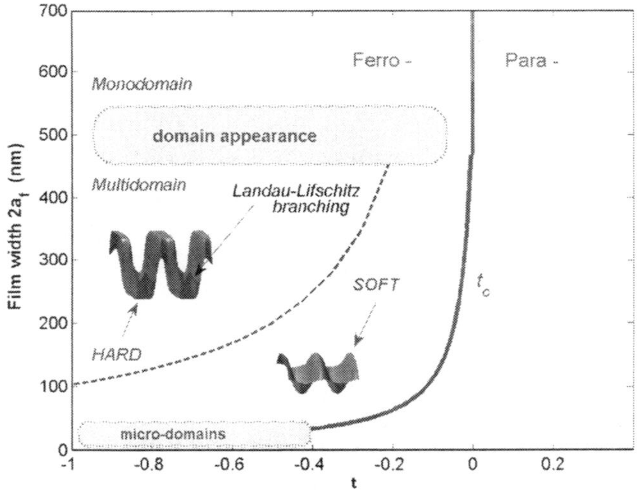

Figure 4. Phase diagram of the domain states in thin film.

To study the crossover from soft to hard domains we plotted in Fig. 5a the temperature dependence of polarization square in the domain center $P_{\max}^2(T)$.

In the soft-profile regime (in vicinity of transition) $P_{\max}^2(T)$ follows the Landau-type behavior ($\sim T - T_c$). In the hard-profile regime the temperature dependence again becomes linear but with the bulk transition temperature T_{c0} as the critical temperature parameter ($P_{\max}^2 \sim T - T_{c0}$). Therefore the thermodynamics of the film with hard domains is the same as that of the bulk short-circuited sample and neither the surface-located depolarization field nor the thin domain walls play substantial role in it.

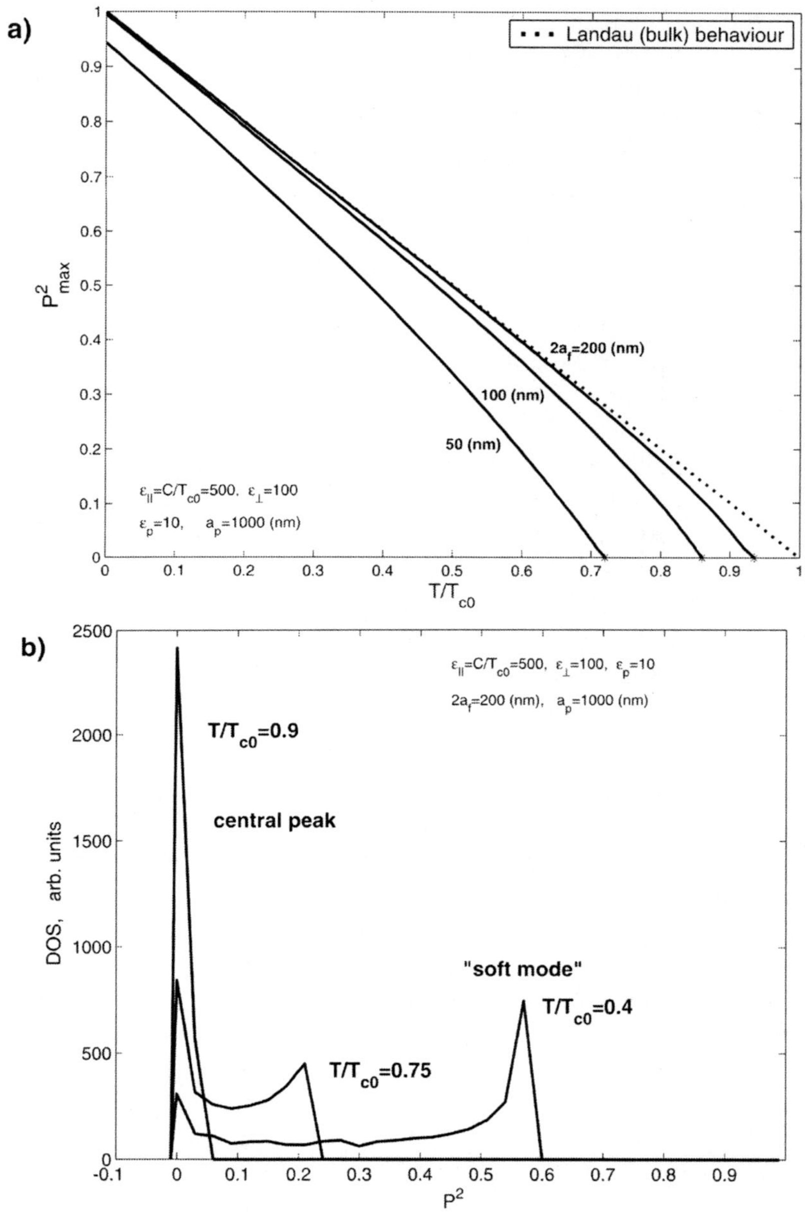

Figure 5. (a) Temperature dependence of the square of polarization in center of domain $P^2_{max}(T)$ for films with different thickness $2a_f$. Dotted line shows the Landau-type behavior for the bulk mono-domain short-circuited sample. (b) Density of states of square of polarization P^2 for the domain structure in film of thickness $2af$ = 200 *nm* at different temperature.

Meanwhile, the depolarization field plays the important role in the polarization profile close to the film surface. As follows from Fig. 4, in the hard regime, it results both to considered[7] surface polarization damping and to intervention of the small surface-located inversely-polarized domains. This phenomenon, known as Landau-Lifshitz domain branching[2] was reproduced in our numerical calculations up to the second branching level. In the soft domains (Fig. 4), polarization always vanishes on the film surface.

As follows from Fig. 5b, the crossover temperature between soft and hard domain states decreases with film thickness in the same way as T_c, i.e. inversely proportional to $2a_f$. The temperature interval of existence of soft domains becomes very large for the thin nanometric films and their physical properties can be governed by the highly nonuniform distribution of polarization across the sample.

The gradual polarization variation in soft domains can be visualized by experimental techniques testing the local distribution of polarization such as Raman and infrared spectroscopy, X-ray diffraction, ESR etc. Figure 5b shows the density of states (DOS) of local values of polarization square P^2 at different temperatures. It appears that in vicinity of T_c the domain walls region pumps over a DOS from the "soft mode" peak at $P_{\downarrow}\mathrm{max}^{\uparrow}2$ into the central peak at $P^2 = 0$ that can result in misinterpretation of soft-mode spectroscopy measurements.

4. Variational analysis

Although the complete set of Equations (5), (6) and (7) can be solved numerically it would be more convenient to have the more general analytical approach permitting generalize the numerical results for the overall parameter region.

Consider first the general properties of this set of equations rewriting them in dimensionless variables z′, x′, t′, P′, φ′ defined by:

$$z = a_f z', \qquad x = \tau^{-1/2}\xi_{0x}x', \qquad t = tt' \tag{13}$$

$$P = \frac{1}{\tau^{n-2}}P_0 P', \qquad \varphi = \frac{1}{\chi_{\|}}\tau^{\frac{n-1}{n-2}}a_f P_0\varphi', \qquad F = \frac{\xi_{0x}a_f}{\chi_{\|}}\tau^{\frac{n-1}{n-2}}P_0^2 F'$$

$$\tau = \left(\frac{\chi_{\|}}{\varepsilon_{\perp}}\right)^{1/2}\frac{\xi_{0x}}{a_f}$$

Neglecting the small terms $\frac{\varepsilon_\perp}{\chi_{||}}\frac{\xi_{0z}}{a_f}(\partial_z P)^2$ and $\frac{\varepsilon_{i||}}{\chi_{||}}\frac{\xi_{0x}}{a_f}(\partial_z \varphi)^2$ we obtain the dimensionless functional

$$F' = \iint\left[4\pi\left(\frac{1}{2}t'P'^{\,2} + \frac{1}{4}P'^{\,n} + \frac{1}{2}(\partial'_x P')^2\right) - \frac{1}{8\pi}(\partial'_x \varphi')^2 + P'\partial'_z \varphi'\right]dx'dz' \qquad (14)$$

that contains only one driving variable – the dimensionless temperature t'. This means that knowledge of the "master" temperature dependence of any physical parameter for some particular film of width $2a_f$ and material constants $\varepsilon_\perp$, $\chi_{||}$, ξ_{0x} allows the re-scaling of this dependence for any other ferroelectric film, characterized by the new set $\varepsilon_\perp$, $\chi_{||}$, ξ_{0x} and $2a_f$.

Taking into account that $2a_f' = 2$ we obtain a new set of equations that is independent on the material parameters and therefore is universal for any ferroelectric film:

$$(t' - \partial_x'^2)P' + P'^{n-1} = -\frac{1}{4\pi}\partial'_z \varphi' \qquad (15)$$

$$\partial_x'^2 \varphi' = 4\pi\partial'_z P' \qquad (16)$$

and the corresponding variational boundary conditions:

$$P' = 0, \quad \varphi' = \varphi'^{(p)} \qquad (17)$$

Scaling properties (Equation (9)) lead to several universal relations for physical properties of multidomain ferroelectric films that will be discussed at the end of this paper.

Now we shall try to analyze the temperature dependence of the polarization profile P(x,z,t) that is the solution of new parameter-independent equations (15–17) presenting it in a variational form of simple x-periodic function

$$P = f(z)sn\left[\frac{4K(m_1)}{2d}x, m_1\right], \; f(0) = f(2) = 0 \qquad (18)$$

that, depending on parameter m_1 ($0 < m_1 < 1$) covers all the domain profiles from the soft ones (at $m_1 \sim 0$) to the hard ones (at $m_1 \sim 1$). Considering m_1 and function f(z) as variational parameters we substitute Equation (18) into functional (Equation (14)) (we have omitted the prime indices and take n = 4) and integrate it over domain structure period 2d.

After some algebra and nothing that according to Equation (16) $\varphi = 4\pi f'(z)\iint P(x'')dx''dx'$ we present the resulting functional as:

$$F = 4\pi \int \frac{1}{2}\left[\alpha(m_1)t + \frac{4K(m_1)\beta(m_1)}{(2d)^2}\right] f^2 + \frac{1}{4}\gamma(m_1)f^4 + \frac{1}{2}\delta(m_1)(2d)^2 f'^2\, dz \quad (19)$$

in which the coefficients are expressed via complete elliptic integrals of the first and second kind K(m), E(m)[8] as:

$$\alpha(m) = \left\langle sn^2(x,m)\right\rangle = \frac{1}{m}\left[1 - \frac{E(m)}{K(m)}\right] \quad (20)$$

$$\beta(m) = 4K(m)\left\langle (sn'u)^2\right\rangle = 4K(m)\frac{1}{3}\left[2 - (1+m)\alpha(m)\right]$$

$$\gamma(m) = \left\langle sn^4(x,m)\right\rangle = \frac{1}{3m}\left[2(1+m)\alpha(m) - 1\right]$$

$$\delta(m) = \frac{\left\langle S^2(x,m)\right\rangle}{\left[4K(m)\right]^2} = \frac{8}{m\left[4K(m)\right]^2}\sum_{l=1,3,5}^{\infty}\left[\frac{1}{l}\frac{q^{l/2}(m)}{1-q^l(m)}\right]^2$$

where $q(m) = \exp\left[-\frac{K(1-m)}{K(m)}\pi\right]$, $S(x,m) = \int sn(u,m)du$ and <...> is the average over the period. Dependencies α(m), β(m), γ(m) and δ(m) are presented in Fig. 6.

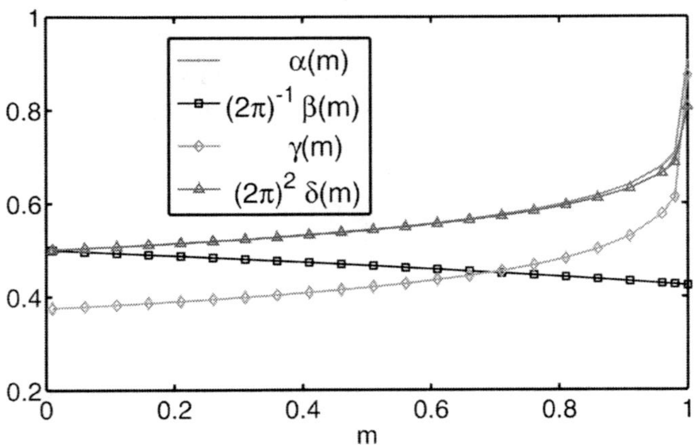

Figure 6. Coefficients α(m), β(m), γ(m) and δ(m) that enter into the variational functional (Equation (20)).

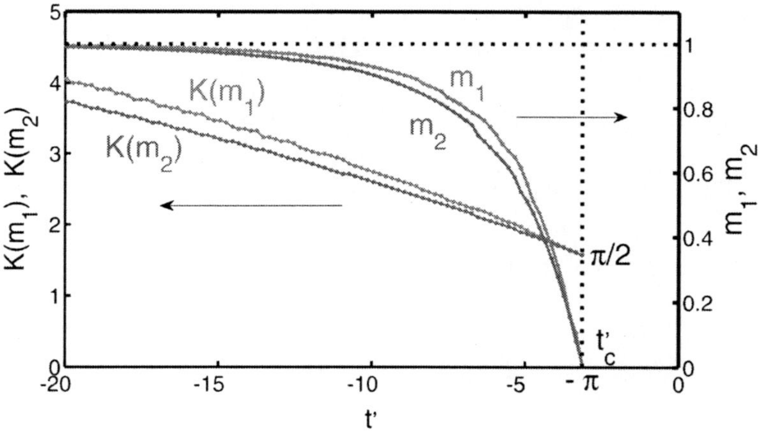

Figure 7. Dimensionless temperature dependences of the elliptic arguments m_1 and m_2 and of the elliptic integrals $K(m_1)$ and $K(m_2)$.

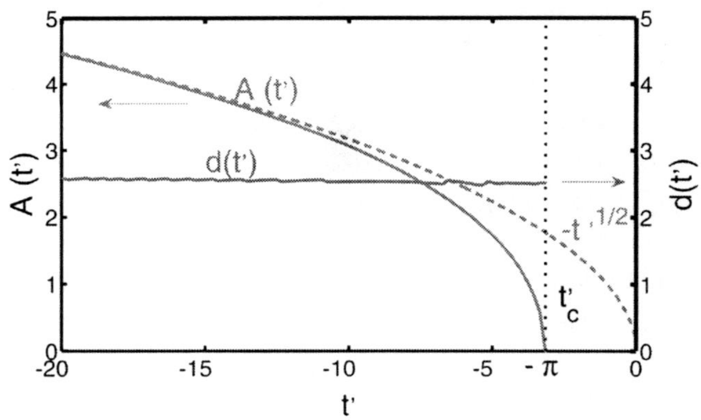

Figure 8. Dimensionless temperature dependences of the domain amplitude A(t') and domain lattice period d(t').

Variational Euler-Lagrange minimum of (Equation (19)) is given by the function:

$$f = A(t, m_1, m_2) sn\left[K(m_2)z, m_2\right] \tag{21}$$

with

$$A(t, m_1, m_2) = 2d(t, m_1, m_2)K(m_2)\sqrt{\left(2\frac{\delta(m_1)}{\gamma(m_1)}m_2\right)} \tag{22}$$

that matches the boundary conditions f(0) = f(2) = 0 providing that dependence $d(t, m_1, m_2)$ is fixed by biquadratic equation:

$$(2d)^4 \delta(m_1)(1+m_2)K^2(m_2) + (2d)^2 \alpha(m_1)t + 4K(m_1)\beta(m_1) = 0 \quad (23)$$

Substituting Equation (17) back into Equation (15) gives:

$$F = -4\pi \frac{1}{4} \gamma(m_1) \int f^4(z)dz \quad (24)$$

$$= -\frac{1}{2} \pi\gamma(m_1) A^4(t, m_1, m_2)\gamma(m_2)$$

Collecting now all the results together, we present the final variational solution as:

$$P(x,z) = A(t, m_1, m_2) sn[K(m_2)z, m_2] sn\left[\frac{4K(m_1)}{2d} x, m_1\right] \quad (25)$$

where the temperature dependencies of m_1 and m_2 and of elliptic integrals $K(m_1)$ and $K(m_2)$ are found by minimization of functional (Equation (21)) over m_2 and m_1 and are presented in Fig. 7. The corresponding temperature dependencies of the amplitude A(t) and domain lattice period d(t) are given in Fig. 8.

Formula (25) gives the approximation for the domain texture at arbitrary temperatures. It recovers all the domain regimes: from the soft one (Equation (12)) close to t_c (when m1, m2 < 0 and sn → sin) to the hard (Kittel-like) one at low t (when m1, m2 < 1 and sn → step-wise function).

By the scaling relation (Equation (13)), expression for *P(x,z)* (Equation (25)) can be expanded for any film width $2a_f$ and for any set of material constants $\varepsilon_\perp$, $\chi_{||}$.

We present several remarkable conclusions about the physical properties of the multi-domain state which can be obtained only from the scaling properties (Equation (13)), without solution of Equations (5) – (7).

(i) Any transverse length parameter scales as $\tau^{-1/2}\xi_{0x}$. This, in particular, justifies the Kittel formula (10) for the domain width d even beyond the flat domain approximation. A convincing demonstration of the validity of this scaling law was reported recently for various ferroelectric and ferromagnetic materials.[9] The temperature dependence d(t) is incorporated into Equation (10) since the numerical factor γ depends on t. Meanwhile, the results shown in Fig. 8 as well as finite-element simulations indicate that the dependence d(t) is very weak and hence one can extend the parameter $\gamma \simeq 3.53$ from Equation (10) to any temperature. This, in particular, implies the low temperature hysteresis related with motion of DW.

(ii) The temperature t scales as τ. Thus, to compare the physical properties of different films (even from different materials) it should be instructive to trace their temperature dependences using the rescaled coordinate t/τ. Another consequence is the $1/2a_f$ scaling for which the evolution of the reduced transition t_c (Equation (12)) and the soft to hard domain crossover t* temperatures is illustrated in Fig. 4. The temperature interval for the existence of soft-domains $\Delta t = t_c - t^*$ growth dramatically with decreasing film thickness and one can expect that for thin films with $2a_f <$ 100 nm only soft domains with a gradual polarization distribution are possible.

(iii) Consider now the effective permittivity of multidomain film, that according to Equation (13) scales as $\bar{\varepsilon} = \dfrac{4\pi\bar{P}}{\bar{\mathrm{E}}} \sim \chi_{||}/\tau$ where $\bar{E} = -\overline{\partial_z \varphi}$ is the average internal field and P is the average induced polarization. Correspondingly, the temperature dependence of multidomain permittivity $\bar{\varepsilon}$(t) (that is negative[11]) can be presented by some universal function f(t′):

$$\bar{\varepsilon} = -\chi_{||}/\tau f(t/\tau) \tag{26}$$

that has the material and nonlinearity-index-n independent character and can be found from plot of $\bar{\varepsilon}$(t) in $\tau/\chi_{||}$ and t/τ coordinates. Function f(t′) can be also calculated theoretically by solution of Equations (5) - (8) in external field. Note however, that for such analysis the film-intrinsic permittivity $\bar{\varepsilon}$(t) should be extracted from the total paraelectric + ferroelectric capacitor-like dielectric constant

$$a\varepsilon_{tot}^{-1} = a_p\varepsilon_p^{-1} + a_f\bar{\varepsilon}^{-1}, \qquad a = a_p + a_f \tag{27}$$

measured[12,13] at different a_p with further linear extrapolation $a_p \to 0$. (The "capacitor" model however is valid[14] when $ap \gg d$)

5. Conclusion

To conclude we have shown that properties of ferroelectric nanometric thin films are substantially governed by the gradual polarization distribution in soft domain texture. At the same time, the width of ferroelectric domains is not sensitive to polarization distribution and is almost temperature independent.

We derived the simple interpolation formula (Equation (25)) that recovers the temperature evolution of polarization profile in the multi-domain state of ferroelectric films. Based on universal scaling relations (Equation (11)) we have demonstrated how the physical properties of the different multi-domain films such as: transition temperature, dielectric permittivity etc., can be compared and mapped onto each other. We hope that such method will give the power tool for analysis and systematization of numerous experimental data for thin ferroelectric films.

ACKNOWLEDGEMENTS

This work was supported by the Region of Picardie, France and by STREP "Multiceral" (NMP3-CT-2006-032616). We thank to Professor M. G. Karkut for the useful discussions.

References

1. M. Dawber, K. M. Rabe and J. F. Scott, Rev. Mod. Phys. **77**, 1083 (2005)
2. L. D. Landau and E. M. Lifshitz, *Electrodynamics of Continuous Media* (Elsevier, New York, 1985)
3. A. M. Bratkovsky and A. P. Levanuk, Phys. Rev. Lett. **84**, 3177 (2000)
4. E. V. Chensky and V. V. Tarasenko, Sov. Phys. JETP **56**, 618 (1982) [Zh. Eksp. Teor. Fiz. 83, 1089 (1982)]
5. V. A. Stephanovich, I. A. Luk'yanchuk and M. G. Karkut, Phys. Rev. Lett. **94**, 047601 (2005)
6. C. Kittel, Phys. Rev. **70**, 965 (1946)
7. Y. G. Wang, W. L. Zhong and P. L. Zhang, Phys. Rev. **51**, 5311 (1995)
8. M. Abramowitz and I. A. Stegun (eds.) *Handbook of Mathematical Functions* (10th ed., NBS, 1972)
9. G. Catalan, J. F. Scott, A. Schilling and J. M. Gregg, J. Phys.: Cond. Matter **19**, 022201 (2007)
10. B. A. Strukov and A. P. Levanyuk, *Ferroelectric Phenomena in Crystals* (Springer, Berlin, 1998)
11. A. M. Bratkovsky and A. P. Levanyuk, Appl. Phys. Lett. **186**, 171 (2006)
12. M. Tyunina and J. Levoska, Phys. Rev. **B63**, 224102 (2001)
13. A. Lookman, R. M. Bowman, J. M. Gregg et al., J. Appl. Phys. **96**, 555 (2004)
14. P. Mokry, A. K. Tagantsev and N. Setter, Phys. Rev. **B70**, 172107 (2004)

Part IV

OTHER FUNCTIONAL MATERIALS

CLAYS AND CLAY MINERALS IN WESTERN HIGH ATLAS: CHARACTERIZATION, GEOLOGICAL SIGNIFICANCE AND INDUSTRIAL USES

LAHCEN DAOUDI[1*], ABELOUAHAB KNIDIRI[1]
AND BENAISSA RHOUTA[2]
[1]*Département des Sciences de la Terre, Faculté des Sciences et Techniques, Université Cadi Ayyad, B.P. 549, Marrakech, Maroc*
[2]*Laboratoire de Matière Condensée et Nanostructures (LMCN), Faculté des Sciences et Techniques Guéliz, Université Cadi Ayyad, B.P. 549, Marrakech, Maroc*

Abstract. In the area of the western High Atlas, the clay formations are very abundant and much diversified. However, the use of clays in industry is still very restricted and the industrial exploitation remains insufficient. The Triassic detrital formations provide the most developed and representative outcrops of clays of the area; they were exploited formerly for the industry of the terra cotta. In the Jurassic-Cretaceous series, the clay assemblage evolution shows a very important geographical and stratigraphical variation. However, because of their low content in the sediments, clays are mainly exploited in the industry of cements. In the Tertiary series, when fibrous minerals characterize the clay fraction of the Palaeocene-Eocene formations, semectites are the most abundant mineral during the Miopliocene. These clay materials have been used since ancient times for cottage industry. The Quaternary continental formations present a clay fraction dominated by the illite and sometimes by kaolinite or smectite. The industrial exploitation of these clays is the most significant in the area. They are especially used in cottage industry and in the industry of the bricks.

Keywords: Clay minerals, Western High Atlas, clay assemblage evolution characterization, geological significance, industrial uses.

*To whom correspondence should be sent: L. Daoudi, Département des Sciences de la Terre, Faculté des Sciences et Techniques, Université Cadi Ayyad, B.P. 549, Marrakech, Maroc: Email: daoudi@fstgmarrakech.ac.ma

I. A. Luk'yanchuk and D. Mezzane (eds.), *Smart Materials for Energy, Communications and Security.*

1. Introduction

The modern society uses clays and clay minerals in increasingly large quantities; research allowed a broad variety of technological applications of clay minerals in industry. They are used in increasingly diversified fields: ceramic industry, industry of cements, metal alloy, therapeutic use, manufacture of absorbing paper…. The improvement of the industrial techniques and the chemical treatments continuous to increase the use of the varieties of clays and clay minerals.

In Morocco, the use of clays in industry is still very restricted. In the area of the High Atlas of Marrakech more particularly, clay materials represent an important economically source for the region. Moreover, except for some exploited deposits in a craft way, the industrial exploitation remains insufficient. This region constitutes one of the areas of Morocco where the clay formations are very abundant and much diversified (Triasic, Cretaceous, Tertiary and Quaternary). During these last 15 years, some detailed and well-documented investigations have been performed on the clay mineralogy of sedimentary series from Westen High Atlas.[5,7–10] The mineralogical and geochemical study, the genesis and the diagenetic transformations undergone by clays were largely approached. The results obtained show that the clay processions of this area are much diversified; all the clay mineral species are represented. The palygorskites and sepiolites), are present sometimes with contents which can currently reach 90%. More recently, several studies are undertaken, the purpose of these studies are to evaluate the economic potential of these clay outcrops and to consider the reserves exploitable.

This article represents a synthesis of the data obtained as 15 years of study of clays of western High Atlas series. Much of the data submitted in this paper were extracted from my thesis, submitted at the Cadi Ayyad University of Marrakech, and from several works published in several revues (see the bibliography). The man objective of the present study is to draw up an assessment on the state of progress of research on clays of the Western High Atlas area likewise on their industrial applications.

2. Materials and methods

The Triassic to Quaternary sediments of the western High Atlas consist chiefly of alternations of sandstone, marl, dolostone, limestone, gypsum and phosphatic bed. Sedimentary facies and clay assemblages evolution of this series are studied on several sections disposed Eastern to Western part of Marrakech High Atlas: the Agadir and Essaouira sections in the western part of the area; the Imin'tanout, the Asni and the Amizmiz sections south of the Marrakech City and Ait Ourir section at the east (Fig. 1). In the western part of this basin, they constitute thick accumulations exceeding several thousand meters; moreover, in the east the sedimentary series have a low thickness.

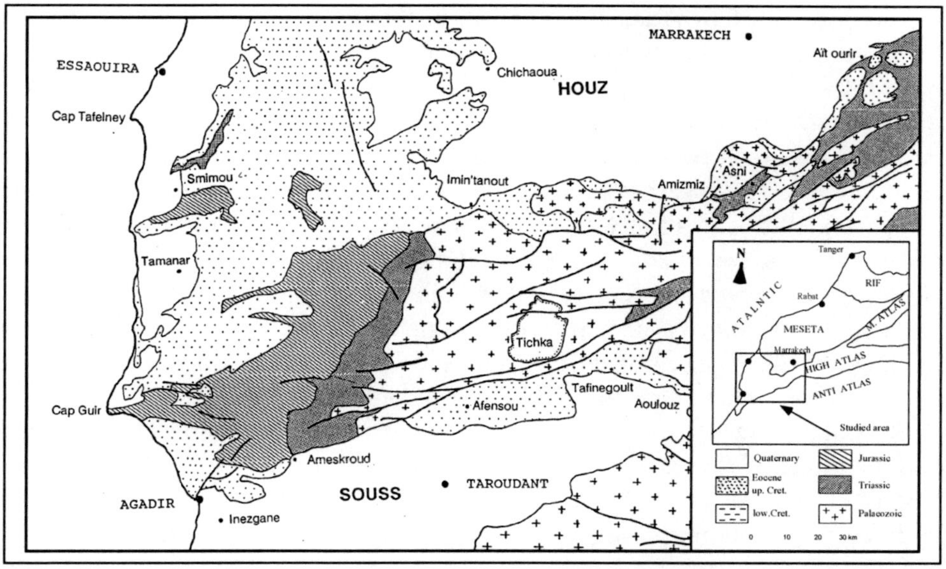

Figure 1. Location map of western High Atlas.

Several hundreds of samples consisting of clays, marls and limestones have been investigated by X-ray diffraction on oriented pastes of the less than 2 μm fraction of decarbonated sediments. Defloculation of clays was done by successive washing with demineralised water after removing carbonates from the crushed rock with 0.2 N HCL. The size-fraction <2 μm, was separated by sedimentation according to the procedures described by Holtzapffel.[12] The XRD analysis was performed using a Philips PW 1730 diffractometer with Cu-Kα radiation and Ni filter. All samples were analysed after air-drying, ethylene glycol solvation and after heating at 490°C for 2 h. The identification of clay minerals was made according to the position of the basal reflections on the three X-ray patterns.[17,21] Semi-quantitative estimations were based on intensity ratios of the 001 reflection of the respective clay species.[12] Transmission and scanning electron microscopic observations have been performed on selected samples, according to the main clay mineralogical assemblages identified by X-ray diffraction.

3. Clays in Triassic rocks

The Triasic basin of the western part of High Atlas represents the classic example for understanding geological events during Triassic stage. It is composed by a very thick detrital sedimentary series covered by Triassic-liassic basalt flow deposition.[3,15] The structuration of Triassic Moroccan margin is mainly controlled by the reactivation of hercynian fault.[19] Clay assemblage of the Triassic sediments of the western High Atlas is organised in three mineralogical zones: (1) the lower zone presents a relatively constant clay assemblage composed mainly by illite (60–75%) and smectite (20–25%). Chlorite and irregularly mixed layer illite-smectite represent less than 10–20% of clay assemblage. Illite cristallinity is also constant in this zone and is about 0,3–0,4°2θ. (2) In the intermediate zone, chlorite (50–90%) is the dominantly clay mineral, associated with illite (20–35%). Illite cristallinity is varied between 0,2°2θ at the top of the zone, and 0,35°2θ at the bottom. Global evolution of clay assemblage in this zone shows a decreasing content of chlorite in aid of corrensite, at the bottom. (3) The upper zone lacated immediately under basalt flow is about 1.5–2 m in thickens. Clay assemblage is composed by corrensite (30–75%), chlorite (0–15%), kaolinite or smectite (10–25 %) and talc (5–40%).

Several assumptions, often contradictory, were advanced to explain the origin and the mode of formation of these clay assemblages: neo-formation or aggradations of minerals in strongly chemical sedimentary environment (Lucas, 1962), hydrothermal metamorphism,[19] epimetamorhism related to the Atlasic rifting,15 heritage of ante-Permian metamorphic minerals on the plate of Meseta[4] and post-sedimentary hyperthermic event in the atlasic domain, accelerating the transformation of clays of surface.[20]

According to our recent investigations,[5] the formation process of clay minerals of the Triasic series surmounted by the basaltic casting can be considered in the following way. During the deposit of the triassic red formation, a contemporary distensive tectonic activity with sedimentation affects the basin. This involves a destabilization of the edges of the basin, which supply the primary mineral at the basin (well crystallized illite and chlorite). The existence of a hot and seasonally humid supports the development of smectites in low altitudes. The contribution of heat and chemical elements, in particular of magnesium, consecutive with the installation of basaltic casting, involves changes of the equilibrium conditions of the various phases. The detrital clay minerals become unstable under these new conditions and the new hydrothermal phases precipitate from fluids of high temperature. The intermediate part of the series is characterised by a partial transformation of the sediment; the smectites are transformed into corrensite then chlorite, and the illites became more crystalline. In the higher part, the hydrothermal fluids rich in magnesium transform completely the sediment. Their migration involves the precipitation of new phases (smectite, corrensite, chlorite and talc). In an ultimate stage, we assist to the precipitation of siliceous and silico-aluminous minerals (quartz and kaolinite). Progressively with accumulation of post-triassic deposits, the effects of burial diagenesis on

Triassic clay minerals take importance. Thus, smectites became instable under these new conditions, and is transformed into illite-smectite irregular mixed layer then in illite.

The Triassic detrital formations provide the most developed and representative outcrops of clays of the area. However, the industrial exploitation remains very insufficient in comparison of the quality and the quantity of clays which offers these formations. Indeed, these clays show characteristics similar to those of the area of Benahmed (Western Meseta), much appreciated by the ceramists of the industrial park Casablanca-Settat. Some clay deposit of the area of Marrakech were exploited formerly for the industry of the terra cotta (pottery, red bricks, tiles), but were quickly abandoned for multiple reasons. One of the major problems which are opposed to the exploitation of these clays is the heterogeneity of the outcrops. Indeed, in some levels of the series, the content and the nature of clays can be very variable within the same outcrop. In addition, the presence of evaporates (salt or gypsum) in the higher part of the series, just as the presence of altered level, make these clays not easily exploitable in industry of the terra cotta. Thus, the choice of the careers must be founded on a very detailed geological investigation. It is the only means which permits to avoid the zones of transitions which are likely to provide clays of variable compositions, as the zones of alterations which would give the clays enriched in carbonates, in evaporates and in hydroxylated minerals.

4. Clays in Jurassic and Cretaceous

The Jurassic-Cretaceous period is represented in the western High Atlas by a very thick sedimentary dominated by marls and calcareous beds with interbedded silty yellowish-green and gypsum. The lithology and fauna indicate fluctuations of water depth from continental and outer littoral to inner shelf depths. The Cenomanian-Turonian period is believed to have been the time of maximum eustatic high.[23] This transgressive period is characterized in the Atlantic Ocean by the deposition in deep sea environments of highly pelagic deposits with extremely low terrigeneous influences. The Turonian shows a very typical facies of hemipelagic; it comprises laminated calcareous marls, cherts and lenticular chert beds. The Jurassic-Cretaceous sequences decrease eastward drastically in thikness, and these facies grade rapidly into dolomites.

The clay assemblage evolution of the Jurassic-Cretaceous series shows a very important geographical and stratigraphical variation.[10] During the Jurassic, clay fraction is usually dominated by illite with high critallinity and sometimes chlorite. The lower Cretaceous is characterised by the apparition of the kaolinite during the Barremian and the development of smectites during the Aptian. During the late Cretaceous, smectite with varying amounts of illite, kaolinite, traces of varied mixed-layer clays and sometimes palygorskite composed the clay fraction. Toward the west of the basin, the clay fraction of the Cenomanian-Turonian period is mainly composed by dioctaedral smectite (beidellite type) and irregular interstratified I/S with varying high proportion of smectite layers.[7] Toward the east, irregular interstratified I/S with lower percentages of smectite layers is the major component of the clay fraction, associated to illite, sometimes palygorskite and to siliclastic influence (quartz, micas).

Clay mineral assemblages of Jurassic and Cretaceous sediments deposited in the Western High Atlas are controlled by the evolution of sedimentary environments through time and by diagenetic processes. Detrital influences on the clay mineral successions of western High Atlas depend mainly on tectonic activity, climate and eustatic fluctuations. Active subsidence on the Atlas Gulf associated with the opening of the Atlantic Ocean is responsible for the abundance of illite in Jurassic and Early Cretaceous sediments. By contrast, the smectite-rich clay sedimentation occurring during late cretaceous, results from the stabilization of the Atlas Gulf margins. The smectitic clay sedimentation is also enhanced in Cenomanian sediments by a decreasing intensity of erosion during that transgressive period. The scarcity of kaolinite suggests arid climatic conditions particularly around the Jurassic/Cretaceous boundary. Burial diagenesis occurs mainly during Jurassic and lower cretaceous in the southern flank of High Atlas. Smectite is progressively replaced with depth, by illite in sediments deposited in an open marine environment, at the west, and by corrensite-like minerals in evaporate sediments, at the east. Neoformations of clay minerals (kaolinite) occur in pore spaces of sandstones and dolomitic formations outcropping in the Northern flank of the High Atlas.

The industrial exploitation of clays and clay minerals of the Jurassic and Cretaceous series is not very developed because of their low content in the sediments. The clays contained in the marls of Cenomanian and Senonian series are exploited industrially much more for their chemical composition (Si, Al and Fe) that for their physico-chemical properties. Indeed, in the area of Agadir, clays of Senonian sediments are exploited in the industry of cements. The career of the cement factory (CIMAR-Agadir) is established in the marls of Coniacien-Santonien. In the area of Marrakech, the CIMAR-Marrakech cement factory, exploits the contemporaneous levels like addition with the taken raw material of the Paleozoic.

5. Clays during Tertiary

The clay minerals commonly recognized in the fraction <2 μm of Palaeocene-Eocene series from the sections studied on the Marrakech High Atlas region comprise illite, smectite, palygorskite, sepiolite and irregular mixed layer. Kaolinite and swelling chlorite occur very locally. Non-clay minerals in the <2 μm fraction include common quartz, feldspars and gypsum and occasional goethite or opal. The Asni section displays very clear vertical changes of the clay composition (Fig. 1) in sedimentary formations whose thickness exceeds 400 m.

In the lower part of the sedimentary system (Palaeocene-Eocene), the clay fraction is composed mainly by illite, and smectite at the west of the studied area (Esaouira basin). Toward the east (Asni), palygorskite and trioctaedral smectite are the more abundant and ubiquitous minerals. Palygorskite constitute usually 30–60% of the clay species, and trioctaedral smectite 30–45%. In the upper member, palygorskite content increases (60–80%) when smectite decreases. At the top of dolomitic marls, it reaches unusually high values, where it represents 90–100% of the clay fraction of some samples. Sepiolite is more abundant in the clayey-gypsum facies of the upper sedimentary member. In the upper part of the sedimentary system (Miocene-Pliocene), the clay fraction is composed by smectite, illite and palygorskite, from wish contents are relatively constant (35–40% for illite, 20–30% for smectite and 20–30% for palygorskite). Sepiolite is associated preferentially to gypsum, it constitute 10–40% of the clay fraction of some samples. Kaolinite and chlorite are identified only in some samples where constitute a few percents of the clay fraction. The random mixed layer contents are relatively constant (5–10%). In the southern flank of the High Atlas, the composition of clay associations displays the same significant trend, with minor differences. The relative abundance of palygorskite tends to increase upwards and toward the east, when the amounts of illite and smectite tend to decrease.

Palygorskite in the Palaeocene-Eocene series of the Marrakech High Atlas formed mainly in a coastal, schizohaline environment, under an arid to semiarid climate with humid intervals. It formed shortly after the deposition and contemporaneous dolomitization of the rocks and reaches a maximum during a transgression in which the dolomitic marls were deposited during the Ypresian. An Mg concentration, in presence of dolomite was adequate for the palygorskite precipitation, with Si and Al supplied by the dissolution of silicates under alkaline conditions.[6] The development of palygorskite in these marginal restricted basins of the Marrakech High Atlas during the Paleocene-Middle Eocene time is contemporaneous with similar occurrences in the Tertiary basins of West Africa and Middle-east. During the Miocene and Pliocene, the sedimentation deposit is characterised by an important terrigenous influence. The smectites became the most abundant clay minerals in the clay assemblages.

From the industrial point of view, Eocene clay materials of western High Atlas are well known and have been used since ancient times for cottage industry. The first mineralogical study of the Eocene clays was carried out at the Moroccan phosphate basin in the Meseta (North of Marrakech City).[2,14] A recent detail study of clay assemblages of the Palaeocene and Eocene successions of the western High Atlas was made by Daoudi.[6] Until now, the industrial use of clays in these series is still very insufficient, in comparison with their quality and their content in the sediments. Indeed, by their physicochemical properties, the palygorskites know a broad range of application in the industrial field (manufacture of drilling muds, recycling of the spent oils, pharmaceutical uses, use like manure in suspensions, manufacture of absorbant paper… etc.). The improvement of these industrial techniques and the chemical treatments continues to increase the use of these varieties of clay minerals. Each year, new or improved uses are discovered. The fibrous minerals being flexible of employment, their physical and chemical properties can be modified to be appropriate for particular applications.

In the Mio-Pliocene, clay deposit of the bordered of the western High Atlas Mountains (south of Marrakech) is rich in smectites. These clays are gray to green in color, soapy to the touch and exhibits wax consistency to the extraction. These texture point out that of smectites of the mined deposit of the rhassoul, the unique one in the word, exploited in the tertiary lacustrine basin of the jbel Rhassoul (north-east Morocco).[1,16] However, semctites clay minerals of the Mio-pliocene series of the western High Atlas do not present the same interest because of their high percentage of lime.

6. Clays of Quaternary

In the area of the High Atlas of Marrakech, the formations of Quaternary arise in various forms. They are developed either in very broad forms of alluvial cone in Piedmont of the chain, or in the form of alluvial terraces in the various valleys, south of Marrakech.[18] The materials composed mainly by conglomerates and clays come from disintegration of materials forming the atlasic chain and from the erosion of soils. The clay fractions of the sediments are dominated by the illite. Other mineral phases can be associate to illite such as kaolinite, smectite or palygorskite.

The industrial exploitation of clays of the quaternary formations is the most significant in the area of the High Atlas of Marrakech, because of the high content of clays in the sediments on the one hand and the easy access to the outcrops on the other hand. Clays and silt deposited in the beds of the various wadis of the area are especially used in cottage industry and in the industry of the bricks.[11,22]

7. Conclusion

In the area of the Western High Atlas, the studies carried out on clay materials of the sedimentary sequences during these 2 last decades, permit to characterize and to identify the composition and the evolution of the clay assemblages since the Triassic age until now. These studies made it possible to establish a good base of sedimentological, mineralogical and geochemical data of the clay fractions. The genesis and the geological significance of these clays are also largely approached. However, the studies relating to the industrial use of these clays are still rare.

The results obtained during these studies show that the clay materials of this area are abundant and diversified. They constitute a considerable potential economic source for the inhabitants of the area. Moreover, until now, the industrial use of clays in these series is still very insufficient, in comparison with their quality and their content in the sediments. Indeed, by their physicochemical properties, several clay minerals (particularly palygorskites and semctites) know a broad range of application in the industrial field (manufacture of drilling muds, recycling of the spent oils, pharmaceutical uses, use like manure in suspensions, manufacture of absorbant paper… etc.). The improvement of the industrial techniques and the chemical treatments continues to increase the use of these varieties of clay minerals. Unfortunately, in the western High Atlas area, these clay minerals are still used in cottage industry.

Studies must be carried out in this direction in order to develop these clay deposits, which can constitute an effective means for the economic development of the area. These studies will make it possible to guide the craftsmen owners of clays towards rough materials more adapted to their industry, to use rough materials with highly specific properties for more adapted industries and more developing, and finally to encourage the industrialists to develop this type of industry which is missing in the area.

ACKNOWLEDGEMENTS

This work was supported by Protars II – P23/04 (Programme thématique d'Appui à la Recherche Scientifique), the PSR 2001 (Programme de Soutien à la Recherche de l'Université Cadi Ayyad) and convention between the CNRST (Morocco) and the GRICES (Portugal). Much of the data submitted in this paper were extracted from my thesis, submitted at the Cadi Ayyad University of Marrakech. Analyses were done at the Cadi Ayyad University of Marrakech, the University of Aveiro (Portugal) and the University of Sciences and Technology of Lille (France).

References

1. A. Barrakad, *Mines et Energie*, Rabat, 49:138–142 (1981).
2. A. Boujo, *Notes Mémoires du Service géologique du Maroc*, 276:82–152 (1986).
3. R.H. Brown, Amer. *Assoc. Petroleum Geol. Bull.*, 64:988–1003 (1980).
4. H. Chamley and P. Debrabant, *C. R. Acad. Sci., Paris*, 296:651–656 (1983).
5. L. Daoudi and J.L. Pot De Vin, *C. R. Géoscience*, 334:463–468 (2002).
6. L. Daoudi, *J. Afr. Earth Sci.*, 39(3–5):353–358 (2004).
7. L. Daoudi, F. Rocha, B. Ouajhain, P. Callapez and D. Chafiki, *EUROCLAY 2007, Aveiro-Portugal*, 126 (2007).
8. L. Daoudi and J.F. Deconinck, *J. Afr. Earth Sci.*, 18:123–134 (1994).
9. L. Daoudi, *Thèse Université Lille I*:196 (1991).
10. L. Daoudi, *Thèse d'Etat Université Marrakech*:247 (1996).
11. A. Knidiri, L. Saadi, M. Waqif and L. Daoudi, *International Meeting on Materials for Electronic Applications, IMMEA-2007-Marrakech*:67 (2007).
12. T. Holtzapffel, *Soc. Géol. Nord.* 12:136 (1985).
13. S. Huon, J.J. Cornee, A. Pique, N. Rais, N. Clauer, N. Liewig and R. Zayane, *Bull. Soc. géol. France*, 164:165–176 (1993).
14. D. Jeannette, A. Moniton, L. Ortelli, and H. Salvan, *20ème Congrès International de Géologie-Mexico, Etude argiles*:53–60 (1956).
15. F. Medina, *Geol. Mag.*, 128:525–536 (1991).
16. G. Millot, *C. R. Acad. Sci., Paris*, 238:257–259 (1954).
17. D. M. Moore and R.C. Reynolds, *Oxford University Press*:322 (1989).
18. A. Nahid, *Rev. C. & G.*, 15(1–2):135–160 (2001)
19. A. Pique and E. Laville, *C. R. Acad. Sci., Paris*, 317:1215–1220 (1995).
20. N. Rais, *Thèse d'état Faculté des Sciences Fes-Saiss*:250 (2002).
21. R.C. Reynolds, In: Brindley G.W., Brown G. (Eds.), *Mineralogical Society, London*: pp. 49–303 (1980).
22. L. Saadi, A. Khalil, M. Waqif, H. Ahamdane and L. Daoudi, *Silicates Industriels, Ceramic Science and Technologie-Belgium* (in press) (2006).
23. P.R. Vail, F. Audemard, P. Eisner and G.A. Perez-Cruz, *In. Cyclic and Event Stratigraphy II, Springer-Verlag, Berlin* (1992).

TEM ANALYSIS OF ADVANCED DEVICES FOR ELECTRONICS OR SPINTRONICS: FROM STRUCTURE TO PROPERTIES

MARIE-JOSE CASANOVE*, CHRISTOPHE GATEL, ANNE PONCHET AND CHRISTIAN ROUCAU
CEMES, CNRS, 29 rue J. Marvig, B.P. 94347, 31055 Toulouse cedex 4, France

Abstract. During the last decades, transmission electron microscopy has become an essential tool for solving problems involving microstructural questions in materials science. Its unique ability to provide direct imaging of particular regions of the material together with structural information made it highly valuable in the study of complex materials including domains as ferroelastic or ferroelectric materials. Moreover, TEM is the dedicated technique for studying defects (dislocations, boundaries, precipitates, interfaces) which can either affect or enhance the desired properties (mechanical, optical or magnetic properties, transport). After a brief review of typical answers given by TEM on materials questions, we discuss the interest of the technique to investigate new devices for energy, communication or security applications. We show in particular that new developments in TEM now allow us to determine various fields in the specimens, in particular strain or magnetic fields, which bring direct information on the physical properties.

Keywords: Transmission electron microscopy, interfaces, nanoparticles, semiconductors, strained layers, magnetism

1. Introduction

Advanced devices for new electronics, security or information storage now include many different components with various shapes and compositions. They are fabricated by means of a wide variety of sophisticated techniques including thin layer deposition, epitaxial growth, lithography, advanced chemical synthesis.... Their properties strongly depend on the characteristics of their

*To whom correspondence should be sent: Email: casanove@cemes.fr

I. A. Luk'yanchuk and D. Mezzane (eds.), *Smart Materials for Energy, Communications and Security.*

different components (size, morphology, structure and chemical composition). With the advent of nanotechnologies, accurate knowledge of these different characteristics require investigations from the micrometer to the atomic scale. While most of the structural techniques provide statistical information, transmission electron microscopy (TEM) offers the main advantage of a very good spatial resolution in image mode. Moreover, its different modes of operation give access to a wide range of scales (from several micrometers to interatomic distances) and combine analyses in direct and reciprocal spaces (imaging and diffraction). Besides, TEM can be fitted with analytical devices as Energy Dispersive X-ray spectrometer (EDX) or Electron Energy Loss Spectroscopy (EELS). Thus, while modern devices can now be controlled at a very fine scale, TEM, and specially its high resolution mode which allows us to image the atomic columns, has become in many fields the only possible characterization method (for instance to image interfaces in heterostructures or to observe specific arrangements of nanomaterials).

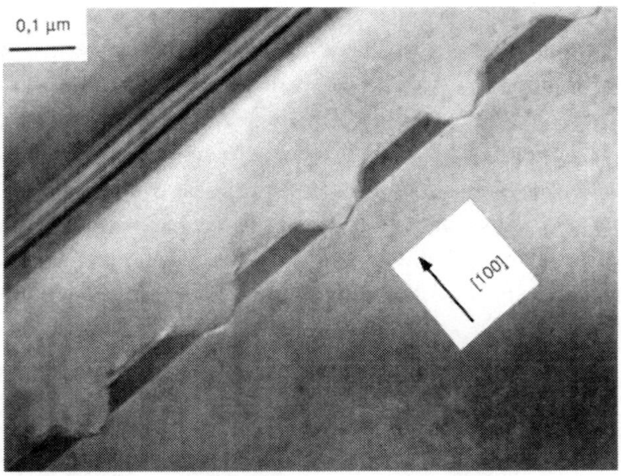

Figure 1. Device for optoelectronics showing multiple quantum wells consisting of quaternary GaInAsP compounds and an optical grating in InP.

Figures 1 and 2 emphasize the interest of TEM investigations in two different devices. Figure 1 presents an example of a typical component for optoelectronics showing an optical grating (center of the figure) grown and patterned on an InP substrate. The multiple GaInAsP quantum wells separated by barriers (at the top of the figure) were grown in a second step. The image was obtained from a cross-section of the device, at medium magnification, in order to observe the regularity of the optical grating and to further analyse the characteristics of the active region of quantum wells. Figure 2 presents ligand-stabilized iron nanoparticles synthesized by a soft chemistry route. Self-assembling of nanoparticles in electronic devices is a bottom-up alternative to lithography for the ever finer devices required by microelectronics and magnetic-storage industry.[1,2] Very few techniques, if not only one, are likely to provide information on such particles deposited on carbon foils and surrounded by ligand molecules. The TEM image clearly show that the ligands (here oleic acid and hexadecylamine) promote self-assembling of the particles in a 2D-lattice (10.3 nm from center to center, mean diameter 7.5 nm). The characteristics of the superlattice are better evidenced in the Fourier transform of the image in inset.

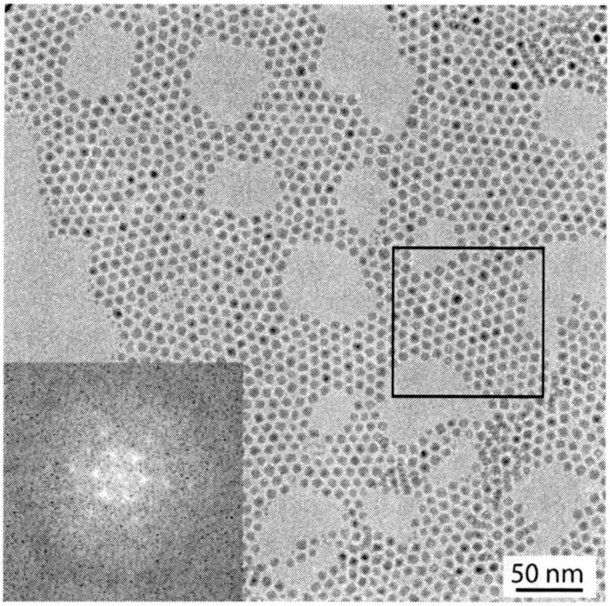

Figure 2. 2D superlattice of self-assembled Fe particles. Fourier transform in inset.

2. Microstructure and properties

It is well known that defects are detrimental to properties. However, properties of materials also rely on their defects if one can carefully control the material microstructure. Analysing defects is then essential to master the behaviour of materials. Study of structural defects, as imaging dislocations in metallic alloys was in fact one of the very first applications of TEM to materials science. Since then, numerous microstructures have been investigated and in particular walls separating domains in smart materials. Well known examples of the relation between properties and microstructure are outlined below.

- Mechanical properties can be analysed through the study of dislocations (nature, mobility, spatial distribution…).

- Transport properties are also tightly linked to structural defects which act as charge carrier recombination centers in semiconductors, pinning centers in superconductors ....
- Domain walls (thickness, nature, mobility) play an essential part in ferroelastic or ferroelectric porperties.

2.1 HOW CAN TEM IMAGE DEFECTS

Different imaging modes in TEM can give evidence for the presence of defects. However, one of the most useful technique to study the microstructure is the so-called bright-field/dark field conventional TEM. In this mode the sample is oriented in such a way that only one diffracted beam is highly excited, which means that a single family of lattice planes is very close to the Bragg conditions for diffraction. When a single diffracted beam (or respectively the transmitted beam) is selected in the objective aperture, a dark field image is formed (respectively bright field image). The contrast in the image results from local variations in the orientation or spacing of the diffracting planes. Defects such as dislocation lines, which induce local deformation of the lattice planes, can thus be imaged owing to the remarkable sensitivity of the technique.

High resolution TEM (HREM) is the privileged technique for imaging atomic columns in the specimen. To form HREM images, the specimen is oriented in such a way that the incident beam is parallel to a family of atomic columns. Several diffracted beams are selected in the objective aperture which is centered on the transmitted beam. A lattice image results from the interference of these different beams. If the specimen is thin enough, the contrast in the image is then proportional to the projected crystal potential. The resolution in the image is limited by the different aberrations of the microscope, the spherical aberration of the objective lens being the most critical.

Details on the different TEM techniques and how to use them can be found for instance in the William and Carter books.[3]

2.2 IMAGING THE MICROSTRUCTURE

2.2.1 Dislocation and domain walls in dark field mode

Figure 3 presents a plane view image of semiconducting islands of GaSb grown on a (001) GaAs substrate. The lattice mismatch between the two crystal lattices is accommodated by a misfit dislocations network. The dark field image presented in the figure gives evidence for the remarkably regular spacing of the dislocation lines. Their spacing reveals the full relaxation of the epitaxial stress.[4,5]

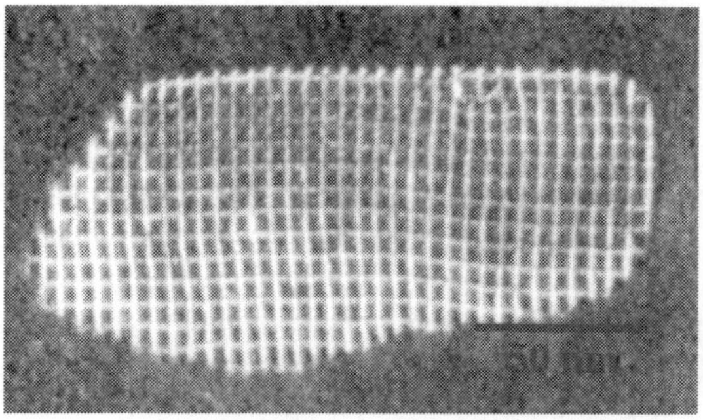

Figure 3. Network of misfit dislocations at the GaSb/GaAs interface. (Courtesy Dr. A. Rocher, CEMES, CNRS, France.).

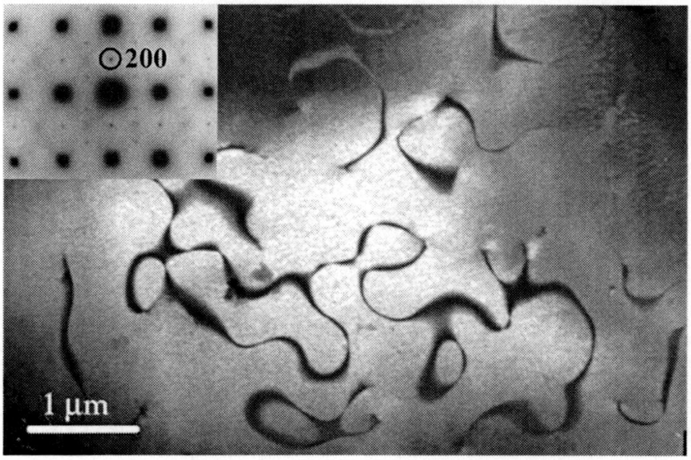

Figure 4. Shape memory alloy-200 dark field image showing the antiphase boundaries separating two B2 ordered alloy domains.

Another example of microstructure is observed in Fig. 4, taken from a specimen of a CuZnAl shape memory alloy.[6] This alloy presents a B2 ordered body-centered cubic phase (austenite) at high temperature. The phase transition between martensite and austenite phases results in the presence of different ordered domains separated by antiphase boundaries. Such boundaries are clearly observed in the dark field image formed by the 200 reflection, which belongs to the B2 structural phase. Indeed the B2 order is lost in the domain walls, which then appear as dark regions in the image.

2.2.2 High resolution imaging

Figure 5 presents twin walls observed at the atomic scale in a [110] oriented gold thin layer. This HREM image has been formed using a spherical aberration corrected TEM microscope operated at 200 kV and fitted with a field emission gun (Cs-corrected TECNAI F20 at the CEMES in Toulouse, France). Such instruments gather the more efficient developments now achievable in TEM. Compensation of the spherical aberration (Cs) of the objective lens has two major effects: enhancement of the point resolution in the image up to the information limit (here 0.12 nm) and even more important in the study of advanced devices, suppression of contrast delocalization effects at interfaces.[7–9]

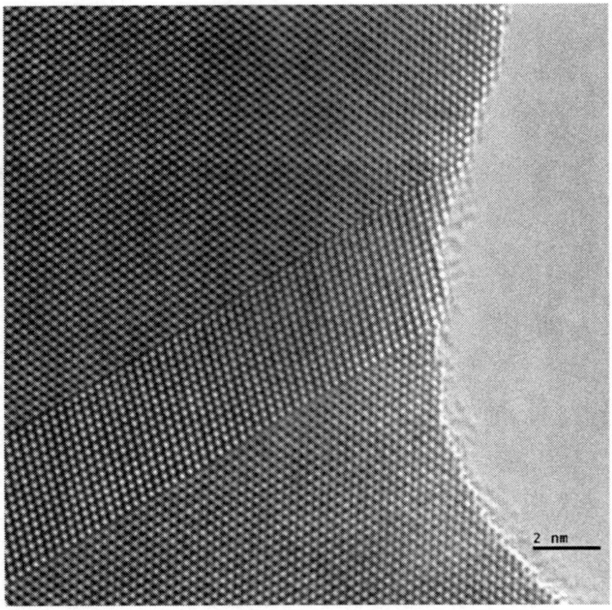

Figure 5. Twin walls in [110] oriented gold film. Cs-corrected HREM image.

3. Inside the nanoworld – how far can we go?

With the advent of nanotechnologies and the use of ultrafine material, structural characterization has to be performed at the ultimate scale. At interfaces for instance, not only the stacking sequence but also the precise composition and position of the different planes must be determined. As a matter of fact, small displacements of the atomic columns, due for instance to epitaxial stress or to charge effects, are likely to influence the properties of ultrafine layers. This is also the case for small variations of the chemical content. The same precision is required in the analysis of other kinds of defects. Besides, because of the fabrication processes, many regions of the devices undergo important stresses, which will alter their properties. Measuring strain or stress in various regions of a component is now the subject of intense work and the most recent developments of TEM will play an important part in this domain. Indeed, recent experiments in all these fields benefit from highly coherent electron sources produced by field emission guns, the absence of contrast delocalization at the interfaces thanks to the use of Cs corrector of the objective lens, which also lead to enhanced point resolution of 0.1 nm now.

We present in the following some of the most recent examples on TEM analysis at the nanoscale of materials involved in the important fields of security, energy or communications. In particular, we emphasize the interest to combine modelling and experiments for analysing the properties of these nanomaterials.

3.1 NANOPARTICLES FOR MAGNETISM

Nanomaterials share the particularity of a strong contribution of surface atoms which can represent more than 50% of the total number of atoms in the so-called nanoparticles. More sensitive to their environment and in particular to surface stresses, these materials adopt structural and chemical orders that can strongly differ from the ones in the bulk material. In parallel, their macroscopic properties are considerably enhanced, or even completely original. Such materials are therefore particularly interesting from a fundamental point of view. More recently, they have been the subject of intense research in specific fields as information-storage and magneto-electronics, which require the development of magnetic materials with still higher technological potential and level of device integration.

An important challenge remains the ability to tune the magnetic properties in order to optimize the devices, and this comes with improved understanding of the underlying physical mechanisms. In this context, particular attention has been paid during the last decade to the structural and chemical organization in nanoparticles of pure metals and of their alloys which display the highest magnetic anisotropy as well as the highest spontaneous magnetization.

One of the major interests of TEM in these studies is to provide combined evidence of spatial distributions, morphologies and structures.

3.1.1 Morphology and chemical composition in Fe nanoparticles

Figure 6 shows a cube-shaped Fe nanoparticle synthesized by a soft chemistry route. Such particles have been observed to form dense 3D lattices.[10] This HREM micrograph of a single particle taken in a 2D superlattice shows that the core of the particle consists in pure iron crystallized in its body-centered cubic (bcc) phase. The larger facets of the particles, and hence the more stable ones, are parallel to (100) planes, which is surprising in a bcc structure. In fact, the surface energy of these facets is indeed lower than the one of the (110) facets

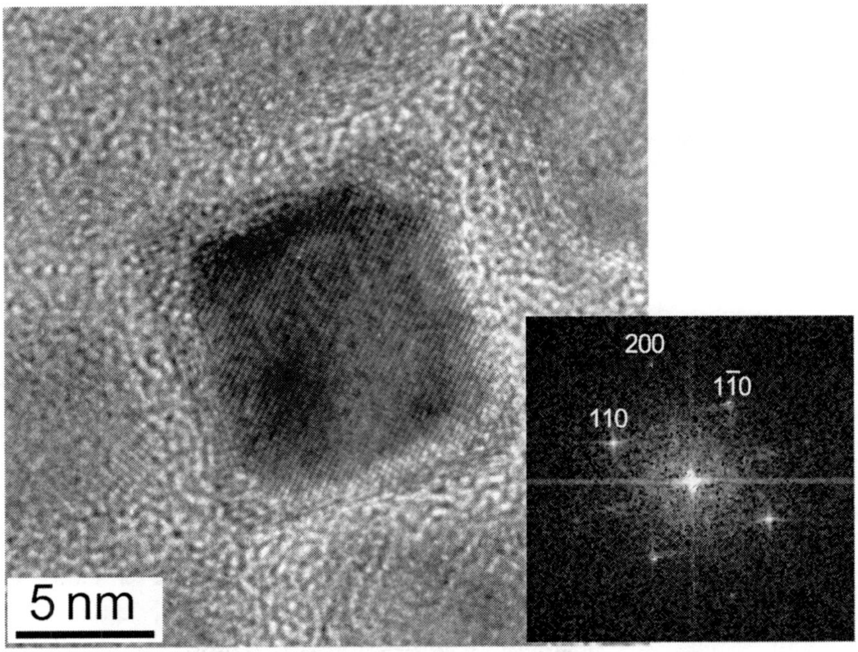

Figure 6. HREM image of a Fe particle with cube morphology – the Fourier transform inset shows that the main facets are parallel to (100) planes.

having higher atomic density because of ferromagnetism. The HREM image gives also evidence for the presence of an oxide layer at the particle surface.

3.1.2 Chemical order in CoRh particles

The way the different species organize inside a particle, and specially core-shell organizations, usually requires the use of element-sensitive techniques. Energy filtering TEM (EFTEM) technique, which requires a post-column filter, offers the advantage to provide chemical maps, highly useful in complex device analyses. Figure 7 presents a chemical map of CoRh nanoparticles resulting from the combination of Co and Rh elemental maps. Element maps were obtained by energy filtering on the Rh-$M_{4,5}$ and on the Co-$L_{2,3}$ edges respectively. It shows that rhodium atoms concentrate in the nuclei while cobalt atoms spread over much larger zones corresponding to the particles size.[11]

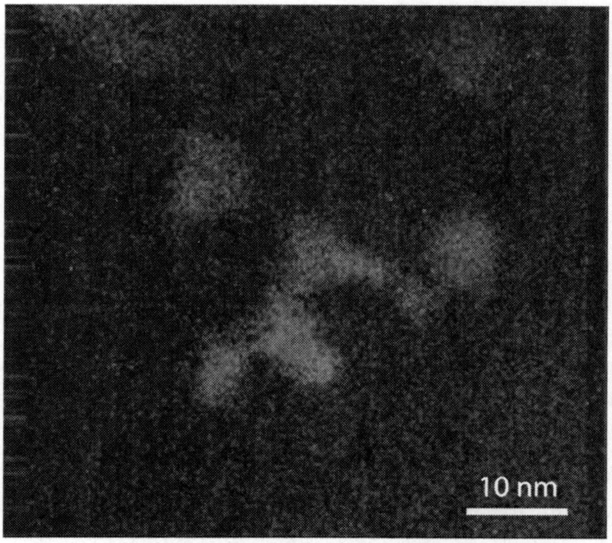

Figure 7. Elemental map of CoRh particles obtained by Energy filtered TEM (EFTEM)-Evidence of a Rh core (blue) and Co shell (red) chemical order.

3.2 STRAIN FIELD DETERMINATION IN SEMICONDUCTING HETEROSTRUCTURES

3.2.1 Internal stress, an essential parameter in advanced devices

Internal stresses result from the physical processes involved in the nanostructuration of crystalline materials. Epitaxial stresses in particular are now fully integrated in the design of opto-electronic, micro-electronic or magnetic devices. Epitaxial growth is a complex out of equilibrium process which involves both thermodynamic (competition between surface energy, interface energy, elastic energy…) and kinetic mechanisms. The actual microstructure of a strained epitaxial device is thus extremely dependant on the growth conditions, which can modify the growth mode (2D or 3D) and the quality of interfaces (nature of chemical bonds, roughness, interdiffusion…). The knowledge of the stress and strain states is then essential for mastering the whole design-elaboration-properties process.

3.2.2 Strain field determination through combined modelling and HREM image processing

Figure 8 presents an InAs quantum well embedded in two $Ga_{0.47}In_{0.53}As$ barriers; the lattice parameter of InAs is 0.605 nm while $Ga_{0.47}In_{0.53}As$ has the same lattice parameter than the InP substrate (0.582 nm).[12] Accommodation of

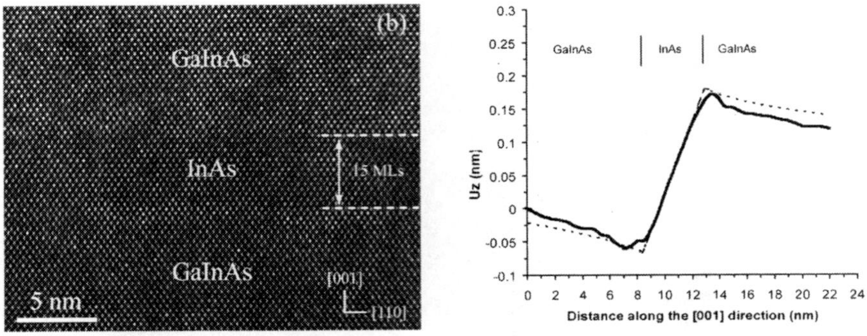

Figure 8. HREM image of 15 monolayers of strained InAs (left); the $Ga_{0.47}In_{0.53}As$ barriers are lattice matched to InP substrate. Displacements of the atomic positions along the growth direction (right).

the lattice mismatch (3.2% in this case) during the epitaxial growth results in an internal strain in the InAs quantum well. The principle of strain field determination consists in analyzing the displacements of the atomic positions in an HREM image with respect to a virtual lattice. This virtual lattice is built from a zone of the crystal considered as an unstrained reference. Displacements are determined by Fourier image processing[13] or peak-finding[14] techniques. Note that, in the vicinity of interfaces, displacements can be affected by contrast delocalization; as previously outlined, this can now be avoided by use of spherical aberration (Cs) corrector of the objective lens. The accuracy of strain determination is limited by the partial relaxation of the internal strain due to sample thinning with respect to the bulk sample.[15] Such relaxation must be taken into account in the analysis. In the example below, the displacements experimentally determined from the Fourier analysis of the image (full line) are well fitted by finite element modelling of a thinned cross-sectional sample (dashed line).

3.2.3 Some other techniques: CBED and TEM curvature

Convergent Beam Electron Diffraction (CBED) is also able to provide information on the strain components with a remarkable sensitivity down to 10^{-4}. Due to the convergence of the incident beam, the transmitted beam forms a disk in the focal plane of the objective. This disk displays fine dark lines (deficiency lines), each of them being associated to a particular diffracting plane. The relative position of these lines is extremely sensitive to the crystal lattice parameters. Strain field gradients have thus been measured in patterned devices with a spatial resolution of about 10 nm.[16,17] As devices often present heavily strained and inhomogeneous regions, much effort is now devoted to the development of reliable methods taking into account inhomogeneous strain fields along the beam direction.[18–20]

Surface relaxation effect, which is unavoidable, can in some cases be exploited. This is done in the curvature method which measures the internal stress of a layer from the curvature of the substrate.[21] Curvatures are usually measured by deflection of a laser beam or by x-ray diffraction. Curvatures are widely amplified if the substrate is thinned like in TEM plane view observations and they can be determined with an high accuracy from extinction bending contours in conventional TEM images.[22] This new method is well adapted to nanometric layers with epitaxial stresses of an order of magnitude around a few GPa.[23] It differs from CBED and HREM techniques in the sense that one uses the substrate deformation (i.e. the curvature) to determine the stress in the layer, while CBED and HREM (being sensitive to atomic positions) give access to strain.

3.3 MAGNETIC NANOLAYERS FOR HIGH DENSITY RECORDING

During the past 50 years, increase of the recording density in hard disk drives has been mainly achieved by decreasing the size of the magnetic grains in the storage layer. However, the recorded data can be erased by thermal fluctuations of the magnetization when the grain volume is highly reduced (the magnetic energy per grain becomes indeed comparable with the thermal energy). This is why hard disk drives include materials with large uniaxial magnetic anisotropy energy.[24,25]

An important evolution happened with perpendicular recording which opened the route to high density data storage.[25] A major issue is to evaluate the ability for a recording head to flip a bit of information, and hence to estimate the stray field to magnetic moment intensity ratio inside the sample.

3.3.1 Imaging magnetic domains by TEM – some basics of Lorentz microscopy and electron holography

Lorentz Transmission Electron Microscopy (LTEM) is one of the techniques enabling analysis of local magnetic properties.[26] The domain structure of magnetic materials can be observed by this technique which was improved over the past few years and also enables dynamical investigations. Classical LTEM relies on the fact that an electron beam will be deflected by a Lorentz force when its trajectory is perpendicular to the magnetic induction in the analyzed sample. The electron beam will be sensitive to the perpendicular magnetic induction averaged along the electron path (sample plus vacuum). Electron Holography (EH) in TEM is another method to study magnetic configurations, well adapted to the nanometer range.[27–29] This technique gives access to both the amplitude and the phase shift of the electron wave after its interaction with the studied material. The phase shift is sensitive to both electric and magnetic fields in the sample.

3.3.2 Analysis of FePd nanolayers

Maps and quantification of magnetic induction are obtained through specific data processing which depends on the technique (LTEM or EH). Magnetic maps can then be compared with micromagnetic simulations. An example of a FePd ordered (L_{10}) alloy layer having a high perpendicular magnetic anisotropy[24] is given in Fig. 9. This layer was epitaxially grown on a chemically disordered $FePd_2$ (or FePd) buffer layer having a vanishing anisotropy.

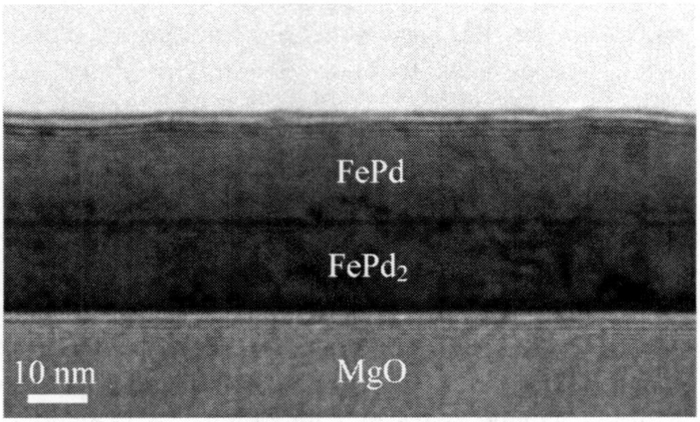

Figure 9. Cross-sectional view in conventional TEM of a FePd ordered layer epitaxially grown on a $FePd_2$ disordered layer.

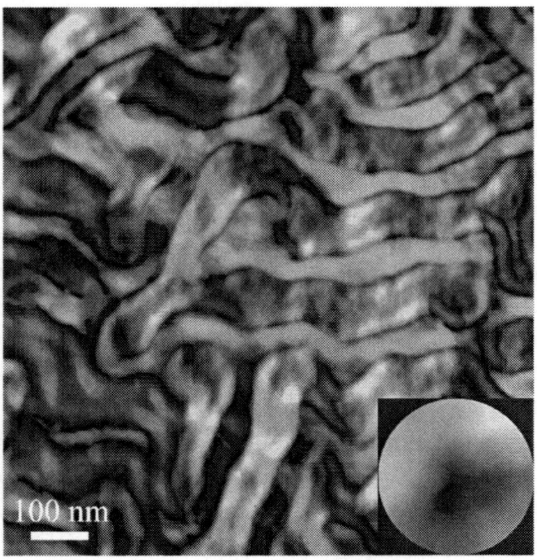

Figure 10. Map of the magnetic induction obtained in plane view by LTEM on a FePd ordered layer epitaxially grown on a $FePd_2$ disordered layer. The inserted color wheel indicates the direction of magnetization (color) and the strength of the magnetic signal (intensity).

Magnetic maps have been obtained in plane view by LTEM (Fig. 10) and in cross-sectional view by EH (Fig. 11) where the "up" and "down" domain configuration is clearly observed with flux closure within the FePd disordered layer.[30,31]

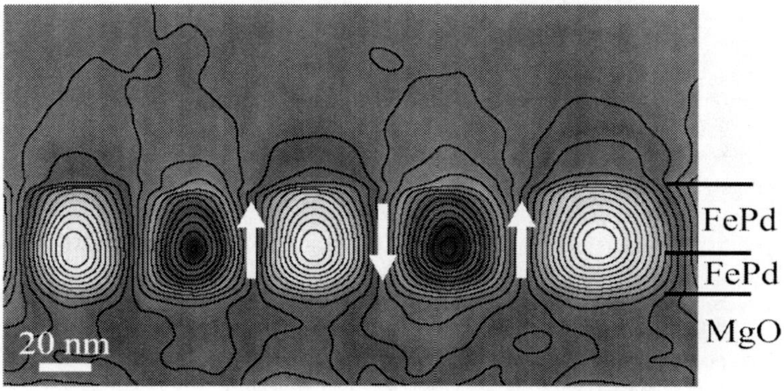

Figure 11. Magnetic contribution to the phase of the electron wave in cross-sectional view by EH. The equi-phase lines reveal the direction of the magnetic induction.

4. Conclusion

Well established in materials science, TEM has recently known very important developments which make it an even more valuable tool for structural and chemical characterization of advanced materials. We have shown in particular that quantitative analyses of HREM images, can now be performed with very high accuracy, for instance to determine strain fields in stressed materials. Magnetic properties can also be investigated at the nanometer scale using image processing of magnetic maps or holograms.

Many unsolved interface problems can now be reconsidered and important fields as security, energy or communication should largely benefit from these new achievements in the next future. For instance, TEM in-situ catalysis experiments on nanoparticles designed for important applications like gas sensors are already in progress.

References

1. K. Soulantica, A. Maisonnat, M.-C. Fromen, M.-J. Casanove, and B. Chaudret, *Angew. Chemie International Edition.* 42:945–1949 (2003)
2. F. Dumestre *et al.*, *Faraday Discussions* 125:265–278 (2004)
3. D.B. Williams and C.B. Carter, Transmission Electron Microscopy: A Textbook for Materials Science. Plenum, New York (1996)

4. A. Rocher, A. Ponchet, S. Blanc, and C. Fontaine, *Applied Surface Science* 188:55 (2002)
5. J. M. Kang, M. Nouaoura, L. Lassabatère, and A. Rocher, *Journal of Crystal Growth* 143:115 (1994)
6. L. Buffard, P. Charbonnier, C. Roucau, and J. Jaud, *Journal-de-Physique-IV* 1(C4):123–128 (1991)
7. M. Haider, H. Rose, S. Uhlemann, E. Schwan, B. Kabius, and K. Urban, *Ultramicroscopy* 75:53 (1998)
8. M. Lentzen *et al.*, *Ultramicroscopy* 92:233 (2002)
9. C.L. Jia, M. Lentzen, and K. Urban, *Science* 299:870 (2002)
10. F. Dumestre, B. Chaudret, C. Amiens, P. Renaud, and P. Fejes, *Science* 303(5659):821–823 (2004)
11. M.-C. Fromen, P. Lecante, M.-J. Casanove *et al.*, *Physical Review B* 69:235416 (2004)
12. C. Gatel and A. Ponchet, *to be published*
13. M.J. Hytch, E. Snoeck, and R. Kilaas, *Ultramicroscopy* 74:131 (1998)
14. R. Bierwolf, M. Hohenstein, F. Phillipp, O. Brandt, G.E. Crook, and K. Ploog, *Ultramicroscopy* 49:273 (1993)
15. M.M. Treacy and J.M. Gibson, *Journal of Vacuum Science and Technology B* 4:1458 (1986)
16. A. Toda, N. Ikarashi, and H. Ono, *J. Crystal Growth* 210:343 (2000)
17. S.L. Toh *et al.*, *Journal of Vacuum Science and Technology* B 23:940 (2005)
18. L. Clément, R. Pantel, L.F.T. Kwakman, and J.L. Rouvière, *Applied Physics Letter* 85:651 (2004)
19. F. Houdellier, C. Roucau, L. Clément, J.-L. Rouvière, and M.-J. Casanove, *Ultramicroscopy* 106:951 (2006)
20. F. Houdellier, A. Altibelli, C. Roucau, and M.J. Casanove, *Ultramicroscopy* 108:426–432 (2008)
21. L.B. Freund, J.A. Floro, and E. Chason, *Applied Physics Letters* 74(14):1987 (1999)
22. A. Ponchet, M. Cabié, and A. Rocher, *European. Physical Journal: Applied Physics* 26:87 (2004)
23. M. Cabié, A. Ponchet, A. Rocher, V. Paillard, and L. Vincent, Applied Physics Letters 84:870 (2004)
24. J.-P. Attane, Y. Samson, A. Marty *et al.*, *Applied Physics Letters* 79(14):794–796 (2001)
25. S. Iwasaki and Y. Nakamura, *IEEE Transactions on Magnetics* 13:1272–1277 (1977)
26. J. Chapman, *Journal of Physics D: Applied Physics* 17:623–647 (1984)
27. D. Gabor, *Nature* 161:777–778 (1948)
28. A. Tonomura, *Advances in Physics* 41:59 (1992)
29. R.E. Dunin-Borkowski, M.R. McCartney, B. Kardynal, S.S.P. Parkin, and D.J. Smith, *Journal of Microscopy* 200:187 (2000)
30. A. Masseboeuf *et al.*, in Quantitative Electron Microscopy for Materials Science, edited by E. Snoeck, R. Dunin-Borkowski, J. Verbeeck, and U. Dahmen (Mater. Res. Soc. Symp. Proc. Volume 1026E, Warrendale, PA. 2008).
31. A. Masseboeuf, C. Gatel, P. Bayle-Guillemaud, A. Marty, and J.-C. Toussaint, *Journal of Applied Physics* (in press)

MICROSTRUCTURAL AND MECHANICAL PROPERTIES OF COPPER PROCESSED BY EQUAL CHANNEL ANGULAR EXTRUSION

RAKIA DALY[1*], MOUHAMED KHITOUNI[1] AND NABIL NJAH[1]

[1]*Laboratoire de Métallurgie Appliquée, Sfax – Tunisia.*

Abstract. The spurt in research activities pertaining to nanotechnology, much interest has arisen in processes involving severe plastic deformation. Equal channel angular pressing (ECAP) in particular has drawn considerable attention due to its potential to produce ultrafine-grained or in some cases nanometer grain size materials in bulk form. It is now well known that materials with such small grain sizes have extraordinary properties and, such as simultaneous ultrahigh strength and high ductility as well as the capability of superplastic forming. They therefore have great potential for technological applications. Equal channel angular extrusion (ECAE) was used to investigate the formation of submicron grain in copper deformed to ultra-high plastic strains by die angle of 90°. The mechanical result was characterized by the use of tensile tests, microhardness measurement Scanning Electron Microscopy (SEM) and Differential Scanning Calorimetry (DSC). ECA extrusions at room temperature result in a significant reduction of grain size due to a fragmentation of the preexisting coarse grains. In TEM, many grains are separated by high-angle boundaries. For the specimen deformed N = 2, A drop of the hardness level was observed at 573 K, indicating that a fully recrystallized state was achieved.

Keywords: Equal Channel Angular Extrusion (ECAE), copper, deformation, hardness, stress–strain

1. Introduction

Severe plastic deformation (SPD) has been arousing great scientific interest, especially due to its capacity of producing materials which display unusual properties.[1] The advantages of the SPD methods is the possibility of achieving

*To whom correspondence should be addressed: Rakia Daly, Phone number: +216 74 274400, Fax number: +216 74274437, Email: Daly.rakia@caramail.com

fabrication rates and costs comparable to those for the production methods of conventional materials, making use of equipments similar to those used in conventional deformation processing.[2] One of the most common SPD methods is Equal Channel Angular Extrusion (ECAE), invented by Segal and his Co-Workers in 1970s and 1980s at an institute in Minsk in the former Soviet Union.[3] ECAE was considered, the most prominent SPD technique due to its capability of producing submicron-grained materials[4–7] and bulk work pieces free of porosity and inclusions.[4] There have been also attempts to adapt this technique to continuous processing, enabling the production of materials in a large scale, efficient and cost effective manner.[5]

One of the aims of grain refinement is to enhance plasticity; the main difficulty is to preserve a small grain size at the forming temperature which is close to. In this range of temperature, diffusive transformations are to be considered such as recrystallization. It is then necessary to dragging recrystallization and grain growth; this can be achieved by introducing second phases like precipitates. However this leads to an increase of the flow stress and a decrease of ductility of the material and therefore a limitation of the amount of deformation which can be introduced. The use of low-purity metals may be an interesting way since they are relatively easy to work and may retain a suitable grain size at sufficiently high temperatures due to the presence of impurities.

In the present work, an electrolytic copper containing low-fraction of oxide was investigated. The aim of the work is the studied of the microstructure and the mechanical properties of copper after ECAE.

2. Experimental part

The starting material for the experiments was copper (99.99% purity) in form of rectangular bars of 10 × 10 mm cross section. The die for ECAE consisted of two channels with square cross section intersecting at an angle of 90°. The ECAE was carried out at room temperature with MoS2 as lubricant. Sample was passed through the die and it was rotated about the extrusion axis by 90°, after the first passed. The process used has been designated as route A (see Fig. 1).

Transmission electron microscopy (TEM): TEM was used to examine the microstructure after ECAE. The transmission electron microscopy used in this study is a JEOL 2000 EX. We used negative photographs for the photographic support.

Tensile testing: Tensile specimens with a rectangular section (5 × 10 × 0.5 mm of width, length and thickness, respectively) were cut along the longitudinal direction of the specimens ECAE at room temperature with an initial strain rate of 1×10^{-3} s^{-1}. Tensile strength and modulus values correspond to the average of three samples.

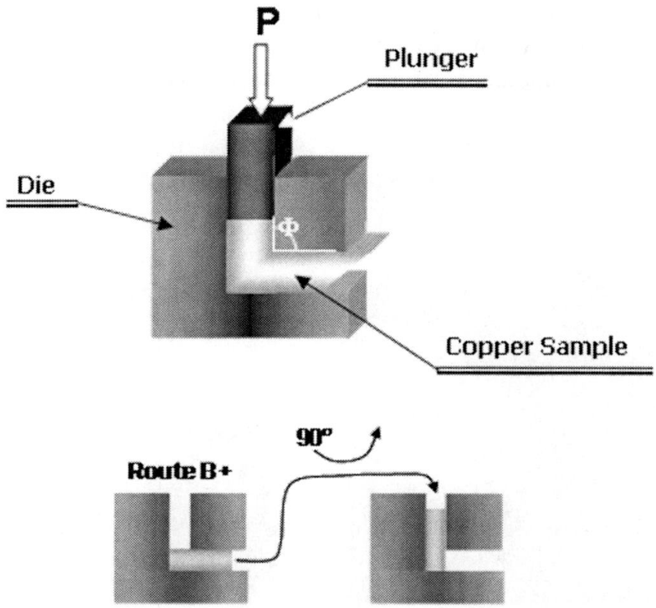

Figure 1. Schematic illustration of the equal-channel angular extrusion (ECAE) technique.

Scanning electron microscopy (SEM): The tensile fracture surfaces of the copper materiel were examined with a scanning electron microscope. Experiments were carried out in a SEM Philips 505 equipped with a PGT SiLi energy dispersive spectrometer (EDS).

Hardness: The Vickers hardness, HV, was measured on samples using shimadzu HVM-2000 microhardness.

Differential scanning calorimetry (DSC): DSC was performed with Perkin–Elmer DSC7 equipment. Samples, around 10 mg, were placed in pressure-tight DSC cells and at least two individual measurements were carried out to ensure perfect reliability of measurements. Each sample was heated from +20°C to +450°C at a heating rate of 10°C/min.[8]

3. Data and result

3.1 MICROSTRUCTURE AFTER ECAE

Figure 2a and b show TEM micrographs of the microstructure obtained by ECAE in two selected magnifications. After two passages through the die, the microstructure consisted of elongated grains which appear to be fragmented into smaller grains. Since no subsequent heating was carried out after deformation, recrystallization should not be expected at room temperature in this material. The splitting of coarse grains is then a pure mechanical feature and it is formed during extrusion. The characteristic features of the given structure are the existence of high internal stresses which is confirmed by the bending extinction contours and the presence of defects mainly dislocations within the grain boundaries. Also, we note that diffraction spots are still resolved on the SAD patterns even through they are slightly elongated; this indicates a weak misorientation between the newly created grains accompanied by an eventual distortion of the lattice.

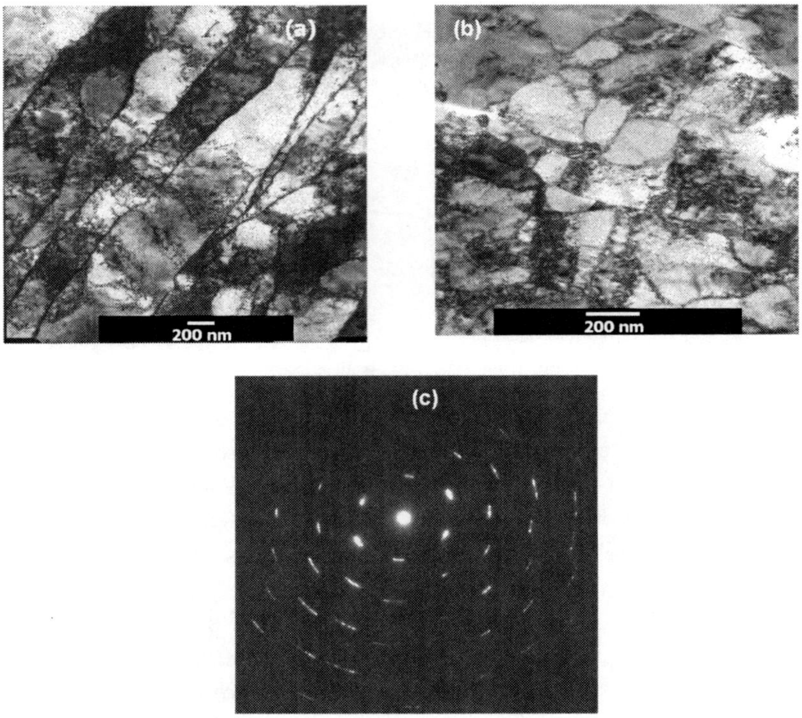

Figure 2. Microstructure of copper prepared by ECAE: (a) region of elongated grains, (b) region of equiaxed grains, (c) selected area diffraction patterns.

3.2 THE STUDIES OF THE MECHANICAL PROPERTIES OBTAINED AFTER ECAE

The effect of the ECAE deformation on the mechanical properties of the copper material was investigated. Figure 3 shows the evolution of microharness as a function of temperature. Between 100°C and 150°C we note a small increase in the microhardness. At the higher temperature of 200°C, we show a beginning diminishing of microhardness, the increasing of temperature around 300°C leads to a drop of hardness which is due to a recrystallization phenomenon.

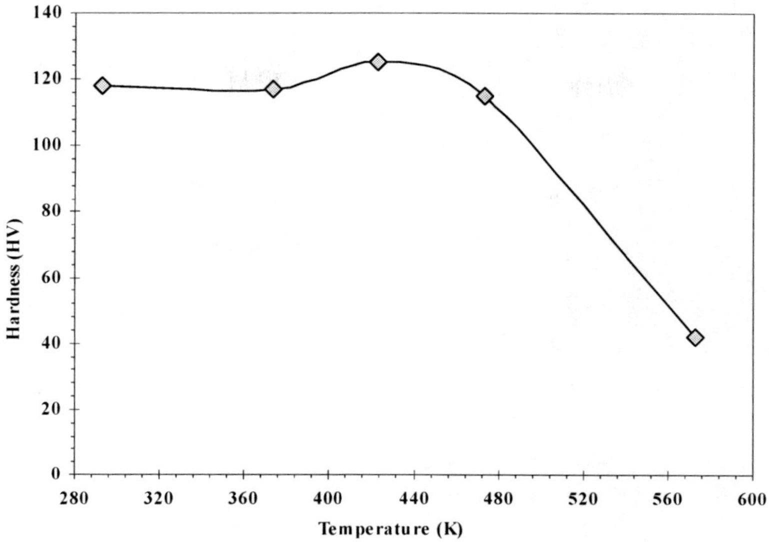

Figure 3. Evolution of the hardness as a function of temperature.

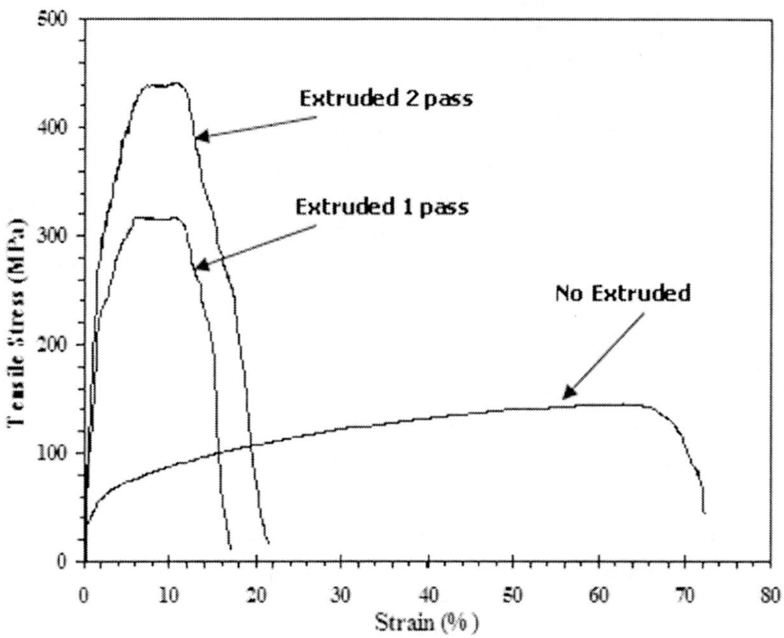

Figure 4. Tensile stress–strain curves for copper material pressed (N = 1 and N = 2 passes) at room temperature.

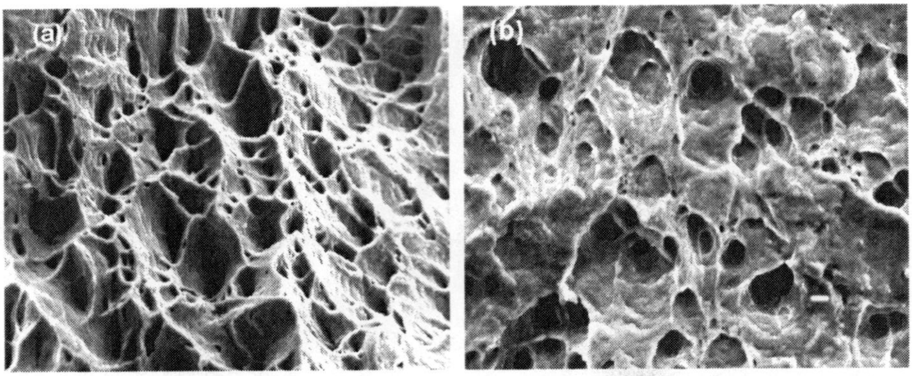

Figure 5. SEM images showing the tensile fractured surfaces of copper material before (a) and after tow pass (b).

Typical true stress–strain curves obtained for extruded copper are shown in Fig. 4. With the passage from the first pass to the second pass, the tensile strength increase, while the elongation decreases. Indeed, the strength increased from 145 MPa for no deformed copper sample up to 310 and 435 MPa for ECAE copper sample after 1 and 2 pass, respectively. These can explain the difference between SEM observations of tensile fractured surfaces of annealed coarse-grained and extruded material sight in Fig. 5a and b respectively. In fact, in the ultra fine grained material, the dimples seem to be shallower in comparison with those observed in non extruded copper which present a very profound holes. This is a direct consequence of the diminish of ductility occurring during ECAE due to the deformation in the ultra-fine grains copper ensured by the displacement of dislocations. When dislocations were obstructed in its displacement, the strength necessary to make it move will be more important, This explain the increasing in the hardening observed during the tensile tests from the extruded samples, related to the development of barriers opposing the displacement of dislocations.

3.3 CALORIMETRIC STUDIES

The thermal behaviors of samples before and after ECAP have been investigated by DSC measurements. The results are shown in Fig. 6. The DSC traces contain a broad exothermic peak occuring over quite a temperature interval, ranging from 300°C to 350°C approximatily, with peaks at 338.37°C, 335.63°C and 335.05°C for the undeformed and deformed samples (N = 1 and N = 2), respectively. A second heating run proved the irreversibility of all the observed transformations. The estimated enthalpies are ΔH = 0.45, 0.561 and 1.298 J g^{-1} for undeformed and deformed samples (N = 1 and N = 2), respectively. This peak corresponds to the energy strain release and microstructure recrystallization. The maximum of the exothermic peak in the DSC curve was slightly shifted to lower temperature. On the basis of the results obtained, we can say that the thermal behavior of deformed samples was highly influenced by ECAP.

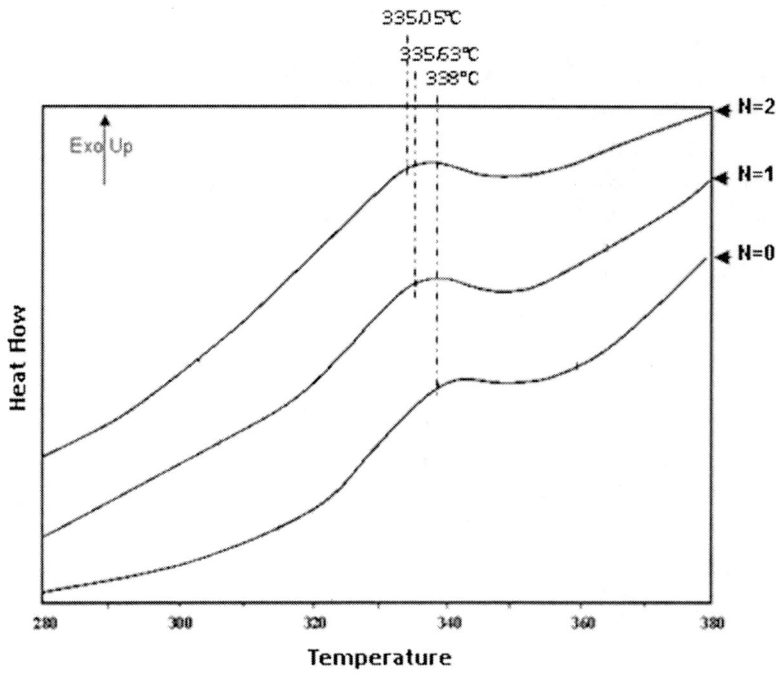

Figure 6. DSC curves for samples after the tow pass of ECAE.

4. Conclusions

ECAP is an effective process of producing ultrafine microstructure in pure copper. The bulk of grain refinement seems to take place in the first pass.

The microstructure of ECA extrusion specimens was investigated by TEM. It was found that in copper the crystallite size decrease significantly even after one ECAP pass, at the same time the dislocation density increases.

The strength in the copper material pressed at room temperature was much higher than that of as-received copper material, but the elongation to failure was significantly smaller, this explain the difference between the fractures surfaces observed by SEM.

The differential scanning calorimetry (DSC) experiments revealed a peak in the heating curve of the extruded material, associated essentially to a recrystallization phenomenon; in addition the thermal behavior of deformed samples was highly influenced by ECAP.

References

1. Z. Horita, T. Fujinami, M. Nemoto, T.G. Langdon, *Journal and Materials Proceeding Technology* 117(3):288–292 (2001).
2. Y.T. Zhu, T.C. Lowe, T.G. Langdon, *Scripta Materialia* 51(8):825–830 (2004).
3. V.M. Segal, V.I. Reznikov, A.E. Drobyshevskiy, V.I. Kopylov, *Russian Metall* 1:99 (1981).
4. J. Wang, M. Furukawa, Z. Horita, M. Nemoto, R.Z. Valiev, T.G. Langdon, *Materials Science Engineering A* 216:41–46 (1996).
5. R.K. Islamgaliev, F. Chmelik, R. Kuzel, *Materials Science Engineering A* 237:43–51 (1997).
6. R.Z. Valiev, *Materials Science Engineering A* 234–236:59 (1997).
7. Y. Iwahashi, Z. Horita, M. Nemoto, T.G. Langdon, *Acta Materialia* 46:3317–3331 (1998).
8. G.J. Raab, R.Z. Valiev, T.C. Lowe, Y.T. Zhu, *Materials Science Engineering A* 382:30–34 (2004).

INDEX

H

I

J

L

M

N

O

P

INDEX OF MATERIALS

Printed in the United Kingdom
by Lightning Source UK Ltd.
135725UK00008BA/134/P

9 781402 087943